Women and the Energy Sector

Natalia Rocha Lawton • Cynthia Forson
Editors

Women and the Energy Sector

Gender Inequality and Sustainability in Production and Consumption

palgrave
macmillan

Editors
Natalia Rocha Lawton
Coventry University
Coventry, UK

Cynthia Forson
Lancaster University
Lancaster, UK

ISBN 978-3-031-43090-9 ISBN 978-3-031-43091-6 (eBook)
https://doi.org/10.1007/978-3-031-43091-6

This Palgrave Macmillan imprint is published by the registered company Springer Nature Switzerland AG.
The registered company address is: Gewerbestrasse 11, 6330 Cham, Switzerland

Paper in this product is recyclable.

Acknowledgments

An edited book such as this necessarily invites much appreciation and gratitude. First and foremost, we would like to thank all the authors for their valuable contribution to the book and also for their engagement with the spirit of the topic of women's inequality in the energy sector. Working on this book together has been stimulating multidisciplinary and collaboratively intellectual experience from beginning to end.

Secondly, special thanks go to our Palgrave production editors Steven Fassioms and Vinitha Jayabalan for their assistance, patience and insights into this project and editors Palgrave Macmillan for their great support and enthusiasm in making this project possible.

Contents

1 **Introduction** 1
Cynthia Forson and Natalia Rocha Lawton

Part I Women and Gender Inequality in Energy Policy 13

2 **Merging Local Content and Women's Economic Empowerment in Tanzania's Extractive Industries** 15
Jesse Salah Ovadia

3 **Women's Empowerment Through Electrification: What Is the Evidence from the Indian Subcontinent?** 37
Debajit Palit, Mini Govindan, and Rashmi Murali

4 **A Feminist Policy Analysis of the Gender and Climate Change Nexus in the Colombian Coal and Energy Sector** 63
Ana Maria Mahecha Groot, Natalia Daza Niño, Silvia Rojas Castro, Kathrin Meyer, María Laura Rojas, and Kerstin Mohr

Part II Women and Gender Inequality in Energy Transition 89

5 **Empowering Women in a Climate-Changing World Through Climate-Resilient Energy Access** 91
Katrin Lammers, Camille Belmin, Zakia Soomauroo, and Martha M. Hoffmann

6 **The Energy Transition and Gender Considerations of the Workforce Transformation: A Critical Review of the Dutch Case** 125
Marielle Feenstra and Maria Kottari

7 **How to Build a Gender-Balanced Solar Sector Workforce in the Brazilian Energy Transition** 149
Aline Cristiane Pan, Kathlen Schneider, Natália Helena Ribeiro Chaves, Larissa Boing, Patrícia Betti, Aline Kirsten Vidal de Oliveira, and Izete Zanesco

Part III Women and Gender Inequality in Energy Communities 175

8 **Gender-Just Energy Communities: A Catalyst for Sustainable and Just Development** 177
Katharina Habersbrunner and Marika Kuschan

9 **Engendering the Energy Transition: Inspiring Examples of Gender-Just Citizen Energy Communities in Europe** 211
Antonia Proka

Part IV Women at the Intersection of Gender and Energy Equality 241

10 **Women in Energy Communities: An Intersectional Analysis of Their Participation** 243
Aurore Dudka

11 Gender and Energy Poverty in Africa: An Intersectional Approach 263
Verena Tandrayen-Ragoobur

12 Participants or Recipients? Negotiating Gender and Energy as Empowerment in the Displaced Setting 297
Alison Halford

13 Complex Intersections: Reflections on Women, Gender Inequality and Energy 315
Natalia Rocha Lawton and Cynthia Forson

Index 333

Notes on Contributors

Camille Belmin is a doctoral researcher in the Future Lab Social Metabolism and Impacts at the Potsdam Institute for Climate Impact Research and at Humboldt University of Berlin (Germany). Her research focuses on the effect of access to modern energy on the fertility transition in low-income countries and on relevant applications for sustainable development policies. Her main fields of scientific interest include energy in relation to gender and poverty, demography and social metabolism. Since 2021, she also works as the main data scientist in the social start-up company SoGreen, whose aim is to finance girls' education through climate finance.

Patrícia Betti is a PhD student and holds a Master's in Environment and Development from the Federal University of Paraná. She has 14 years of experience in project development and public policies in the socio-environmental area. Since 2016 she has been working with gender mainstreaming in projects of the German Cooperation for Sustainable Development.

Larissa Boing has a Master's in Regional Development and Environment from the State University of *Santa Cruz* and works with the management of social and environmental programs and projects aimed especially at the female audience since 2008. She is a member of the Centre for Gender and Sexuality Studies—GENI, of the Federal Institute of Education, Science and Technology Baiano—IFBaiano, a nucleus focused on education on the diversity of gender and sexuality in educational institutions.

She is also a consultant in the development of gender analysis in projects, especially focused on renewable energy and facilitator of courses on the subject.

Silvia Rojas Castro is a legal advisor and South-North specialist at the European Centre for Constitutional and Human Rights previously with Polis180 in Berlin. As a Colombian-trained lawyer with MA in Public Economics, Law and Politics, she has ample experience in human rights, strategic litigation, transitional justice, gender and climate justice.

Natália Helena Ribeiro Chaves is an environmental scientist specialised in Renewable Energy Management, with emphasis on research and studies about the solar energy sector. She works on developing projects about sustainability, with a focus on Energy Transition and Professional Training at the Chamber of Commerce and Industry Brazil-Germany (AHK-RIO). With knowledge and experience in the photovoltaic sector, she is one of the co-founders of Brazilian Women's Network in Solar Energy (MESol). In addition, she takes part in the International Global Woman's Network for Energy Transition (GWNet) as a participant in the Mentorship Program (2019–2020).

Natalia Daza Niño holds a bachelor's degree in political science from Universidad de Los Andes, with minors in gender studies and philosophy. Over the past years, she has worked as a human rights researcher and advocate, with a focus on the nexus between human rights, climate justice, and gender justice. Currently, Natalia works as a researcher at the Natural Resources and Just Transitions area at the Business and Human Rights Resource Centre. She also served as the junior research associate in Colombia for the "Colombia, Gender and Climate: Women as Actors for Transformation (Colombia_GC)" project, where she co-authored the policy paper "Gender-Responsive Climate Policy: A Case Study of the Colombian Coal Sector" in October 2020.

Aline Kirsten Vidal de Oliveira is an electrical engineer graduated at *Universidade Federal de Santa Catarina* (UFSC). She has experience with microgrids and did an exchange program at *Technische Universität München* (TUM), in Germany, where she specialized in the integration of renewable energies and electromobility. Since 2017, she is a researcher at the Solar Energy Research Laboratory at UFSC where she obtained her doctoral degree with a focus in PV plants inspection. She is also president of the Brazilian Association of Solar Energy (ABENS) and co-founder of Brazilian's Network of Women in Solar Energy.

Aurore Dudka holds a PhD on energy communities in fostering energy democracy and justice. Dudka is currently a researcher at the University of Trento and is involved in various European projects, focusing especially on women's inclusion in renewable energy projects. Dudka has led consultancy activities on the development of energy communities.

Marielle Feenstra is co-founder and scientific director of the NGO 75inQ, an expertise centre on gender diversity in the energy sector. She works on gender and inclusive research and innovation as a senior researcher at the Delft University of Technology, the Netherlands. She has been working on gender and energy since 2000. After 15 years as a policy advisor for municipalities in the Netherlands, she returned to academia with her PhD project "Gender Just Energy Policy: Engendering the Energy Transition in Europe" graduated in 2021 at University Twente, the Netherlands.

Cynthia Forson is an associate professor and Deputy Provost at Lancaster University Ghana. Her research interests cover equality, diversity and inclusion in the labour market and organisations. She focuses particularly on gender, ethnicity, class and migrant status and their intersectional influences in the lives and careers of women, particularly African women on the continent and in Diaspora. In that regard, she has published research reports, edited books, book chapters and journal articles in *British Journal of Management; Equality, Diversity and Inclusion; Gender Work and Organisation; International Journal of Entrepreneurial Behaviour and Research, Journal of Management Development*; *Journal of Small Business Management;* and *Work, Employment and Society*, among others.

Mini Govindan is a senior fellow at TERI. She has a PhD in Development Studies and around 20 years of experience working in various projects on gender, energy and sustainable development. She is involved in grassroots institutions' development, gender inclusive approaches and capacity building. She has widely written on gender mainstreaming and the energy-gender-poverty nexus and is a sought-after speaker on gender and sustainability issues.

Katharina Habersbrunner is a board member and team leader at WECF. Her professional expertise is decentralized citizen's energy projects, local energy markets and advocating for socially sound, gender-just climate solutions on national and international levels. Increasing gender equality is central to the energy and climate projects with gender mainstreaming as one of the core tools. Katharina is also board member of *Bündnis Bürgerenergie BEEn*, the umbrella association for citizen's energy in Germany, advocating for an ambitious decentralized energy transition. Katharina holds a diploma in mathematics, a master's degree in education and psychology and an MBA in sustainability management with focus on renewable energy.

Alison Halford is an assistant professor (Research) and Impact Lead at the Centre for Computational Science and Mathematical Modelling, Coventry University, UK. As a transdisciplinary scholar, her work aims to dismantle knowledge silos between different disciplines and across sectorial boundaries to develop new ways of thinking about real-world problems. Her current research is looking to improve the design of e-solar cooking systems in Rwanda through meaningful participation with women that amplifies their voice at all levels of energy decision-making.

Martha M. Hoffmann is writing a doctoral thesis on the integration of justice into energy planning. She receives a scholarship from the Reiner Lemoine Foundation (RLS) and works with the RLS Graduate School Energy System Transition. Her research is supervised by Prof. Dr. Pao-Yu Oei from the University of Flensburg. In 2023 she was able to discuss the different perspectives on justice in the energy transition with researchers from the University of California/Berkeley in the scope of a research exchange organised by the Reiner Lemoine Institut (RLI) and funded by RLI and the Federal Ministry for Economic Affairs and Climate Action of Germany (BMWK). From 2019 to 2022, Martha worked as a researcher and project leader in the research unit Off-Grid Systems at the RLI. She developed and applied the simulation and optimization tool Multi-Vector Simulator for local sector-coupled energy systems in the scope of the European H2020 research project E-LAND. Additionally, she investigated issues of electrification and developed Offgridders, a tool for the optimization of electrification options. Accompanying this, Martha published and presented her work in different journals and at conferences, including studies related to electrification, blackouts, micro grid planning and sector coupling. She holds a master's degree in renewable energy systems from the Berlin Institute of Technology and gained on-hand experience in electrification planning during an internship in India at Oorja Development Solutions.

Maria Kottari is a research associate at the School of Transnational Governance of the European University Institute (Climate). She holds a PhD in International, European & Regional Studies with a focus on Energy Politics and her expertise and network relate to the EU and international politics and governance, energy/climate and environmental policies, sustainability and social challenges. In 2015, she conducted postdoc research at ADA University (Baku, Azerbaijan) on the topic of Green Economy & Energy Efficiency in the wider Caucasus region. Maria worked for three years at the European Commission's Joint Research Centre, focusing on energy transition and innovation.

Marika Kuschan is an energy and gender expert, graduated with a Diploma in Physics, currently serves as project manager for the Energy, Climate and Gender team at WECF Germany. She is certified in academic adult education, has experience in social entrepreneurship as well as in leading project management in technical research projects. Thus, her women power is mainly focused on R&D projects, capacity building and the implementation of innovation, strongly advocating for social solidarity, equality and diversity in the energy sector and promoting decentralised renewable energies.

Katrin Lammers is an expert in off-grid electrification and climate change resilient mini-grid design. In her doctoral theses, she analysed climate change–induced risks on Southeast Asian islands and developed an approach for climate resilient island electrification. She applied and evaluated this approach to three case study islands in the Southeast Asian region. Katrin holds a master's degree from the Technical University and Mining Academy Freiberg in Environmental Engineering focusing on decentralised energy systems and thermal insulation. Within the Off-Grid Systems team at RLI, Katrin specialises on rural electrification (energy system modelling and demand estimation), climate resilience of energy systems, adaptation potential arising from energy access and international perspectives of green hydrogen partnerships. From 2015 to 2018, Katrin worked at GIZ in Bangkok, where she supported renewable energy reference project development on community level as well as within the commercial and industrial sector, managed a project on rural electrification on

Thai islands and developed electrification concepts and technical system designs for two pilot sites together with a project consortium. She also supported a project funded by the International Climate Initiative (IKI) for improved provincial energy planning. While working on her thesis at the Fraunhofer THM Institute and Schott Solar, she gained additional expertise in degradation effects and thermal losses of solar panels.

Ana Maria Mahecha Groot is a social and environmental researcher with a PhD in geography. She has a background in women and environmental issues, climate change adaptation, environmental education and management of strategic ecosystems. She has worked with various organisations such as the National University of Colombia, Rutgers University, United Nations and NGOs. She directed the "Gender and Climate: Women as Actors for Transformation" project in Colombia and co-authored a policy paper on Gender-Responsive Climate Policy, when she was in Transforma. Currently, she works at *Mundo Común* as an interweaver of worlds.

Kathrin Meyer is working as an advisor for the German Development Cooperation in Costa Rica, promoting knowledge management in the topics of mitigation, decarbonization and NDC implementation. She studied Political Science, Environmental Science and Latin American Studies in Germany, Cuba and Colombia.

Kerstin Mohr is a policy advisor on international climate justice and energy policy who specializes in just transition and gender. Previously part of Polis180, she currently works for a Berlin-based think tank. She studied Governance and Public Policy and Political Science in Germany, Brazil and Colombia and obtained her PhD in Ecological Economics.

Rashmi Murali is an associate fellow with the Centre for Social Research and CSR at TERI. Since joining TERI, she has worked in the fields of renewable energy, gender, water-energy-food nexus, waste, coastal communities and eco-tourism. Her key expertise lies in conducting primary and secondary research, qualitative analysis, community engagement and documentation. She has co-authored several publications on gender and energy. Her academic qualifications include master's degrees in physics and renewable energy technology.

Jesse Salah Ovadia is an associate professor at the University of Windsor, Canada. His research focuses on the political economy of resource extraction and development in Africa as well as sustainable energy transitions. He is the author of *The Petro-Developmental State in Africa* (Hurst 2016). He has consulted for organisations such as the World Bank, African Development Bank, African Climate Foundation and others. His academic research has been published in journals such as *Third World Quarterly, New Political Economy, Resources Policy, Journal of Modern African Studies, Journal of Agrarian Change* and *the Extractive Industries and Society*.

Debajit Palit has 25 years of experience working in the domain of renewable energy, energy access, rural electrification, micro-grids, energy transition and gender-energy-livelihood nexus. He is featured in the Top 2% World's Scientists ranking by Stanford University for the last three years. Dr Palit is currently Professor of Energy at the NTPC School of Business, India. Dr. Palit was earlier associated with The Energy and Resources Institute as a director and senior fellow and led its Rural Energy & Livelihood Division for almost a decade. Dr Palit holds a master's degree in physics and PhD in Energy Policy.

Aline Cristiane Pan graduated in Physics from *Pontifícia Universidade Católica do Rio Grande do Sul* (PUCRS) (2002), has a Master's in Materials Engineering from PUCRS (2004) and a PhD in Photovoltaic Solar Energy from the Polytechnic University of Madrid (2009). Currently, she is Professor in Energy Management Engineering at the Federal University of Rio Grande do Sul (UFRGS)/Campus Litoral Norte; member/guest of the Working Group "Gender Equity" promoted by CAPES (2018); and co-founder of the Brazilian Network of Women in Solar Energy (Rede MESol).

Antonia Proka works at REScoop.eu, the European federation of citizen energy cooperatives, focusing on community organising and capacity building for the development of gender-just energy communities. She also coordinates the Gender Power working group of REScoop.eu, which enables the exchange of good practices and tools. Antonia coordinated the European Citizen Energy Academy, supported the development of the European Commission's Clean energy for EU islands secretariat and facilitates the collaboration between energy cooperatives and local authorities. She holds a PhD from the Erasmus University Rotterdam (and DRIFT) on the (self-) organisation of renewable energy initiatives and their impact on the energy transition.

Natalia Rocha Lawton is an assistant professor at Coventry University. She holds a PhD in Sociology from the University of Warwick. Prior to joining academia, she worked at the Secretariat of Energy for the Mexican Government. Her research has focused on the social implications of economic deregulation of the energy sector on employment relations, gender and diversity. She has been a co-investigator on a research project on gender and economic inequality as well as policy associate for the Energy Policy Fellowships. She has been stream convenor at the Gender Work and Organisation Conference on topics of energy, inequality and intersectionality and guest editor for books, journals and author for SAGE, Routledge and Cambridge University Press.

María Laura Rojas is a lawyer with a Master's in Environment and Sustainable Development. She has experience with climate change, gender, international negotiations and diplomacy. Beyond her legal expertise, Maria Laura is an enthusiastic dancer and diver with a passion for animals and feminist political ecology. She is drawn to environmental issues as a means of illuminating the complexities of development and underscoring our interconnectedness.

Kathlen Schneider has bachelor's and master's degrees in civil engineering from *Universidade Federal de Santa Catarina* (UFSC); has experience in PV systems, energy efficiency in buildings simulation; distributed generation regulation context with a focus on energy community initiatives; and participated as a mentee at the Mentorship Program from the Global Women's Network for Energy Transition (GWNET) in 2019. Currently, Kathlen is a researcher from the Solar Energy Research Lab (Fotovoltaica-UFSC), director of the Institute for the Development of Alternative Energies in Latin America (IDEAL), and co-founder and coordinator of the Brazilian Women's Network in Solar Energy (MESol Network).

Zakia Soomauroo is a doctoral candidate affiliated with the Reiner Lemoine Institute and the Technical University of Berlin. Within the framework of her dissertation, she analyses potential decarbonisation pathways through a socio-technical-economic lens with a focus on untangling the carbon lock-ins and power dynamics of the current energy and transport system. Zakia holds a graduate degree in Industrial and Network Economics from the Technical University of Berlin. Within the Off-Grid System Team at RLI, Zakia is the project leader of the FAIRberlin project, exploring a feminist and inclusive approach to the sustainable mobility transition in Berlin.

Verena Tandrayen-Ragoobur is Associate Professor in Economics, University of Mauritius. Her research areas are in development economics, labour economics, international economics, poverty and gender issues. She is currently heading the Pole of Research Excellence in Applied Social and Economic Research Analysis and is the co-chair of the WTO Chairs Programme at the University of Mauritius.

Izete Zanesco is currently a professor of the School of Technology and of the Postgraduate Program in Materials Engineering and Technology (PGETEMA), and she coordinates the Solar Energy Technology Nucleus of the School of Technology, at the Pontifical Catholic University of Rio Grande do Sul (PUCRS). Izete holds a PhD degree in photovoltaic engineering from Polytechnic University of Madrid, Spain. She has been a member of Brazilian Association of Solar Energy (ABENS) and of the International Solar Energy Society (ISES) and is co-founder of the Brazilian Network of Women in Solar Energy (MESol).

List of Figures

Fig. 5.1	Linkages between climate change impacts, women's empowerment, and energy access	93
Fig. 5.2	Resilience framework, based on DFID (2011)	95
Fig. 5.3	Economic activities with strong energy input links, based on Stuart (2017)	104
Fig. 5.4	Climate change impacts on technical equipment. Solar panels destroyed by hail and storm (*left*), house destroyed by a storm (*middle*), electricity lines destroyed (*right*), Bulon Don island, Thailand. (Taken by Katrin Lammers, 2017)	112
Fig. 6.1	Women's share in employment in selected economic activities and women's employment in the same sectors, EU-28, 2018 (Czako V., 2020)	132
Fig. 6.2	Percentage of occupations relevant to the energy transition and female participation, Credits: 75inQ based on CBS figures (CBS, 2019)	137
Fig. 6.3	Visual representation of gender-responsive energy policy roadmap. (Authors' own visualisation)	141
Fig. 8.1	Gender dimensions in the energy sector, WECF 2021	190
Fig. 9.1	Zuiderlicht: energy (and coffee) with a positive story ©Zuiderlicht	224
Fig. 9.2	Aylin from the scrolly telling of BBEn ©BBEn	225
Fig. 9.3	Charter for Gender Equality Cooperatives Europe ©Cooperatives Europe	232
Fig. 10.1	Interaction between gender and occupation for joining the cooperative	251
Fig. 10.2	Interaction between gender and income for joining the cooperative	252

Fig. 10.3 Repartition of women by income in the cooperative 253
Fig. 10.4 Interaction between gender and income for investing in the cooperative 254
Fig. 10.5 Interaction between gender and seniority for investing in the cooperative 255
Fig. 11.1 Conceptual framework—Modelling gender energy poverty from an intersectionality perspective. (Source: Author's adaptation and extension on various models namely Großmann and Kahlheber (2017) and Gonzalez Pijuan (2018)) 271

List of Tables

Table 3.1	Framework for analysis of electricity–empowerment linkages	44
Table 5.1	Overview over climate change impacts on women	97
Table 5.2	Overview of technical adaptation measures to mitigate climate risks of (off-grid) energy systems based on Lammers et al. (2020)	113
Table 11.1	Electricity access by region from 2000 to 2019	275
Table 11.2	Socio-demographics of women surveyed across 20 SSA countries from 2010 to 2015	276
Table 11.3	Micro-level dynamics: Individual characteristics and family dynamics of African women	278
Table 11.4	Micro-level dynamics: Housing conditions of African women surveyed	284
Table 11.5	Meso-level dynamics of African women surveyed	285
Table 11.6	Macro-level dynamics of African women surveyed	288
Table 12.1	Overview of data collected, interventions, and handover of interventions	305

CHAPTER 1

Introduction

Cynthia Forson and *Natalia Rocha Lawton*

This book, *Women and the Energy Sector: Gender Inequality and Sustainability in Production and Consumption*, builds on conference stream papers presented at the Gender Work and Organisation Conference, Sydney, 2018. It collates reflections developed on the papers presented in the 'energy, gender and inequality' stream that identified and reflected on different sources of gender inequality in the energy sector in diverse settings and geographic contexts.

The purpose of this book is to go beyond the initial stream contributions, and its focus is to examine the significant roles different dynamics within the energy landscape play in understanding the inequality women face in energy systems. Patriarchy is a persistent, prevalent and resilient feature in societies worldwide and not just an issue for the North or South. It is marbled in private and public life, intersecting with other strands of inequality to shape women's lived experiences, including access to clean

C. Forson (✉)
Lancaster University, Lancaster, UK
e-mail: c.forson@lancaster.ac.uk

N. R. Lawton
Coventry University, Coventry, UK
e-mail: natalia.rochalawton@coventry.ac.uk

N. Rocha Lawton, C. Forson (eds.), *Women and the Energy Sector*,
https://doi.org/10.1007/978-3-031-43091-6_1

affordable energy. The book seeks to highlight these experiences in relation to energy inequality, giving voice to women in different parts of the world. As such, in this book, we use the concept of gender in binary terms—as relationships between men and women.

The discussions in the chapters examine concepts of gendered energy, unequal gender relations and inequality in the access to and production, reproduction and sustainable consumption of energy. Furthermore, the analyses aim to address the dominant discourse on energy capitalism that has obscured gender as an important factor in the, by and large, homogeneous perceptions of energy equality, reinforced by traditional literature on energy that assumes the neutrality of gender (Ryan, 2014). Further, the conversations in this book seek to show that the policy focus on energy has also been based on a homogeneous view of energy consumption that needs deep critical academic intervention, reflected in, for example, the disconnect between energy policy and gender policy in many jurisdictions. In this sense, the chapters base their analysis on theoretical and empirical data that show that 'woman is a heterogeneous category, and as such energy systems and technologies can be analysed in layers of identity, structures, institutions, and representations' (Sovacool, 2014, p. 15).

Energy inequality and gender inequality in energy systems are not only a Southern (African, South Asian and Latin American) phenomenon, although the North has a higher level of access to modern energy sources. However, inequality is a feature present in poor sustainable development initiatives, restricted access to affordable energy and limited political participation in organisations and their communities.

Gender, inequality and energy relationships are discussed in growing empirical and theoretical research from North to South (Arctic, Africa, Latin America, Australia, Europe) (Anditi et al., 2022; Feenstra & Özerol, 2021). Most of the discussion in the literature on inequality in the energy sector has been addressed by diverse disciplines, but the focus has been to give a broad idea of sustainable development and the impact of technological change on energy access without considering the 'gendered nature of energy' (Lieu et al., 2020, p. 1) and its consequences on the daily lives of women and the communities in which they live.

Indeed, research on gender and energy over the last twenty years has assumed that the way to achieve gender inclusion is by ensuring access to sustainable modern energy (SDG7). This perspective assumes that helping women cover their material necessities will contribute to fostering gender equality. However, 'in practice, this has typically led to gendered energy approaches that focus more on technological fixes rather than providing appropriate energy services and on meeting women's immediate needs, rather than addressing the broader cultural, socio-economic and political contexts important for attaining genuine gender equity' (Johnson et al., 2019, p. 169).

Against this trend in energy studies, the literature has also suggested that gender inequalities, gendered energy and the relevance of diversity at work have become significantly critical within the energy sector (Feenstra & Özerol, 2021; Johnson et al., 2020; European Institute for Gender Equality, 2012; Ernest & Young, 2015; McCright & Xiao, 2014). Also, there have been attempts in the gender and energy literature to recognise that intersectionality could be a tool that would contribute to unveiling the complex relationships between women and energy (Johnson et al., 2020; Søraa et al., 2020; Gray et al., 2019; Winther et al., 2018; Baruah, 2015).

The chapters are supported mainly by theoretical and empirical research developed in a variety of regions from the Global North and Global South. The multidisciplinary contributions from authors affiliated to different backgrounds, such as policymaking and practitioners' networks, provided diverse theoretical perspectives and methodological approaches to understand and explain the complexity of the inequalities women face in the access, production, reproduction and consumption of energies.

Chapter contributions have come from authors from different disciplines including political economy, humanitarian engineering, sociology, geography, environmental and energy studies, political science and political economy, public policy, sustainability and business and organisational studies. Their arguments are based on the gender implications of different forms of energy, an approach that has contributed to the understanding of the complexities behind gender inequalities in the energy sector. Some of the authors identified or showed the inadequacy of theoretical and conceptual tools that tend to take gender for granted, and some of them suggest some ideas which may open up a fresh discussion on the complexities of gender inequality in the energy sector.

The discussions centre around the identification of factors that contribute to creating and reproducing gender inequality in the energy sector and different energy communities in the Global North and South. The book has been designed to show that gender inequality is not only a global Southern (African, South Asian and Latin America) phenomenon, although the North has a higher level of access to modern energy sources;

inequality is a feature present in poor sustainable development initiatives, restricted access to affordable energy and limited political participation in organisations and their communities. Gender, inequality and energy relationships are discussed in empirical and theoretical research from the Global North to the Global South (Africa, Asia, Europe and Latin America).

The book includes empirical and theoretical contributions with regard to sustainable consumption and production. Some of the chapters address the consequences of the green deal in Africa (Chap. 11), South America (Chaps. 4 and 7) and South Asia (Chap. 3). Contributions from different geographies (Chaps. 7, 8 and 10) provide insights into different dimensions at individual, community and organisational levels (Chaps. 3, 4, 5, 6, 7, 9, 11 and 12). Other chapters take an intersectional approach to the analysis of energy gender inequality.

Energy is an essential resource that intersects all aspects of men's and women's lives; it is one of the most important components which drives economies and sustains societies (UN). Energy is a fundamental resource in terms of economic and human development and has a variety of issues associated with its production and consumption. For instance, in environmental terms, ecological degradation of the atmosphere (Chap. 5), national security and other social considerations that imply that the study of energy has to go beyond technical and financial considerations (Chap. 12) and be sustained by theoretical perspectives from social sciences. Chapter 11 explores the complexities of energy transitions in diverse African countries from an intersectional approach.

Structure of This Book

The book is organised in four main parts:

Part I—Women and gender inequality in energy policy
Part II—Women and gender inequality in energy transition
Part III—Women and gender inequality in energy communities
Part IV—Women at the intersection of gender and energy equality

These are followed by a final reflection on the analyses of the complexities of gender inequality in the energy sector that builds on debates on women in the energy system.

Part I focuses on women in energy policy. Examining policy from three different continents, the chapters that appear under Part I consider ways in which gender is constituted and enacted in energy policy and the need for an integrated approach to gender equality and energy equality policy development and implementation.

Jesse Ovadia (Chap. 2) begins the section with an analysis of Tanzanian policy that seeks to increase the participation of Tanzanians in the extractive industries in the context of the promotion of economic opportunities for women within the sector. Ovadia argues that local content regulations erroneously assume that women will benefit automatically from policies aimed at all Tanzanians, ignoring the many important ways in which women have different characteristics, face different problems and have different needs from men. The chapter challenges the dominant discourse on energy and gender in the relationship between Tanzanian local content policy and women's empowerment in the mining sector. It highlights the gender blindness that leads to a failure to see the relationship between local content and women's economic empowerment (WEE) policies.

Building on this argument, **Debajit Palit**, **Mini Govindan** and **Rashmi Murali** (Chap. 3) provide a useful analysis of the links between gender-responsive policy and women's social and political empowerment, in the context of electrification projects in India and Nepal. Analysing the policy-programme nexus and employing Winther et al.'s (2017) electricity-empowerment framework, they examine the impact of access to electricity on women's rights, resource access and agency. The policy review revealed that while, in the main, the supply-driven policies were gender-neutral, their assumed mutual benefit for men and women was not always realised due to the inherent differential impacts of sociocultural norms on men and women. Furthermore, the policies focused on traditional perspectives of the role of women and failed to engage with alternative and complimentary roles in the energy supply chain, including entrepreneurship. Off-grid programmes seemed to offer better evidence of positive, yet sociopolitically situated, outcomes for women.

Examining the intersection of climate change and gender policy, **Ana María Mahecha Groot**, **Natalia Daza Niño**, **Silvia Rojas Castro**, **Kathrin Meyer**, **María Laura Rojas** and **Kerstin Mohr** (Chap. 4) present an interesting gender analysis on the energy transition of the Colombian energy sector and coal sub-sector and its implications for social justice—showing once again the disconnect between gender policy and energy policy. The authors reference the accepted need for the incorporation of gender perspectives in climate policies yet question how gender is perceived and incorporated in climate and coal policies and how current policies can be strengthened to promote gender equality and respond more efficiently to the climate crisis. The findings suggest that climate and energy policies are currently not gender-responsive and incorporate stereotypical understandings of gender. Secondly, gender policies have not sufficiently incorporated concerns about the climate crisis. More specifically, they provide very limited opportunities for communities to participate in these policies, and therefore the policies fail to engage with community understandings of the environment. The authors call for an urgent need to formulate a gender-responsive climate policy based on intersectional methodologies that account for the peculiarities of the context in which the policies are to be implemented.

Part II explores the realities of women in energy transition shaped by climate change impacts, gender inequality and energy systems in a range of contexts. The section shows how gendered energy inequalities work in multi-level contexts—the broader society, the energy system and the energy labour market. In this book, the multi-level impacts of gendered energy inequalities are evident in a variety of contexts with the dynamics within each space producing different forms and degrees of inequality. We begin with the triangular dynamic between climate change impacts, gender equality and energy access.

Katrin Lammers, Camille Belmin, Zakia Soomauroo and **Martha Hoffmann** (Chap. 5) open the section by examining the specific role that energy access plays in reducing gender inequalities in the context of climate change. A review of the literature on (i) the adverse effects of climate change on women, (ii) the role of energy access on climate resilience and gender equality and (iii) climate-resilient energy systems sets the scene for the analysis. In the context of the impacts of climate change, women benefit from energy access in three main ways: improved economic well-being, improved health and agency. They argue that in a climate-changing world, climate-resilient energy systems are critical in promoting gender equality and call for a holistic approach to energy planning that considers technical solutions while integrating social and gender aspects to enable the building resilient and equitable communities.

Mariëlle Feenstra and **Maria Kottari** (Chap. 6) turn their attention to Europe and women in the Dutch transition. From technical, economic and social perspectives, the chapter provides a comprehensive overview of the role of women in the labour market in energy transition. They argue that the availability of a well-qualified diverse workforce is an essential precondition for solving current and future climate and energy issues and that this requires a significant transformation of the labour market due to the energy and climate imperatives. This transformation, they posit, will provide opportunities for women. The chapter then goes on to outline the status of women working within the Dutch energy transition and critically assesses the objectives, priorities, strategies and policies developed by the Dutch public and private sectors at the intersection of gender, energy transition and the climate change combat. The authors point out the failure of the current labour market to adequately reflect the diversity in society and therefore misses the opportunity to develop policies and design products that align with the needs of energy consumers. They argue for a multi-level gender-responsive energy policy roadmap that starts with the ambition to resolve pressing climate issues. This requires a strategy at the heart of which is a human capital agenda that prioritises gender mainstreaming, diversity and inclusion leading to policy that addresses the diverse needs to groups of women in the energy labour market.

Aline Cristiane Pan, Kathlen Schneider, Natália Helena Ribeiro Chaves, Larissa Boing, Patrícia Betti, Aline Kirsten Vidal de Oliveira and **Izete Zanesco** (Chap. 7) focus on the dynamics of building a gender-balanced solar sector workforce in the Brazilian energy transition. In the context of the male-dominated Brazilian energy sector, the paper seeks to identify and discuss actions that will enable the entry, progression, inclusion and retention of women in solar energy in Brazil and throughout its value chain. Discussing four key areas, (i) education, training and access to information, (ii) economic opportunities, (iii) health and well-being and (iv) participation and leadership, it highlights the opportunities as well as challenges faced by women in the sector and outlines practical solutions, in order to make the sector more attractive to women and thus promote gender equity within and alongside the energy diversification and transition process in Brazil.

Part III turns the focus on women in energy communities. In this part, the book looks at the dynamics of energy communities and their importance in energy transition. As citizen-driven collectives, energy communities offer more opportunities for people to engage in the energy market and promote a decentralised energy transition. As 'bottom-up' organisations, energy communities have been hailed as egalitarian democratic spaces that carry the promise of open participation, inclusivity and gender justice. Yet there are variations in the levels of political and financial support among European and other nations. The two chapters in this section interrogate these claims and ask several questions: To what extent do energy communities live up to their promise of energy democracy? Who participates in energy communities? How do they participate? Who makes decisions? How can energy communities be structured and managed to make them more inclusive, for women in particular?

Katharina Habersbrunner and **Marika Kuschan** (Chap. 8) question whether energy communities are gender-just environments that afford women safe spaces to engage in the energy market. The chapter is based on a literature review, outlining the background, structures, categories, functions and key characteristics of energy communities in Spain, Germany, Turkey and Uganda. An analysis of energy communities in these countries using a gender dimensions framework—productive work, reproductive work, power and decision-making power, public resources and infrastructure, body and health, as well as institutionalised male-centeredness—concludes that, in the main, energy communities do not meet the egalitarian claim that they enable open participation and membership for all parts of society and that they are, in reality, gender blind and their structures and processes are gendered. The authors make recommendations for policymakers and other actors in the energy system, with the view to making energy communities a feasible and viable alternative to hierarchical, individualistic organisation around energy production and consumption, one that incorporates a gender lens into community actions and strategies, opening up a new space for innovative debates on an inclusive and sustainable development model of energy transition.

The discussion on energy communities continues with **Antonia Proka** (Chap. 9). She acknowledges that most energy communities fail to live up to their characterisation as being gender-just, inclusive and participatory spaces and that, despite the progress made, a gender gap still exists concerning female participation, visibility and representation in management and top-level positions in the energy field. Proka, however, brings the reader's attention to the fact that gender-just energy communities are possible and a work in progress by highlighting those energy communities that are exemplars in the sector. Her contribution focuses on inspiring examples from the European energy community context, such as female-led citizen energy cooperatives, gender-just citizen energy alliances and female-driven citizen solar photovoltaic installers, who illustrate alternative cultures, structures and practices that have the potential to radically transform the energy system.

In **Part IV**, we turn our attention to women's experiences of energy inequality at the intersection of gender and other strands of social difference. Gender inequality is mutually constructed with other forms of social differences in terms of ethnicity, race, socioeconomic class, sexual orientation, age as well as other axes of social power and oppression. These factors tend to exacerbate the gender energy gap, yet access to energy remains an important tool to improve women's livelihoods. The chapters in this part of the book examine women's experiences in the energy sector at the intersection of gender and other variables. Using intersectional analysis as a tool, they deepen the conversation about the notion that energy communities are sites that perpetuate and exacerbate gender inequality.

Aurore Jeanne Stanislava Dudka (Chap. 10) continues the focus on the potential of energy communities to foster energy democracy and gender justice but argues that a lack of adequate data as a basis for analysing women's experiences in the energy sector makes it difficult to make informed decisions. Dudka contends that energy justice without gender justice could lead to the production and reproduction of new gender inequalities in the energy world. Dudka highlights the need to consider the multidimensionality and complexity of gender inequalities and be able to drive concrete actions to foster women's participation in the energy world. She begins the section by taking an intersectional approach to examining women's participation in energy communities in Belgium and discusses the different ways in which gender can intersect with other stands of inequality such as income, education, class, marital status and parenthood to deepen disadvantage for women's participation in the sector.

Chapter 11 by **Verena Tandrayen-Ragoobur** adopts an intersectional approach to the gender-energy-poverty nexus across 20 sub-Saharan African countries over the period 2010–2015, using the Demographic Health Surveys. In doing so, the chapter disentangles the mechanisms of multiple discriminations that arise at the same time and are particularly relevant in explaining the gender-energy disparity. The chapter groups the different axes of inequality into micro-, meso- and macro-level dynamisms that influence the gender-energy poverty link. Tandrayen-Ragoobur shows how education, class and geography intersect with gender so that African women with low education levels, living in poverty and those having difficult access to health and transport services tend to be energy poorer than other women in the sub-region. Similarly, those women who spend more time in unpaid care and domestic work and those victims of intimate partner violence are more likely to be energy deprived.

Alison Halford (Chap. 12) closes the section with a comparative look at women at the nexus of gender, refugee status and geographical location. Despite progress on measures to improve gender equality, women and young girls are disproportionately impacted by humanitarian crises. In addition to increased exposure and risk of sexual violence, female refugees have challenges accessing information, being involved in decision-making and engaging in economic opportunities at local and governmental levels. Energy interventions can be instrumental in reducing inequities, but refugee camps are sites of highly gendered energy access and use. Drawing upon empirical research of the experiences of Congolese refugees in Rwanda and internally displaced people in Nepal, Halford explores the extent to which improved energy access benefits men and women equally in those settings, how women and young girls negotiate gender and energy and ways in the narrative regarding refugees (particularly women) can be shifted to recognise women's agency.

Contribution of This Book

The main contribution of this book is that the chapters focus on the analysis of the complexities of inequality in the energy sector. The chapters posit their analysis from a gendered energy approach and from different multidisciplinary perspectives. Their enquiry is based on the recognition of concepts such as patriarchy and masculinity in the environment. Gendered democratic participation in the energy communities is also addressed. However, one of the aspects that is relevant to highlight is the dearth of theoretical and conceptual tools that contribute to understanding and explaining gender inequality in the energy sector. As such one of the central contributions of this book is the inclusion/application of intersectionality as an analytical tool to explain the social dimensions of gender inequality in the energy sector (Chaps. 4, 7, 8, 10, 11 and 12).

Along these lines, this book presents a new view of the different elements that may help to explain, and provide suggest solutions for, inequality in the energy sector.

References

Anditi, Ch., Musango, J., Smit, S., & Ceschin, F. (2022, June). Addressing gender dimension in energy innovation: A gender analysis framework for informal urban settlements in Africa, *Energy Research & Social Science,* 88, 102476. https://doi.org/10.1016/j.erss.2021.102476

Baruah, B. (2015). Creating opportunities for women in the renewable energy sector: Findings from India. *Feminist Econmics, 21*(2015), 53–76.

Ernest & Young. (2015). Women in Power and utilities Index 2015. Retrieved September 2, 2018, from http://www.ey.com/Publication/vwLUAssets/EY-women-in-power-and-utilities-index-2015/$FILEY/EY-women-in-power-and utilities-index-2015.pdf

European Institute for Gender Equality. (2012). Review of the implementation in the EU of area K of the Beijing platform for action: Women and the environment. In *Gender Equality and Climate Change Report*. Vilnius: EIGE.

Feenstra, M., & Özerol, G. (2021). Energy justice as a search light for gender-energy nexus: Towards conceptual framework. *Renewable and Sustainable Energy Reviews, 138*, 1106668.

Gray, G., Boyle, A., Francks, E., & Yu, V. (2019). The power of small-scale solar: Gender, energy poverty, and entrepreneurship in Tanzania. *Development in Practice, 29*, 26–39.

Johnson, W. O., Gerber, V., & Muhoza, C. (2019). Gender, culture and energy transitions in rural Africa. *Energy Research & Social Science, 49*(2019), 169–179.

Johnson, W. O., Yi-Chen, J., Knight, A., Mortensen, S., Aung, M., Boyland, M., & Resurection, B. (2020). Intersectionality and gender transitions: A review of gender, social equity and low carbon energy. *Energy Research & Social Science, 70*, 101774.

Lieu, J. et al. (2020, April 1). Three sides to every story: Gender perspectives in energy transition pathways in Canada, Kenia and Spain. *Energy Research &Social Science*, Elsevier. Retrieved October 24, 2021, from https://reader.elsevier.com/reader/sd/pii/S2214629620301262?token=189A264D89F1D590DA7CFB2B8C3C4F2EA44FA9DA4357D89799B435E91D974E882C8064FE0D3E5E772B73C1765969739C&originRegion=eu-west-1&originCreation=20211024141612

McCright, A., & Xiao, C. (2014). Gender and environmental concern: Insights from recent work and for future research. *Society & Natural Resources, 27*, 1109–1113.

Ryan, S. (2014). Rethinking gender and identity in energy studies. *Energy Research & Social Science, 1*, 96–105.

Søraa, R., Anfinsen, M., Foulds, C. H., Korsnes, M., Lagesen, V., Robinson, R., & Ryghaug, M. (2020). Diversifying diversity: Inclusive engagement, intersectionality, and gender identity in a European Social Sciences and Humanities Energy research project. *Energy Research & Social Science, 62*. https://doi.org/10.1016/j.erss.2019.101380

Sovacool, B. (2014). What are we doing here? Analysing fifteen years of energy scholarship and proposing a social science research agenda. *Energy Research & Social Science, 1*, 1–29.

Winther, T., Matinga, M. N., Ulsrud, K., & Standal, K. (2017). Women's empowerment through electricity access: Scoping study and proposal for a framework of analysis. *Journal of Development Effectiveness, 9*(3), 389–417. https://doi.org/10.1080/19439342.2017.1343368.

Winther, T., Ulsrud, K., & Saini, A. (2018). Solar powered electricity access: Implications for women's empowerment in rural Kenya. *Energy Research & Social Science, 44*, 61–74.

PART I

Women and Gender Inequality in Energy Policy

CHAPTER 2

Merging Local Content and Women's Economic Empowerment in Tanzania's Extractive Industries

Jesse Salah Ovadia

Introduction

With the passage of the Mineral Policy (URT, 2009), the Mining Act (URT, 2010) and the Petroleum Act (URT, 2015), Tanzania signalled a renewed interest in local content in its oil, gas and mining industries. This culminated with new local content regulations in petroleum and mining as well as three further laws in 2017, including most prominently the Permanent Sovereignty Act, which makes any agreement that does not fully secure the interests of Tanzanian citizens and the Tanzanian state illegal. However, in designing local content in oil, gas and mining, Tanzania has not made sufficient effort to design policies that will benefit all citizens. In doing so, it has continued a long-running global trend of ignoring the gender dimension of local content.

J. S. Ovadia (✉)
Department of Political Science, Windsor University, Windsor, ON, Canada
e-mail: Jesse.Ovadia@uwindsor.ca

N. Rocha Lawton, C. Forson (eds.), *Women and the Energy Sector*,
https://doi.org/10.1007/978-3-031-43091-6_2

This new approach to resource nationalism (Jacob & Pedersen, 2018; Poncian, 2021) is intended to address decades of natural resource extraction that has not benefitted the country or its citizens. Large-scale oil, gas and mining do not offer significant direct employment opportunities. Formal employment in oil, gas and mining extraction is traditionally male dominated. In Tanzania, formal employment in these sectors is estimated to be over 80 per cent male (NBS, 2014, p. 46; VETA, 2016, p. 20). However, the negative impacts of these industries, especially in terms of social disruption and dislocation, environmental degradation and loss of livelihood, are more likely to be felt by women (Eftimie et al., 2009; Keenan & Kemp, 2014; Oxfam, 2017; Scott et al., 2013). According to some, employment of women in large-scale mines may be even lower—closer to 10 per cent (Daley et al., 2018, p. 16, citing Nayopa, 2015 and UN Women, 2016b). Women may be more involved in artisanal and small-scale mining (ASM), representing a 'feminisation' of mining (Lahiri-Dutt, 2015). However, this sector, like large-scale mining, remains male dominated.

The African Mining Vision (AMV), formulated by the African Union, African Development Bank and United Nations Economic Commission for Africa, specifically calls for the empowerment of women in order to ensure gender equality in line with Goal 5 of the United Nations Sustainable Development Goals (SDGs) (AU, 2009). Progress towards gender equity and the empowerment of women is also included in the tentative framework for action to implement the AMV (ILO, 2021, p. 15). This chapter goes beyond the headline developments in local content development to explore the gender dimensions of economic empowerment and the promotion of new opportunities for women in Tanzania's oil, gas and mining industries. Local content and economic empowerment are often seen as separate areas of public policy in Tanzania; however, they can be brought together to provide the skills and experience necessary to participate in the industry directly and through the participation of women-led micro, small and medium-sized enterprises (MSMEs) in the extractive industries' value chains.

Direct employment and opportunities for MSMEs to supply goods and services to the extractive industries must involve not only economic empowerment policies but also the mandating and setting of targets for women's participation in Tanzania's emerging legal regime for local content promotion and the country's strategy for offering Technical and

Vocational Education and Training (TVET) and MSME development training to women.

While there is a mention of gender as a crosscutting issue in the Mineral Policy of 2009 and of gender mainstreaming in the National Energy Policy of 2015, there has not been a concerted effort to include gender considerations in recent legislation and regulations. The Mining Act of 2010 and Mining (Local Content) Regulations of 2018 (URT, 2018) similarly fail to mention gender or women. The Petroleum Act of 2015 mentions gender in relation to the programmes that companies must submit regarding the recruitment and training of Tanzanians but does not set targets or quotas for direct employment or procurement from women-led enterprises. The Petroleum (Local Content) Regulations of 2017 (URT, 2017b) also fail to mention women or gender. The local content framework overall remains silent on the question of women's economic empowerment (WEE). Therefore, based on desk review of relevant literature and the author's personal experience as a consultant working with the government of Tanzania, this chapter argues that the government must pursue a holistic approach to gender equality in legislation, regulation, policy, education and training in its emerging oil, gas and mining sectors in order to maximise the benefit from the extractive industries.

Gender in Tanzania's Extractive Industries

Research on gender and the extractive industries confirms that the benefits and risks of resource extraction are distributed unequally among different segments of the community. Men have greater access to benefits such as employment and income. Meanwhile, women are more vulnerable to the impacts of oil, gas and mining projects (Eftimie et al., 2009; IGF, 2021).

Hill et al.'s (2017) work shows that the impacts of mining, oil and gas projects are not gender-neutral. Extractive companies often fail in terms of adequate consultation and compensation for the loss of land and land-based livelihoods. Additionally, pollution can lead to chronic illnesses, increasing the women's unpaid care work. Finally, they write, forcing people off their land 'can increase workloads for women in providing for their families where women are traditionally responsible for meeting the subsistence needs of families and are no longer able to do so' (Hill et al., 2017, p. 4).

Few countries have laws governing resource extraction that fully mainstream gender equality. In fact, many fail to acknowledge women as active participants in resource extraction or to include women's participation—particularly in the formal workforce (IGF, 2021, p. 1). For this reason, the Intergovernmental Forum on Mining, Minerals, Metals and Sustainable Development (IGF) has called for the creation of national action plans to address gender equality in the mining sector, for governments to collect, track and analyse gender-disaggregated data, for the promotion of gender equality in supply chains through local content policies that explicitly address gender-based discrimination and to 'make sure that local content laws are sensitive to the plight of women as a marginalised segment of society,' which 'may involve measures such as quotas or support programmes targeting women entrepreneurs' (IGF, 2021, pp. 3–8). Collection of gender-disaggregated data as well as contract transparency are also seen as keys to supporting women in mining by the Natural Resource Governance Institute (NRGI) (Menard & Moses, 2021).

Beyond the gendered impacts of the extractive industries, there are what Bradshaw et al. describe as 'patriarchal power relations between men and women' and how they are 'produced and reproduced within extractive industries' (2017, p. 439). They look at 'extreme and exaggerated gender roles and relations' within the industry and especially at sites of extraction. They then contrast this with the 'supernormal profits' to be made, which they suggest 'encourage the development of "supernormal patriarchy"' (Bradshaw et al., 2017, p. 439).

These insights call attention to what Lahiri-Dutt calls the 'masculinist discourses of mining' (2012, p. 193). In response, she argues that there is a need to 'engender' mining. While mining evokes large machines, mechanised processes such as drilling and blasting, and what might be thought of as men's work, Lahiri-Dutt pushes back against the association of the tools and machinery of mining with masculine characteristics and gendered meanings (2012, p. 194). The same is true—perhaps truer—of oil and gas, particularly offshore oil and gas extraction.

There are numerous cultural issues that manifest at mining sites and lead to the employment of very few women in mines as well as in the broader fields of geology, mining, plant processing, metallurgy and engineering (Lahiri-Dutt & Macintyre, 2006). As a result, in countries like Tanzania, the main jobs for women are sweeping, cleaning, catering and office work, with few women employed in technical, professional or managerial positions (ILO, 2006, p. 695).

Pugliese argues:

> Mining companies of the past and present have promoted specific roles for men and women through their management practices. In DR Congo, first colonial and then state-owned companies naturalised the role of men as employees and breadwinners. At the same time, women were assigned responsibility for reproduction and were understood as being financially dependent on men's salary, either that of their fathers or husbands. By contrast, some LSM (large-scale mining) companies today support gender equality programmes, mainly to improve their corporate reputation. (Pugliese, 2021, p. 1)

The situation may be somewhat better when it comes to ASM given that Lahiri-Dutt has suggested that a 'feminisation of mining' is beginning to occur as more women enter mining with 'an expanded notion of mining as a livelihood' (2015, p. 523). She argues that these initiatives are occurring as industrialised mining mainstreams gender and civil society pushes for change with new policy initiatives. However, when it comes to oil and gas, there is no evidence that any such transition has begun. According to Lange and Wyndham, as enlightened a company as Norway's Equinor has failed in Tanzania to make its investment in the country's gas sector gender sensitive or to alter the fact that the main beneficiaries of that investment are men (2021, p. 1).

Pushing back against Lahiri-Dutt, Bradshaw et al. contend:

> In extractive industries, including small-scale informal artisanal mining, the dangerous contexts in which miners operate promote a particular set of male–male relations among those working the mines. These are based on both co-operation and conflict. Emphasising the (hyper-)masculine characteristics of work undertaken in dangerous contexts such as construction or mining may be a way for men to build collective self-esteem about their work (Saucedoa & Morales, 2010). The solidarity that this creates protects men in dangerous work conditions. It also gives men power over women because it is built on devaluing the contributions of others, in contexts where men themselves may have little other power. (2017, p. 445)

Neither the involvement of a company that comes from a country that prides itself on promoting gender equality nor the formal gender equality enshrined in Article 9 of Tanzania's constitution (URT, 1995, cited in Daley et al., 2018, p. 11) is sufficient on its own to create better outcomes without an approach that focuses on interventions specifically targeted at the extractive industries. In fact, Daley et al. note that Tanzania has measures to prohibit discrimination based on sex, gender or marital status and to mainstream gender perspectives within policy-making in Development Vision 2025, the National Women and Gender Development Policy and National Strategy for Gender Development of 2000, the 2004 Employment and Labour Relations Act and National Employment Policy of 2008 and the Mkukuta II (2018, p. 11). There is then something important in understanding what Bradshaw et al. call 'supernormal patriarchal relations' and addressing the production and reproduction of these relations in order to 'promote gender equality and natural resource justice, as part of an agenda to redistribute wealth gains from natural resource extraction' (2017, p. 439).

Such an approach would at a minimum require revising the policies that govern the extractive industries. However, as will be demonstrated below, creating better outcomes goes well beyond revising legislation. The remainder of this chapter will look at opportunities to merge local content and economic empowerment in Tanzania, with a focus on TVET, MSME development, direct employment in the gas sector and indirect employment through policies to encourage local procurement in the extractive industries.

Local Content in Tanzania

Local content policies (LCPs) are meant to increase the utilisation of national human and material resources in the domestic resource extraction and to domicile in-country extractive industry-related economic activity that was previously located abroad (Ovadia, 2014). These policies have gained considerable currency as a strategy to stimulate natural resource-based economic development (Hilson & Ovadia, 2020). LCPs promote the participation of local companies in economies otherwise geared for the export of raw materials and the employment of national citizens in industries that typically rely heavily on expatriate labour. Encouraging international firms to procure goods and services from local companies, or to work in partnership with local companies, allows for the development of backward, forward and sideways linkages along resource value chains.

Local content has existed in different forms since Tanzania's independence. Local content is a form of industrial policy similar to what many countries have used to foster industrial development. During the period of *ujamaa* (African socialism), the mining sector was largely under the direct control of the state. However, the 1979 Mining Act did require applicants to present a plan for local procurement of goods and services (Lange & Kinyondo, 2016, p. 1097). The Mining Act of 2010 had few concrete clauses related to local content, although it did reserve small-scale mining operations exclusively for Tanzanian citizens and companies (URT, 2010, Article 8).

The 2017 Miscellaneous Amendments Act (URT, 2017a) amended the Mining Act of 2010 to put in place some local content provisions and require mining firms seeking new licenses to submit a local content plan and to give preference to goods produced or available in Tanzania and to services provided by Tanzanian citizens or local companies (Woodroffe et al., 2017, p. 11). In 2018, the Mining (Local Content) Regulations Act was passed. This act expanded on the same provisions, explaining in more detail what was required for the local content plan, employment and training sub-plan, succession plan and research and development sub-plan. It also dealt with the promotion of local insurance services, legal services and financial services for the mining industry. Finally, it addressed monitoring and evaluation and the creation of a common qualification system for local suppliers. In an attached schedule, the regulations specify minimum local content levels in goods and services.

In the gas sector, Tanzania's approach to local content has evolved since the government produced a draft local content policy in 2014. Initially there were few concrete or 'hard' regulations. Instead, there was a more 'soft' or voluntary approach (Ovadia, 2016a, pp. 27–28). Although the Petroleum Act of 2015 as well as the provisions of the Production Sharing Agreements contained some more concrete requirements, this approach really only changed after 2016 with the publication of The Petroleum (Local Content) Regulations of 2017, which set out minimum local content targets for employment of Tanzanians as well as for the use of Tanzanian goods and services. As with the local content regulations for mining, these levels were set in a schedule at the end of the regulations. As Scurfield et al. note, however, the definitions of local company and local goods and services as well as the power granted to the Minister of Energy and Minerals to grant exemptions weaken their impact (2017, p. 7).

While in some ways it is too early to know what impact the new legal regime for local content in petroleum and mining has had, especially with the complications caused by the impact of COVID-19. In the case of coal mining, a key difference from oil and gas was that the implementation of a ban on coal imports was also meant to enhance linkages and promote local coal producers. However, the ban had numerous unintended consequences that may have ultimately harmed industrial development more than it assisted it (Jacob, 2020b). Additionally, there has been somewhat of a 'retreat' from the resource nationalist approach to local content in recent years with some watering down of what it means to be a Tanzanian company (Jacob, 2020a, p. 205).

Recognising that oil, gas and mining are never going to be significant direct employers, successful LCPs encourage linkages to the service sector and beyond. This can also help diversify economies that are dependent on extraction and help prepare for a sustainable energy transition in a post-carbon future. For this reason, LCPs overlap with economic empowerment policies, as implemented in many countries in southern Africa, including Tanzania.

Economic Empowerment in Tanzania

Economic empowerment in Tanzania is a response to the country's historical context of colonialism and experience with socialist development. The main objective of empowerment policies is to help citizens of Tanzania access opportunities to participate effectively in economic activities in all sectors of the economy. Tanzania's National Economic Empowerment Act (URT, 2004) was written to promote and facilitate economic initiatives aimed at empowering Tanzanians by providing them economic opportunities to 'generate wealth' and 'boost the small and medium enterprise sector' in order to 'bring about sustainable affirmative action and facilitate genuine and positive economic empowerment to the population of Tanzanians' (2004, Preface). Thus, economic empowerment is defined in such a way as to include the promotion of local participation in the economy.

Fisher (2007) contends that there are existing inequalities and forms of discrimination or social exclusion ignored by the Tanzanian state. The institutionalisation of mineral titles conceals social and power relations that perpetuate highly unequal access to resources. Institutionalisation defines who gets involved in mining and who is marginalised. Women, to the greater extent, are discriminated against and marginalised. In Tanzania, artisanal mining is also an area in which women are able to transgress boundaries and accrue wealth. However, due to historic discrimination and cultural norms, women generally do not own pits, but they may be co-workers with their husbands. Production and revenue from diamond mining are sporadic and are, in principle, divided equally between members of a pit (although a woman's share may be taken by her husband because men traditionally sell the diamonds). Another area women are able to accrue wealth is their success in the service industry where they have the ability to finance mining operations.

The National Economic Empowerment Act established the National Economic Empowerment Council (NEEC) to carry out empowerment initiatives and implement the National Economic Empowerment Policy. NEEC's Empowerment and Facilitation Department carries out a number of initiatives, such as its programme for women working in ASM, which is a joint project with the Tanzania Women Miners Association (TAWOMA). Programmes for youth, such as the Young Graduate Program, Youth Business Tanzania and My Business, provide entrepreneurship training and coaching, as well as access to finance. Many of the department's activities are in Lindi and Mtwara, close to Tanzania's gas resources, as well as the mining areas of Mwanza and Geita. A separate department, the Empowerment Funds Department, also operates programmes for women and youth (Ovadia, 2016b, pp. 26–27).

In 2015, near the end of President Kikwete's administration, NEEC was given authority to regulate local content. However, NEEC's work on gender and economic empowerment was never integrated into its work on local content. The two aspects of the organisation's work are largely separate because local content in oil, gas and mining is not seen to be the same thing as economic empowerment.[1] When the local content regulations for petroleum and mining were finalised, NEEC's role was not formally addressed, perhaps because local content and economic empowerment are seen to be separate. NEEC's exclusion in my view creates additional confusion in terms of how the various agencies involved in LCPs are meant to interact and coordinate their efforts—something that was already a problem in Tanzania (Ovadia, 2017).

Other international donors work on WEE through the programmes they are running on skills development for oil and gas. In this way, they actively support and encourage opportunities for women. Both the government of Tanzania (through the Ministry of Health, Community Development, Gender, Elders and Children as well as NEEC) and international donors (especially the Enhancing Employability through Vocational Training (EEVT), Skills for Oil and Gas Africa (SOGA) and

[1] The author worked as a consultant with the NEEC in 2017. This insight is based on his personal experience.

Tanzania Local Enterprise Development (TLED) projects) are actively promoting opportunities for women. These initiatives emerged because these organisations have a mandate of gender mainstreaming for all their initiatives. Yet when local content experts advise the government on local content regulations, they often overlook these initiatives and rarely mention the gender dimension. These initiatives, in turn, are largely unaware of LCPs in the oil, gas and mining sector as well as their connection to WEE in the extractive industries. As a result, in creating the legal framework for local content, Tanzania largely missed out on the opportunity to bring in a gender dimension to LCPs, and in creating these programmes, donors missed out on the opportunity to support local content development.

Merging Empowerment and Local Content in Tanzania

Given the overlap between local content and economic empowerment, LCPs should consider economic empowerment in general and WEE in particular in order to maximise the potential developmental impact of the policies and of resource extraction overall. Noting that men tend to benefit more than women from oil and gas industries, a 2017 report from the African Natural Resources Centre (ANRC) of the African Development Bank on WEE argues, '… there are many potential benefits in promoting women's economic empowerment within O&G industries, such as more inclusive development, improved welfare for families, greater diversity and productivity, better corporate images of firms as gender champions, as well as more wealth and jobs staying in the host country' (ANRC, 2017, p. 7). Increasing the local participation, or local content, of women in oil, gas and mining involves not just encouraging direct employment but a comprehensive and 'hard' approach to local content and economic empowerment involving quotas and targets for women's participation in the sector.

Such an approach is not without precedent. Tanzania already pursues a 'hard' approach to local content overall. Additionally, as Daley et al. (2018) note, mandatory female representation and quotas for female representation are already in place in Tanzania. These provisions include requiring that 30 per cent of all seats in Parliament must be held by women, that at least three of the seven members of each Village Land Council shall be women, that land adjudication committees should include at least four female members out of nine and that at least 25 per cent of seats on village councils be held by women (Daley et al., 2018, p. 11). While in practice meeting these targets is currently a challenge in many village institutions, the precedent exists for this approach to WEE.

As the experience with unmet quotas on village councils suggests, mandatory targets require a broader set of conditions to be in place that allow the targets to be met. This is where TVET can be helpful. As Fox argues, 'When females have equal access to education and training, *they can compete for higher-productivity jobs and enhance the pool of skilled labour available*' (Fox, 2016, p. 13 emphasis in original).

TVET was historically thought of as education for the purposes of work. Ashipala shows that, in colonial Namibia, a policy of vocational education was initially advanced by the German colonial administration in order to solidify German domination and the idea that natives of the country are servants whose virtue is understood in terms of their capacity to work for the colonial economy (Ashipala, 2020, p. 2, citing Zollman, 2020). Tanzania's skills sector is also a source of conflict over different objectives, resource misallocation in the public sector and tax avoidance in the private sector (Andreoni et al., 2021). However, TVET is nevertheless an important approach to improving the chances of both males and females entering formal employment, reducing their probability of working in the informal economy or in agriculture and increasing their overall earnings. In fact, the earnings returns 'are substantially higher and statistically significant for females than males' (Joseph & Leyaro, 2019, p. 1).

Another important initiative is the government's National Skills Development Strategy (NSDS). The NSDS seeks to increase the supply of skills for industries with high potential for growth and job creation, including the energy sector. The strategy is supported by the World Bank under the Education and Skills for Productive Jobs (ESPJ) Programme (VETA, 2016). Unfortunately, initiatives from various ministries to address health, education, community development or other areas generally lack coordination among themselves and especially with government officials involved in energy and minerals. Daley et al. suggest that this is because gender is often equated with 'women's issues' (2018, p. 11), leading to initiatives related to promoting gender equality being ignored by those not directly involved. Additionally, there are numerous institutions and agencies involved in both TVET and MSME development. These include the Ministry of Education, Vocational Education and Training Authority

(VETA), NEEC, Ministry of Energy and Minerals, Ministry of Industry, Small Industries Development Organisation (SIDO), Tanzania Entrepreneurship and Competitiveness Centre (TECC), Tanzania Investment Centre (TIC), Tanzania Industrial Research Development Organisation (TIRDO), Tanzania Engineering and Manufacturing Design Organisation (TEMDO), Tanzania Chamber of Commerce, Tanzania Bureau of Standards (TBS), Industry and Agriculture (TCCIA), Tanzania Trade Development Authority (TANTRADE), Tanzania Private Sector Foundation (TPSF), Centre for Agricultural Mechanisation Rural Technology (CAMARTEC) and more (Anderson, 2017; Ovadia, 2017). There are also numerous international agencies and donors.

Joseph and Leyaro argue, using data from the 2014 Integrated Labour Force Survey (ILFS), that the effect of TVET for women is greater because there are more women who are either unemployed or employed in the informal sector (2019, p. 1). The gender divide at more senior levels of formal employment is also quite stark. Overall, over 82 per cent of legislators and administrators as well as 83 per cent of company directors and corporate managers are male (NBS, 2014, p. 46). Such disparity can only be addressed through policy intervention.

Improvements in TVET must occur in concert with reforms in higher education. In the case of South Africa, Cruise (2011) shows there has been tremendous change in the degree to which mining engineering has undergone a gender and racial transformation due to empowerment policies. Universities in South Africa have made great strides over an extremely short period of time in promoting gender transformation. This has produced a notable increase in the number of female mining engineering graduates. Over a four-year period, an average of 20 per cent of all mining engineering graduates at the University of Pretoria were women. Over the same period at the University of the Witwatersrand, 35 per cent of all mining engineering graduates are women. While Tanzania's tertiary education system is not as well funded as South Africa's, institutional change to address women's participation in petroleum and mining-related fields is important.

Job creation in the extractive industries requires not only TVET and higher education to empower women to compete for direct employment but also skills development and effective supplier development programmes. Many of the donor-supported and NEEC-supported programmes mentioned above such as EEVT, SOGA and TLED already connect TVET with enterprise development. According to United Nations Economic Commission for Africa (UNECA), such programmes 'further entrench the participation of micro, small and medium-sized enterprises along the value chains' and allow for non-resource sectors to promote further industrial growth and technological advancements in other sectors (UNECA, 2020, p. 53, citing SADC, 2019, p. 60).

MSME support can have a major impact on women in their everyday lives. Kotsadam and Tolonen (2016) employ a multi-country quantitative analysis of local employment impacts for men and women in large-scale mining in the African continent with a focus on women's labour market opportunities. They contrast their findings with the effects for men. According to them, access to employment when a mine opens significantly improves women's lives and opportunities. Overall, the impacts on employment from industrial mining are not as clear-cut. While some scholars argue that extractive industries may have a negative impact on women, these arguments may be less applicable in the African context (Kotsadam & Tolonen, 2016, p. 325). More importantly, the issue of what impact LCPs have on women's employment and livelihoods is not well studied.

Another understudied area would be the link between the feminisation of mining in general and MSME development. In recent years, both male and female ASM operators are increasingly being taken seriously as business people. According to Kinyondo and Huggins, 'many formalisation efforts aim to facilitate ASM to become more productive and invest surplus profit in various forms of capital (human, technological, financial, etc.) to "grow" businesses and move towards MSM operations' (2020, p. 759).

Esteves argues:

> To ensure that women benefit from enterprise development opportunities arising from company operations, management approaches need to be examined from a gender-aware development perspective. Potentially, major resources companies provide a significant opportunity to stimulate the participation of women in local economies—by integrating SMEs with female ownership and employment into the corporate supply chain, and building the capacity of these SMEs to become competitive. However, economic growth does not affect men and women equally. Genderless approaches to SME development will ignore the complexities that underlie exchanges between women and men. (2011, p. 136, citing Srinivasan & Mehta, 2003)

According to Anderson, MSMEs employ between 3 and 4 million people in Tanzania, accounting for 20–30 per cent of the total labour force and contributing between 35 and 45 per cent of the country's gross domestic product (2017, p. 1). As UNECA notes, these enterprises need assistance to supply to oil, gas and mining firms: 'As a result of the capital intensity and technical nature of the mineral beneficiation value chains, structured interventions from government in support of micro, small and medium-sized enterprises are critical to enable their participation' (2020, p. 51).

While the UNECA report doesn't specifically mention empowering women-led MSMEs, the ANRC does suggest that specific policies are also needed to ensure, as they put it:

> … that women benefit equally from compensation, community programs and royalties (social investments) from these industries; promoting women's direct, waged employment in the (oil and gas) sector; encouraging or mandating suppliers and subcontractors to employ women (indirect employment); and also enabling women entrepreneurs to gain access to (oil and gas) industries as suppliers and subcontractors. A prerequisite for all these measures is to enable women and girls' equal access to education, including STEM subjects, skills training and capacity development, along with their male counterparts. (ANRC, 2017, p. 7)

According to the ANRC, these policies may take the form of and mandatory requirements for employers and subcontractors to employ both women and women-owned firms and incentives for such initiatives. The promotion of procurement from women-owned or women-led firms is another important aspect of approaching value chains from a gender perspective (ANRC, 2017, p. 8). Such an approach suggests the possibility of targets and incentives being written into local content legislation and regulations.

In its flagship report on women in mining, the ILO now recommends a series of policies designed to encourage direct employment of women in large-scale and artisanal and small-scale mining. They write:

> Governments play a key role in ensuring that SMEs in the mining sector can continue to generate opportunities for decent and productive jobs for both women and men in the future. This requires a combination of local procurement policies and well-designed, gender-sensitive and inclusive SME policies in alignment with national circumstances and the characteristics of the

> mining sector in the country. Such policies should increasingly focus on enhancing the capabilities of SMEs to innovate, adopt new technologies, and develop new products and services for the mining industry in order to address social, economic and environmental goals. (ILO, 2021, p. 29)

Although direct employment in oil and gas is not going to be significant for Tanzania, any skilled jobs that do go to citizens are well-paid and can potentially support numerous others. In the mining industry, there is more potential for significant employment as ASM operations are formalised and if mining firms embrace women across all job types. According to Jenkins, over one-third of all mine workers globally are women (2014, p. 331). A key aspect of creating good jobs is making sure women are employed on equal terms with men. Bashwira et al. note that in the mining industry, laws and policies relating to women's employment tend to focus on health and safety 'protection' for women, rather than on gender equality (2014, p. 112). Yet it is not enough to legislate equal pay and equal access to employment if more isn't done to address the masculine culture of mining and other factors that also lead to women's exclusion.

Writing with regard to the Democratic Republic of Congo (DRC), Pugliese argues that more research is needed on women in large-scale mining. While there are many studies on women in ASM, few studies focus on the formal employment of women in large-scale mining (Pugliese, 2021, p. 2). In the case of DRC, Pugliese finds mining companies hinder female work, often unintentionally, 'through recruitment processes, subcontracted work, and masculine workplace policies that discourage women's career and favour men in the role of the family's main provider.' In other cases, discrimination persists because of the representation of the mining context as a dangerous working environment where women must be protected, cannot perform even safe mechanised tasks and are seen as weak, fragile and 'naturally lacking technical skills' (Pugliese, 2021, p. 7).

The stereotypes and masculine culture of oil, gas and mining operations present a formidable challenge. However, it is a challenge that can be overcome with the right policy and genuine engagement from mining firms. Moving beyond direct employment, it is ultimately in the provision of goods and services that there is the most potential for women's economic empowerment and increasing the participation of women in the extractive industries in Tanzania. It is for this reason that empowerment must be merged with 'hard' local content targets and interventions.

Fortunately, when it comes to Tanzania's legal regime for local content, there is precedent for increasing women's participation in procurement in the Public Procurement Act, which was amended in 2016 to require procuring entities to select procurement methods that achieve certain social objectives, including setting a specific percentage of procurement from special groups including women, youth, elderly and persons with disabilities. However, the language is not very strong and is difficult to operationalise in its current form (URT, 2016). There is an ongoing movement to specify that 30 per cent of all contracts be given to women-led companies.

Further, the 2016 amendments require unbundling of works to enable higher levels of local participation in the bid. The Public Procurement Regulatory Authority (PPRA) is granted wide powers to carry out investigations, request relevant information and documents, call witnesses and enforce the Act with stiff penalties. Nevertheless, the precedent exists to set targets and quotas for the employment of women in all areas of resource extraction and for procurement contracts with women-owned or women-led Tanzanian companies. In fact, TAWOMA has lobbied along similar lines for quotas of mining land to be dedicated to small-scale women miners (Eftimie et al., 2009, p. 28).

Conclusion

A disconnect between new frameworks put in place to regulate local content in the extractive industries and already-existing programmes and policies to promote women's economic empowerment has exposed a blind spot for government officials in Tanzania. As a result, the country's LCPs do not sufficiently address TVET or MSME development, even though these are crucial areas when it comes to encouraging local participation through direct employment and the provision of goods and services to the industry.

The disconnect impacts the participation of all Tanzanians in the oil, gas and mining sector; however, the impact is greater for women given their increased vulnerability to the risks of resource extraction and unequal share of the benefits due to patriarchal social relations and the masculinised culture of resource extracting industries.

Revisions are required to Tanzania's legal framework for local content as well as to the approach responsible government officials take to regulating oil, gas and mining to include gender equality as a priority area. Merging local content and WEE may require hard targets and quotas for women's participation in these industries. These targets can apply to the number of people trained and employed by these industries as well as to the number of women-led MSMEs that are offered entrepreneurial support as well as the number and value of contracts for supplying goods and services. It will also require coordination between the various government agencies and donor institutions that work in these areas.

Women have taken a more prominent role in ASM; however, increasing the participation of women in large-scale mining, in oil and gas and in procurement is an overlooked area when it comes to maximising the impact of local content and aligning the extractive sectors with the priorities outlined in the AMV and SDG 5.

Large oil, gas and mining firms also have a role to play in applying the principles of gender equality not only in their hiring practices but also to their training, enterprise development and procurement practices. Lahiri-Dutt suggests that gender equality can be applied by mining firms both vertically and horizontally across the spectrum of their business activities (2011, p. 7). A vertical approach involves the creation of a separate unit or department to deal with gender equity issues. Most firms in Tanzania have already adopted such an approach. However, the horizontal approach requires the application of a 'gender lens' to every aspect of an organisation's work, including procurement.

Taken together, more robust government regulations and a concerted effort by extractive firms can promote gender equality in Tanzania's extractive industries. Given the growing importance of these sectors to the country's economy, merging the programmes and policies of both the government and private firms when it comes to LCPs and WEE will greatly increase their positive developmental impact, offering new opportunities for economic transformation and natural resource-based development.

'Portions of this chapter were first published in the article Addressing gender inequality through employment and procurement: Local content in Tanzania's emerging gas industry, The Extractive Industries and Society, 9, 2002, 101028.'

References

African Natural Resources Centre (ANRC). (2017). *Women's economic empowerment in oil and gas industries in Africa*. African Development Bank.

African Union (AU). (2009). *Africa mining vision*. African Union.

Anderson, W. (2017). Factors affecting Small & Medium Enterprises (SMEs) start-up and growth in Tanzania. *The Pan-African Journal of Business Management, 1*(1), 1–26.

Andreoni, A., van Huellen, S., Katera, L., & Jahari, C. (2021). *Breaking the impasse: Aligning incentives to address corruption in Tanzania's skills sector. Anti-corruption evidence briefing paper 014.* SOAS University of London.

Ashipala, S. N. (2020). Technical and vocational education and the place of indigenous labour in the mining industry of Namibia, 1970–1990. *Journal of Southern African Studies, 47*(1), 127–142.

Bashwira, M., Cuvelier, J., Hilhorst, S., & van der Haar, G. (2014). Not only a man's world: Women's involvement in artisanal mining in eastern DRC. *Resources Policy, 40*, 109–116.

Bradshaw, S., Linneker, B., & Overton, L. (2017). Extractive industries as sites of supernormal profits and supernormal patriarchy? *Gender & Development, 25*(3), 439–454.

Cruise, J. A. (2011). The gender and racial transformation of mining engineering in South Africa. *Journal of the Southern African Institute of Mining and Metallurgy, 111*(4), 217–224.

Daley, E., Lanz, K., Mhinda, A., Driscoll, Z., Ndakaru, J., Grabham, J., Kereri, E., & Mbise, E. (2018). *Gender, land and mining in pastoralist Tanzania.* WOLTS Research Report No. 2. Mokoro Ltd & HakiMadini.

Eftimie, A., Heller, K., & Strongman, J. (2009). *Gender dimensions of the extractive industries: Mining for equity* (Extractive Industries and Development Series #8). The World Bank.

Esteves, A. M. (2011). Women-owned SMEs in supply chains of the corporate resources sector. In K. Lahiri-Dutt (Ed.), *Gendering the field: Towards sustainable livelihoods for mining communities* (pp. 133–143). ANU Press.

Fisher, E. (2007). Occupying the margins: Labour integration and social exclusion in artisanal mining in Tanzania. *Development and Change, 38*(4), 735–760.

Fox, L. (2016). *Gender, economic transformation and women's economic empowerment in Tanzania.* Overseas Development Institute.

Hill, C., Madden, C., & Collins, N. (2017). *A guide to gender impact assessment for the extractive industries.* Oxfam Australia.

Hilson, A. E., & Ovadia, J. S. (2020). Local content in developing and middle-income countries: Towards a more holistic strategy. *The Extractive Industries and Society, 7*, 253–262.

IGF. (2021). *Gender in mining governance: Opportunities for policy-makers.* Intergovernmental Forum on Mining, Minerals, Metals and Sustainable Development.

ILO. (2006). *National report for promoting the linkage between women and employment and the reduction of child labour*. Dar es Salaam: ILO Gender Promotion Programme.

ILO. (2021). *Women in mining: Towards gender equality*. International Labour Organization.

Jacob, T. (2020a). A political economy of resource nationalism and revived state-owned enterprises in Tanzania's coal sector. PhD Dissertation, Roskilde University.

Jacob, T. (2020b). When good intentions turn bad: The unintended consequences of the 2016 Tanzanian coal import ban. *The Extractive Industries and Society, 7*, 337–340.

Jenkins, K. (2014). Women mining and development: An emerging research agenda. *The Extractive Industries and Society, 1*(2), 329–339. 10.1016/j.exis.2014.08.004

Jacob, T., & Pedersen, R. H. (2018). New resource nationalism? Continuity and change in Tanzania's extractive industries. *The Extractive Industries and Society, 5*, 287–292.

Joseph, C., & Leyaro, V. (2019). *Gender differential effects of technical and vocational training: Empirical evidence for Tanzania* (CREDIT Research Paper, No. 19/04). The University of Nottingham, Centre for Research in Economic Development and International Trade.

Keenan, J. C., & Kemp, D. L. (2014). *Mining and local-level development: Examining the gender dimensions of agreements between companies and communities*. Centre for Social Responsibility in Mining, The University of Queensland.

Kinyondo, A., & Huggins, C. (2020). 'Centres of excellence' for artisanal and small-scale gold mining in Tanzania: Assumptions around artisanal entrepreneurship and formalization. *The Extractive Industries and Society, 7*, 758–766.

Kotsadam, A., & Tolonen, A. (2016). African mining, gender, and local employment. *World Development, 83*, 325–339.

Lahiri-Dutt, K. (2011). Introduction: Gendering the Masculine Field of Mining for Sustainable Community Livelihoods. In K. Lahiri-Dutt (Ed.), *Gendering the field: Towards sustainable livelihoods for mining communities* (pp. 1–19). ANU Press.

Lahiri-Dutt, K. (2012). Digging women: Towards a new agenda for feminist critiques of mining. *Gender, Place & Culture, 19*(2), 193–212.

Lahiri-Dutt, K. (2015). The feminisation of mining. *Geography Compass, 9*(9), 523–541.

Lahiri-Dutt, K., & Macintyre, M. (2006). Where life is in the pits (and else where) and gendered. In K. Lahiri-Dutt & M. Macintyre (Eds.), *Women miners in developing countries: Pit women and others* (pp. 1–24). Routledge.

Lange, S., & Kinyondo, A. (2016). Resource nationalism and local content in Tanzania: Experiences from mining and consequences for the petroleum sector. *The Extractive Industries and Society, 3*, 1095–1104.

Lange, S., & Wyndham, V. (2021). Gender regulation and corporate social responsibility in the extractive sector: The case of Equinor's social investments in Tanzania. *Women's Studies International Forum*. 84102434-10.1016/j.wsif.2020.102434

Menard, A., & Moses, E. (2021). *Gender and extractive governance: Lessons from existing legal and policy frameworks*. Natural Resource Governance Institute.

NBS. (2014). *Tanzania integrated labour force survey 2014*. National Bureau of Statistics.

Ovadia, J. S. (2014). Local content and natural resource governance: The cases of Angola and Nigeria. *The Extractive Industries and Society, 1*, 137–146.

Ovadia, J. S. (2016a). Local content policies and petro-development in Sub-Saharan Africa: A comparative analysis. *Resources Policy, 49*, 20–30.

Ovadia, J. S. (2016b). *Local content situational analysis in Tanzania in the mining and gas sectors*. African Development Bank.

Ovadia, J. S. (2017). *Local content in Tanzania's gas and minerals sectors: Who regulates? Briefing*. Chr. Michelsen Institute.

Oxfam. (2017). *Position paper on gender justice and the extractive industries*. Oxfam America.

Poncian, J. (2021). Resource nationalism and community engagement in extractive resource governance: Insights from Tanzania. *Review of African Political Economy* (online).

Pugliese, F. (2021). Mining companies and gender(ed) policies: The women of the Congolese Copperbelt, past and present. *The Extractive Industries and Society, 8*(3).

Scott, J., Dakin, R., Heller, K., & Eftimie, A. (2013). *A survey and analysis of the gendered impacts of onshore oil and gas production in three developing countries*. The World Bank.

Scurfield, T., Woodroffe, N., & Olan'g, S. (2017). *Localizing Tanzania's gas sector: Determining optimal policies for an emerging producer*. National Resource Governance Institute.

UNECA. (2020). *Strategies and policies for the integration of micro, small and medium-sized enterprises into the industrialization process in Southern Africa*. United National Economic Commission for Africa.

URT. (2004). *National economic empowerment act*. United Republic of Tanzania.

URT. (2009). *The mineral policy*. United Republic of Tanzania.

URT. (2010). *The Mining Act*. United Republic of Tanzania.

URT. (2015). *The Petroleum Act*. United Republic of Tanzania.

URT. (2016). *The Public Procurement (Amendment) Act*. United Republic of Tanzania.

URT. (2017a). *The Permanent Sovereignty Act.* United Republic of Tanzania.
URT. (2017b). *The petroleum (local content) regulations.* United Republic of Tanzania.
URT. (2018). *The mining (local content) regulations.* United Republic of Tanzania.
Vocational Education and Training Authority (VETA). (2016). Vocational skills needs survey in oil and gas supply chain. Vocational Education and Training Authority and Transport Resource Centre Limited. Draft. September, 2016.
Woodroffe, N., Genasci, M., & Scurfield, T. (2017). *Tanzania's new natural resources legislation: What will change?* National Resources Governance Institute.

CHAPTER 3

Women's Empowerment Through Electrification: What Is the Evidence from the Indian Subcontinent?

Debajit Palit, Mini Govindan, and Rashmi Murali

Introduction

Access to modern energy is a critical enabler for economic and human development. Of particular importance is the growing recognition of the role energy plays as an enabler to improve gender equality and social inclusion. The gender–energy nexus plays out broadly in two ways—the life-enhancing aspects of energy and the disproportionate effects of its lack, as has been validated by literature (Dutta et al., 2017; Wilhite, 2017; Blackden & Wodon, 2006). Reduced drudgery, decreased poverty, increased income, enhanced livelihood opportunities, better entertainment and communication facilities, and enabling conditions for children

D. Palit (✉)
NTPC School of Business, Noida, India
e-mail: debajit.palit@nsb.ac.in

M. Govindan • R. Murali
The Energy & Resources Institute (TERI), New Delhi, India
e-mail: gmini@teri.res.in; rashmi.murali@teri.res.in

N. Rocha Lawton, C. Forson (eds.), *Women and the Energy Sector*,
https://doi.org/10.1007/978-3-031-43091-6_3

to study are some of the positive outcomes of reliable access to electricity in people's lives, particularly on rural women (Dinkelman, 2011; Grogan & Sadanand, 2013; Standal & Winther, 2016; Sovacool et al., 2013). On the other hand, energy poverty affects rural people adversely, particularly rural women. Without adequate light or modern appliances, women have to endure drudgery and issues related to safety and mobility, among others (Rewald, 2017). Reliable electricity service delivery also enriches the productive uses of energy and benefits both women and men for more equitable economic growth and employment opportunities (Cecelski, 2000; Clancy et al., 2012). It also contributes to quality public services such as potable water, quality health and education, and access to information, communication and entertainment (Winther, 2008; Cecelski, 2006).

Several energy projects, programmes and policies are increasingly realising the importance of operationalising gender approaches for improved performance and increased benefits for both women and men (ADB, 2015; ILO, 2010; PIDG, 2018). A new dimension in the gender–energy paradigm and women's empowerment are their role as change agents along the energy value chain. Women's energy entrepreneurship is being recognised as an effective way of introducing contextually appropriate energy products and services to underserved and unserved communities (Dutta, 2018; SEforAll., 2018). Evidence from the literature suggests that women's involvement in the energy value chain has led to better adoption of technologies, better management, customer outreach at different stages of the energy value chain and higher sales (Smith & Dutta, 2011; Kohlin, Sills, et al., 2011b; Shankar et al., 2018; Glemarec et al., 2016). For instance, an Indian energy company, owned by a women entrepreneur, developed a women entrepreneurship model in the state of Rajasthan where more than 12,000 rural women have been trained as *Saral Jeevan Sahelis* or last-mile delivery agents for selling off grid solar products (Frontier Markets, n.d.). In yet another case, a not-for-profit organisation, through its global campaign on women entrepreneurship, trained rural women from various countries to be barefoot solar engineers and have been promoting off-grid solar home systems (Barefoot College, 2016). Similarly, in Bangladesh, a large not-for-profit organisation with the goal to promote and supply renewable energy technologies to rural households, involved women in the installation and servicing of the solar energy systems (Arthur, 2010; Palit, 2013). They also involved women in the assembly of solar products, a job with flexible working hours to help women earn money on the side while managing their caregiving roles.

There are also other similar examples from developing Asia and sub-Saharan Africa that have promoted women as change agents (Dutta, 2018; UNFCCC, n.d.). Furthermore, SEforALL's global initiative, 'People-Centred Accelerator on Advancing Gender Equality, Social Inclusion, and Women's Empowerment in the Sustainable Energy Sector', has one of its focus areas directed towards empowering women involved in energy service delivery, in terms of autonomy and more significant role in decision-making (SEforALL, 2017). Historical studies and the scriptures from the Indian subcontinent also indicate that women were not excluded in any field of work and enjoyed a comparatively high status in ancient India, surpassing contemporary civilisations in ancient Greece and Rome (Pande, 2014; Ramani, 2015).

While there is no doubt that electricity improves the lives of women and girls, electricity's effect on women's decision-making power, an important requisite for empowerment, is not clearly established (Palit & Govindan, 2016). On the other hand, men's involvement with electricity is linked to their status as the owners of the house, gender norms on decision-making and their higher incomes compared with women (Winther et al., 2020). Studies suggest gender-specific barriers and discrimination, such as lack of access to socio-political networks and finance, restrictive regulatory environment and cognitive inhibitions, contribute to female energy entrepreneurs not achieving their potential in a predominantly male-dominated sector (IRENA, 2019; Kelley et al., 2016). Another study, which attempted to empirically analyse women's decision-making power in India, Nepal and Kenya, observes that women have limited decision-making power than men with respect to electricity and appliances, stemming largely from their subordinated position in the socio-material contexts (Winther et al., 2020). Literature on energy policies also suggests very few energy policies have included gender mainstreaming in their frameworks and guidelines (Oparaocha & Dutta, 2011; Kohlin, Pattanayak, & Wilfong, 2011a; Govindan, Palit, et al., 2020b). The lack of gender-disaggregated data to inform energy policies is often cited as a significant reason for the near absence of gender considerations in the energy discourse (Clancy et al., 2012).

In India, historically, energy planning and policies have primarily donned the hat of a vehicle to push welfare objectives, particularly universal electrification, with provisions to provide access to the poor, marginalised and disadvantaged. For instance, Govindan, Palit, et al. (2020b) observe that gender has had a limited presence in electricity policies in

India, making an appearance only recently. All these aspects clearly indicate pathways to women's empowerment are multifaceted and continue to be elusive and debatable. Thus, one is compelled to revisit the definition of empowerment and there is a need to look beyond the usual economic aspects to social and political empowerment especially electricity's effect on women's decision-making power.

In this background, this paper attempts to take a deep dive to understand the extent to which electricity acts as a pathway for women's empowerment in the Indian subcontinent, focussing on evidence from India and Nepal. To reflect deeply on different facets of empowerment through electricity access, the chapter uses a framework developed by Winther et al. (2017). Applying the framework to cases from India and Nepal, this paper attempts to provide a new way of understanding the gender–energy–empowerment linkages through a combination of analysis of policies and processes of implementation. The analysis will add to the existing body of literature on gender inclusion in the electricity sector and the resultant outcome, in the context of the Indian subcontinent as well as other parts of the World, with possible policy implications. Specifically, the paper attempts to:

1. understand how and to what extent electrification may create a spur in women's empowerment applying a coherent analytical tool
2. provide policy inputs for the electricity sector to create pathways for women's empowerment

Framework for Embedding Women's Empowerment Through Electrification

Several studies have employed and customised different frameworks and methodologies that directly or indirectly attempted to coalesce aspects of gender, empowerment, energy, socio-technical systems and practice theory, among others. While there is no universally agreed definition of women's empowerment nor is there a consensus on the best way to empower them, the United Nations broadly defines women empowerment as the process by which women take control and ownership of their lives through the expansion of their choices. In general, women empowerment indicates an increase in economic, social, spiritual and political strength, boosting their self-esteem, enlarging their decision-making power and allowing them better access to resources (Palit & Govindan, 2016).

Kabeer (1999) conceptualised the idea of women's empowerment as a dynamic process through measurable parameters and proposed a framework focussing on the ability to exercise choice, seen through the lens of three interrelated dimensions—resources (pre-conditions), agency (process) and achievements (outcomes). While resources are classified into three types—human, social and material, Kabeer captures agency as people's ability to decide on issues that concern them so as to fulfil their own aspirations. Kabeer also broadens the scope of agency, moving beyond just 'decision-making' to include bargaining and negotiation, deception and manipulation, subversion and resistance, among others. She elaborates on 'first-order life choices' as major decisions of one's life such as freedom to marry, choice of livelihood and freedom of movement, while 'second-order choices' are less consequential, which may be important for the quality of one's life but do not constitute its defining parameters.

Studies by Bijker and Law (1994) and Hughes (1983) dwell on the interactions and complex relationship between technologies and society through various actors and institutions, including women. The socio-technical model studies the influences of technology design and systems, including electrical systems, on consumers' choices and acceptance. More recently, a systematic analysis of 262 articles on socio-technical perspectives on energy by Sovacool et al. (2020) provided a framework for incorporating important future themes for socio-technical perspectives in energy research, which include conditions of systematic change; embedded agency; justice, power, identity and politics; imaginaries and discourses; and public engagement and governance.

Literature on practice theory talks about how social practices impact gendered outcomes of electrification, including individual agency. A theoretical outline by Bourdieu (1977) on practice introduces the concept of habitus created by structures constitutive of a particular environment, consisting of dispositions that act as principles for generating practices that can be regulated. Practice theories also provide linkages to existing socio-technical systems in the context of the social disposition of people (such as their practical knowledge), which builds the case for energy policies to be grounded on a more genuinely political and democratic approach, where people get agency through active involvement in decision-making (Labanca & Bertoldi, 2018). In the realm of gender and energy, practice theory relies on the repetitive and habitual practices of everyday energy use (such as the use of energy appliances) and the potential to be shaped by changing social contexts including gender norms and power relations (Wilhite, 2008).

Framework of Analysis for the Present Study

The present study employs electricity–empowerment framework developed by Winther et al. (2017), which encompasses key learnings from several studies and frameworks on gender, socio-technical designs and systems, and practice theory including those discussed in the previous subsection. The framework is rooted in the energy justice discourse concerning the distribution of benefits and costs, decision-making (process) and rights/gender norms (recognition). Winther et al.'s framework, influenced by Kabeer's work, especially in terms of contextualising the preconditions and process for empowerment, defines empowerment on the pillars of rights, resources and agency. These dimensions, all influencing one another, propose measurable criteria not just for project-related analysis of the impacts of electricity systems, but other contextual issues as well. Hence, it encompasses all critical issues that enhance or constrain empowerment pathways in the realm of electricity.

Dimension 1: Overarching issues:	These include women and men's rights according to juridical and customary law and practice, women's and men's social positions and gender norms.
Dimension 2: Gendered access to and control over resources:	Resources include material, social and human resources. Material resources comprise assets like money and property including appliances, and other physical assets. Social resources considered are associated notion of social capital and participation in social networks. Human resources include welfare aspects such as knowledge and education, health, safety, security and reduction in drudgery.
Dimension 3: Agency:	Agency is measured in terms of (1) influence over life decisions and (2) influence over everyday decisions.

In addition, the framework emphasises analysing any adverse impacts of an intervention (here electricity) on any of the three dimensions. Finally, the framework attempts to capture agency—(1) women's ability to influence decisions on electricity at the household level, and (2) women's agency within the intervention as such (programme/project/system of supply).

To understand the electricity–empowerment linkage, the framework has proposed four conditional factors that could potentially influence electricity's impact on women's empowerment. These are,

(a) The material and socio-cultural context
(b) The socio-technical design, ownership and management of the system of supply
(c) The gendered organisation of supply and process of implementation
(d) The role of policy, regulations and international actors

Table 3.1 summarises the framework, including empowerment dimensions and conditional factors.

Methodology

The study adopts a two-stage approach mainly through the analysis of policy and practice, complemented by limited review of literature, to examine the gendered outcomes of electricity access—both grid and off-grid options—and build a case for potential pathways for women's empowerment through electricity in the Indian subcontinent and globally.

The study analysed electricity policies from India and Nepal to understand how gender has been acknowledged in electricity policies and programmes. Inputs from a recent study by Govindan, Palit, et al. (2020b) that reviewed electricity policies in India, Kenya and Nepal reflected on how gender elements are depicted and prioritised at the policy level in the context of welfare objectives of the state. The study helped in nuanced analysis of existing energy policies and their status with respect to gender inclusiveness thereby imploring policymakers to relook the policy framing from a gender lens. The present study has aimed to broaden the earlier analysis discussed above, going beyond gender inclusion aspects in energy policies, to also cover barriers and challenges in its effective implementation by also examining key findings from the qualitative fieldwork,

Table 3.1 Framework for analysis of electricity–empowerment linkages

Dimension	*Sub dimensions*	*Specifications*	*Conditional factors*
Empowerment 1 Overarching issues	Women and men's rights	Women and men's respective rights according to judicial and customary law & practice.	1. The material and socio-cultural context 2. The socio-technical design of the system of supply (intervention) 3. The gendered organisation of supply and process of implementation 4. The role of policies, regulations and international actors
	Gender ideologies & norms	How women and men are expected to be behave, their roles and responsibilities	
	Social positions	How women and men of various ages and classes are valued	
Empowerment 2 Resources	Material opportunities	Access to and control over assets in the short term	
	Material endowments	Access to and control over investments, saving, long-term financial security, ownership and accumulation of assets. Includes economic empowerment	
	Social resources	Access to communication and social networks, social inclusion	
	Human resources	Access to information, education, knowledge, degree of drudgery (time use), comfort and convenience, health, safety, includes psychological power	
Empowerment 3 Agency	Influence over life decisions	A person or group's ability to influence what they conceive as significant in their lives, includes political power	
	Influence over everyday decisions	Ability to influence decision-making in everyday life which in sum affects a person's group's autonomy and power to influence matters that affect or concern them	
Negative events	Negative effects	Signs of deterioration in any of the above dimensions	
Agency in the realm of the intervention (electricity)	Influence over decisions regarding electricity access	Ability to influence decisions regarding the household's subscription, installation, use and payment of electricity and acquisition and use of appliances	
	Involvement in & influence over system of supply	Involvement in and ability to influence electricity supply (governing, planning, sociotechnical design, implementation, management and operation)	

Source: Winther et al. (2017)

undertaken in India and Nepal, and analysing cases of gender mainstreaming in the electricity domain.

Documentation of these cases was earlier done as part of two studies. The first titled, 'Exploring Factors that Enhance and restrict Women's Empowerment through Electrification' involved examining the gender–electricity–empowerment nexus in three countries—India, Kenya and Nepal (Winther et al., 2020). The second study was done for the United Nations Climate Technology Centre and Network to document cases from India and Nepal of gender mainstreaming in the energy supply chain (Govindan, Murali, & Dholakia, 2020a). The electricity–empowerment framework is applied to these selected cases to draw insights into the factors that play a significant role in the empowerment of women through electricity and the areas that are currently posing challenges.

Experience from Policy: Electricity Policies of India and Nepal

A recent review of electricity policies of India and Nepal, shed light on the way gender and women are characterised in policy parlance, in a way highlighting the extent to which emphasis is laid on gender specific needs and concerns (Govindan, Palit, et al., 2020b). A total of 14 Indian policies were analysed of which only four showed explicit gender-related terminology or objectives. On the positive side, the review highlights an increase in gender considerations being reflected in recent electricity policies. One of the initiatives that explicitly attempted to mainstream gender is the Indian national rural electrification programme, which identified women or their groups as potential electricity franchisees in charge of metering, billing, providing connections and running local networks. Other policies and programmes that also attempted to recognise women's issues and their participation in the sector, albeit minimalist, include the Rural Electrification Policy 2006, the Integrated Energy Policy 2006 and the PM SAUBHAGYA 2017. However, most policies are found to address beneficiaries in a gender-neutral manner rather than considering any affirmative action for equitable gendered outcomes.

Only 6 of the 23 Nepalese plans and policy documents analysed can be relatively considered as 'gender-responsive' because the policies included gender-inclusive terms such as 'gender mainstreaming', 'women's involvement', 'gender-integrating', 'participation' and 'representation', as guiding principles. For instance, women's role in energy planning and decision-making processes are acknowledged in the Tenth FYP (2002–2007), the National Rural and Renewable Energy Programme (2012) and the National Energy Strategy (2013). The most overtly gender-inclusive guidelines are the Gender Equality and Social Inclusion (GESI) guidelines of the Alternative Energy Promotion Council (AEPC), the nodal agency for renewable energy development in Nepal (Govindan, Palit, et al., 2020b). GESI guidelines lay out detailed procedures for gender inclusion in renewable energy projects and programmes, to ensure women and men get equal access and benefits. The Rural Energy Policy (2006) emphasised on women's involvement in formulating programmes for women's development. The Subsidy Policy for Renewable Energy (2013) promoted rural employment for women in addition to the renewable energy service delivery and productive use. The policy also aided single women-headed households through additional subsidy benefits.

Overall, the policy review in India and Nepal also reveals that none of the electricity sector policies have addressed the gender-differentiated level of access to resources and benefits from electricity-supported development (Govindan, Palit, et al., 2020b). Mostly the policies focus on (1) expanding electricity access to the remote population, (2) making electricity affordable to the poor segments of the society and (3) enhancing the quality and reliability of the services. These policies have missed understanding the linkages between access to electricity and empowerment and thus have been slow to integrate gender into policymaking comprehensively.

This shows that gender mainstreamed policies have been relatively sparse and piece-meal in approach. It is not an isolated trend in the Indian subcontinent and is also observed in other developing nations. The International Union for Conservation of Nature (IUCN) reports that only one-third of the 192 national energy frameworks reviewed contained references to women and/or gender (IUCN, 2017). An IUCN report found around three-fourths of the 45 reviewed documents from 29 Sub-Saharan African countries partially considered gender within their principles, objectives and strategies (IUCN, 2018). However, whether the policies and programmes have achieved what they set out to do on the ground and to what extent policies have made themselves relevant as tools to promote women's empowerment is a matter of debate. This could be attributed to the limited research data available that document their performance over time. Both the policymakers and practitioners have not envisioned a national-level analysis of the existing gender provisions.

Moreover, researchers also argue that translating policy to practice requires a robust mechanism to monitor implementation, investment, track progress on gendered targets and their expected outcomes and document opportunities and challenges (Govindan & Murali, 2020). All these aspects should eventually feed back into the policymaking regime.

Experience from Practice: Gender–Electricity Cases from India and Nepal

India

JEEViKA - Lighting a Billion Lives Initiative in Bihar

The Government of Bihar, through the Bihar Rural Livelihoods Promotion Society (BRLPS) is spearheading the JEEViKA (i.e. livelihood) initiative, with the objective of social and economic empowerment of the rural poor in the state. The main focus has been to develop women's Self-Help Groups (SHGs), wherein women from underprivileged households in the villages are mobilised into groups, their capacities enhanced and are encouraged to save and be thrift. The SHGs are provided financial linkages, enabling them access to credit and thus helping them to augment their incomes. More than anything, SHGs provide women a platform to socialise, express themselves freely, discuss their problems, look for common solutions and take out time for their personal growth. JEEViKA has been successful in creating a much-desired impact on the ground (Hoffmann et al., 2017) in Bihar, a state considered economically backward for a long time since independence from the colonial regime.

As electricity access remained poor in the state's interiors, JEEViKA started enhancing electricity access in 2012 by collaborating with The Energy and Resources Institute (TERI), a global think tank that was steering the Lighting a Billion Lives (LaBL) initiative. Under the JEEViKA–LaBL collaborative initiative, solar home systems (SHSs) were provided to women SHG member households with limited or no access to electricity. The stand-alone energy package consisted of two LED lamps of 2–3-Watt each and a mobile charging point powered by a 20–25 Wp solar panel with battery back-up. The total cost of an SHS was INR 3500 (approximately around US$ 55), where the households were required to pay an upfront cost of at least INR 500 (US$ 8) and the balance amount could be paid in monthly instalments at an interest rate of 2%.

Women, burdened with household and agriculture-based duties, were hardly finding time for recreation or self-development. But with the lighting solutions, they got 'extra hours' in the evening allowing them to use it as per their wishes, which could be socialising, watching TV, monitoring their children's studies or even hosting celebrations and ceremonies. Being able to charge mobile phones at home greatly helped women to stay connected to their friends and family members, especially in cases where men had migrated to other places for employment. While the system delivered on its promised benefits of being a life-enhancing tool for women by reducing drudgery, it also came with a sense of ownership and responsibility for the women (Palit et al., 2015; Chandana, 2017).

The unique socio-technical design of the initiative allowed women SHG members to become not just beneficiaries but owners of the systems. Most importantly, the programme gave women the opportunity to take independent decisions regarding the energy system, even though they consult men for most decisions in the household. This initiative is a prime example of contesting the existing energy-patriarchy, stemming from the overall patriarchal construct of the society, where men are the primary bread-winners making them owners of assets. While the impacts on reducing women's reproductive role such as care giving was very positive, the existing local socio-cultural context prevented most women from taking up opportunities to be trained as technicians and to be engaged in installation of the systems due to the perceived masculine label attached to electricity-related technical jobs. Another aspect to be noted is though the presence of electricity at home enabled the children (both boys and girls) to complete their homework and pay attention to academics, yet girls pursuing higher education are not encouraged and the priority for parents was to get the girls married by the earliest legal age of marriage. Hence, merely owning of SHS by women may not necessarily empower women to change certain entrenched social conditions, which may need more systemic changes in social norms and policies.

Vidyut Sahayaks of Mahavitaran in Maharashtra

An example of policy-level intervention for gender mainstreaming in the electricity sector is demonstrated successfully by the case of the 'Line Women of Mahavitaran'. As the largest electricity distribution utility in India, Maharashtra State Electricity Distribution Company Limited (called as Mahavitaran) distributes electricity in the state of Maharashtra (except Mumbai) and has a consumer base of more than 2.5 million and employs around 81,000 people. Increasing the participation of women in the electricity sector was taken up as one of the key agendas by Mahavitaran. A key policy measure in the form of 30% horizontal reservation for women was thus introduced (Govt. of Maharashtra, 2001), and women were started to be recruited and trained as line staff, operators and vigilance squad amongst other technical field jobs.

Women and men selected as *Vidyut Sahayaks* (electricity helpers), were trained as line staff to work on electrical equipment such as transformers, poles and live wires, which are part of the power infrastructure, in addition to detecting faults and theft, and collecting dues from consumers. Mahavitaran also created an all-women vigilance squad called *Damini Pathak*, where women are responsible for addressing consumer grievances, mostly involving meter readings and tampering. They also undertake the distribution of electricity bills. This squad play a major role in the reduction of electricity theft. To safeguard the interests of women line staff, a committee is formed under the leadership of a senior lady officer, known as *Tejaswini Samiti,* which supports women and helps in resolving difficulties/grievances at the workplace.

An interesting aspect of the process is that the selected candidates undergo training and induction programmes without any gender discrimination. It is understood that women were initially hesitant to undertake jobs that are traditionally done by men, which include activities like climbing up poles and handling live electricity equipment. Thus, women were counselled to change their mind-set about the profession. In fact, their uniforms have been kept similar to those of men to feel equivalent in status and also support them to deal with live wires and climb poles.

Maharashtra is the first state in India to have a policy to formally induct women has been a game changer in providing opportunities to tap the potential of women. Numerous studies have indicated that integrating women throughout the energy supply chain not only brings about greater productivity, efficiency and customer satisfaction, it also promotes gender equality and empowerment (Sovacool et al., 2013). However, in the initial period, women endured structural barriers such as limited access to childcare facilities, lack of flexible work hours. Other cultural constraints included donning a man's clothes, climbing electric poles and many site visits to distant locations, which seemed to be an uncomfortable and uphill task. Moreover, in addition to their own inhibitions regarding the job, women faced reluctance from their families too; especially they feared it might affect marriage prospects of girls employed as technicians. Women also experienced initial opposition from their peers since their jobs seemed too technical and risky, due to which they initially preferred desk-based jobs. Some of the other challenges included long working hours and travelling for site visits to far off locations, which often required approval from men in the household.

With continued complementary support from Mahavitaran, in the form of childcare services and sanitation facilities, and the prospects of having a secured government job, women were able to overcome many of these challenges and do their duties successfully. They are now looked upon as role models for girls in their villages. Around 2000 women have been recruited and trained as line women till 2020, a 100% increase during the last eight years (Singh, 2020). Taking cue from this initiative, another state electricity distribution company in India, Madhya Pradesh West Power Distribution Company (MPWPDC), started a similar pilot initiative in one of the state's district, Indore, in 2018 called 'Pink Electricity Zone' comprising of all-women technical staff, consisting of assistant engineers, junior engineers, line supervisors, line persons, meter readers, accountants and computer operators.

Women-Managed Solar Mini-Grids in Jharkhand
In the remote tribal areas of Jharkhand, the pace of development has been slow and the interventions by government agencies would often not reach the disadvantaged communities owing to their remoteness. Thus, a social organisation, PRADAN, started working for the tribal societies in the state in 1998 and initiated several development projects. Initially, they mainly focussed on agricultural productivity, but from 2012 their focus shifted to a more holistic form of development, when they started supporting the vulnerable people, including women, with livelihoods opportunities facilitating access to government schemes. In the process, based on demand for electricity by the women SHG members, PRADAN partnered with Gram Oorja, a social enterprise working on rural energy technology solutions, to develop a community-managed solar mini-grid project for providing electricity for lighting and productive use in the villages. From its inception, this process involved the community, especially the women SHG members, and the decision was taken with consensus.

Before the actual implementation, the community was acquainted with solar PV technology by Gram Oorja, and community members, including women, were exposed to a mini-grid project for awareness and learning. The implementation process, though carried out mostly by the social enterprise, had the support of the community, in the form of help in civil works, where women also participated. A Solar Energy Committee was constituted after the installation of the technology system, where women held all the key leadership positions. The monitoring was also the responsibility of the women members, trained by PRADAN.

The socio-technical design of the system created rights for women from the project conceptualisation to holding important positions in managing the project. The continuing efforts of PRADAN led to the women gaining confidence and exposure, and a sense of awareness about different government schemes, helping them to make informed decisions and paving the way for the solar mini-grid project. Women's perseverance along with PRADAN's mentoring enhanced their capability in being able to oversee and manage a modern technology without hesitation.

However, ensuring active participation of women in SHGs was not an easy task in a region mostly inhabited by a tribal community due to restrictive norms, which defined the role of women as care givers and encouraged them only to assist the men in farm related activities. Women also required permission from their spouses to be part of the SHGs. Nevertheless, the mini-grid project clearly established that through such initiatives, external actors can mobilise women, lower the barriers for collective action and achieve collective objectives through community action. Moreover, the physical and conceptual space created by the SHGs provided opportunity for women to be involved in the supply chain and generate a sense of ownership and empowerment.

Nepal

Baglung-Bhuji Khola Community-Based Micro Hydro Power Plant

In order to strengthen energy access in rural areas through decentralised renewable energy (DRE), small-scale, community-led Micro Hydro Power (MHP) plants are being promoted by Government of Nepal (GoN). The project 32 kW micro hydro plant on river Bhuji at Bowang village in Baglung district of central Nepal is a community-based MHP plant that was implemented by the Alternative Energy Promotion Centre (AEPC) along with the Ministry of Energy and Ministry of Water Resources and Irrigation. The project electrified all 245 households in the village. A micro hydro User's Group Committee, mobilised by a local grassroots civil society organisation, Dhaulagiri Community Resource Development

Centre (DCRDC), was constituted for construction and operation of the MHP plant. While it is mandated by Nepalese policy to have women members in such projects, the common practice is to induct women just to fulfil the mandatory requirement, without giving them any specific responsibility or decision-making power. In the Bhuji Khola community-based MHP project, the village women were actively involved since its inception. In fact, the project was the first in Nepal to have a female User Group Committee president.

The role of the User Group Committee primarily included supporting technical staff and DCRDC during the project period and operation and management of the power plant after commissioning. The '*Amma Samuha*', a local group of women actively working on development activities in the village, was instrumental in generating interest and mobilising the women for community contribution as well as for other decisions regarding electricity. Further, they also worked as labour for civil works including digging of pits for poles and carrying material for construction to the sites. Women members were also active in the process of tariff fixation, collection of contribution and other committee-related activities.

Several challenges, however, were faced by women in making inroads into a predominantly male domain. At first, women underestimated their ability to take responsibilities outside the household. Their reluctance also emerged from the anticipation of society's comments and opinions on their conduct and thus chose to remain low profile when it came to decisions on energy and new technologies. However, the active role of *Amma Samuha* inspired a sense of ownership amongst the women of the village and encouraged their participation. Involvement and leadership of women in the User Group Committee established faith in women's capabilities as managers of power systems. It also gave confidence to women to participate in similar projects and take responsibilities at par with men.

This community-based MHP plant is a typical example, which proves that just the presence of policy guidelines to have women representatives in the User's Group Committee will not serve the purpose unless specific responsibilities are not delineated. The women's active participation was catalysed by the critical role played by an external actor—*Amma Samuha*—in creating the women committee and generating interest amongst women to participate in the decision-making processes. The Bhuji Khola Community, which already had a strong base of women collectives, exhibited a favourable socio-cultural context for the formation and participation of women in the MHP User's Group Committee. Management of electricity supply by the women also enabled women beneficiaries to use rice cookers due to demonstration organised by the committee members. This resonates with findings from another empirical study in Nepal that reveals women who have better control of material assets are most likely to have electricity access and acquire appliances of their choosing (Winther et al., 2020).

Community Rural Electrification Programme in Udaypur

The Community Rural Electrification Programme (CREP) was launched by the Government of Nepal in 2003–2004 with the objective to improve grid connectivity in remote areas. An implementation body known as the Community Rural Electrification Entity (CREE) was created under the programme, with a mandate to own, operate and manage electrification for a specific area/region. The CREE buys electricity in bulk from the Nepal Electricity Authority at competitive rates to be sold to the domestic consumers and rural enterprises through the existing or new distribution networks. In order to accommodate the costs of managing and maintain the grid infrastructure, taking into account transformer losses and making the entity profitable, the charges to the consumers are set a little higher than the power purchase cost.

In one of the Wards of Katari Municipality, Udaypur district a local women's group known as Samudayeek Balbikash Tatha Mahila Jagaran, working on education and adult literacy, safe-motherhood programme and income generation, registered as a CREE in 2006–2007. The group aimed to enhance electricity access to women and children and foster women's economic empowerment. A seven-member executive committee was formed to oversee the electrification process, as per the mandate of the Social Welfare Act 2060 (2003) (Social Welfare Council 1992), under which the CREE was registered. Though guidelines did not exist for the number of women representatives to be inducted in the executive committee, it was decided to have majority of women (5 women, 2 men) in the decision-making positions. The women members of the Samudayeek Balbikash Tatha Mahila Jagaran had previously taken up leadership and mentorship roles successfully and this influenced their decision to have the women members in the executive committee.

The CREE adopted a flexible approach towards bill payments by allowing people to pay bills from their villages, without having to travel to the electricity office. This helped women, especially those whose male family members had moved out of the village. There was an improvement in the collection of utility charges. The CREE, supported by the government, conducted a 15-day technical training for women on house wiring, to create livelihood opportunities for women, and generated an additional source of income for women by promoting productive use of electricity.

CREE which has a clear policy mandate of gender and social inclusion got effectively implemented on ground by Samudayeek Balbikash Tatha Mahila Jagaran reaffirming to the established understanding of capitalising on the social capital to create and reinforce advantage. Hence, policies with support from other factors like networks and leadership have the potential for achieving diverse common objectives. Also, the active role of women in the management of the supply system has put them in a position of authority in their communities. For instance, the members of CREE have represented their community at national level federations and in other committees.

Discussion and Conclusions

Gender-Blind Policies and Programmes to Gender-Responsive Outcomes and Impacts

It is clear from the review of electricity policies of India and Nepal that while policies aim to be gender-neutral in providing equal opportunities to men and women, the outcomes for men and women may not always be equal because of the inherent differences in baseline. Thus, most of the policies could be considered as 'gender-blind' since they do not explicitly acknowledge the gendered differences needed in creating equitable outcomes. A plausible reason could be that the electricity policies and programmes assumed that the benefits would trickle down equally to benefit all. These policies have also primarily focused on women's domestic role rather than their role in the supply chain or entrepreneurship. Off-grid programmes, however, appear to offer better anecdotal evidence of women's role in the supply chain, gendered outcomes and empowerment. As off-grid projects are usually designed following 'bottom-up' approaches, better participation of women and the marginalised groups is ensured.

Further, while a well-structured gender mainstreaming framework remains largely absent in the overall electricity policy domain, the case studies analysed do reveal attempts to address gender inclusivity in some of the policy provisions and recognise the need for addressing gender inequalities while promoting universal access to modern energy. So, the problem does not entirely rest on the non-inclusion of gender elements in policy and programme formulation, rather its conversion into gender responsive outcomes and impacts to advance women empowerment. Our analysis shows that policies as a standalone component can trigger empowerment, but it is also important to hinge on associated support structures and existing social capital to fully realise the gendered outcomes. It is thus important to understand and build on the processes and conditions that translate the well-intended gender-inclusive policies into visible results and leapfrogging towards empowerment. There is also a need to focus on gender-equitable outcomes and more affirmative actions in the energy policies and programmes, going beyond their gender-neutral characters. If they remain supply-driven, policies will continue to be gender-blind, but if they are formulated based on demand, the differentiated gendered needs will be better accommodated for equitable outcomes.

Women's Participation to Decision-Making in All Spheres

The case studies analysed bring to the fore that specific programmes targeted at women in the supply chain have certainly given women a platform to voice their needs, and concerns, which are steppingstones towards ensuring gender equality. The programmes also contested the existing energy-patriarchy and attempted to make women the decision-makers to operate, manage or own their energy system. However, a deeper look at the impact reveals that while it has enhanced women's social lives, facilitated children's studies and improved their decision-making role in certain realms (e.g. the possibility to get credit or earning from the working as *Vidyut Sahayaks*), it has not necessarily resulted in empowering them from the context of the conditional factors discussed in Table 3.1. There continues to be different challenges in fully achieving the conditions of empowerment where major decisions within the households and the community are otherwise still taken by men.

Despite the small but positive steps of women's empowerment resulting from a few policies and other supporting factors, most policies and programmes hardly associate women with energy supply and productive use of energy. Often, the policies still look at easing the care giving roles as gendered impact rather than designing interventions to pave the way for women's more active role in the organisation of supply and enterprises. Some interventions could be women-friendly loan products and financial instruments and promoting home-based micro-enterprises or designing technology options that can be handled with ease by women. For instance, home-based assembly of solar products or smaller harvesting machines and post-harvest technologies such as motorised grain thresher, ground-nut stripper and other things where women could be potential beneficiaries.

While the challenges to an extent have been addressed in the urban set up with increasing number of women coming forward to establish energy enterprises or enter the energy supply chain, the situation in rural areas is different and needs focussed attention in broadening the horizons of gender inclusion. Specialised women-targeted skill development, both technical and managerial, and affirmative measures will ensure enhanced participation, endowments and opportunities leading to empowerment.

Gendered Electricity Policies Along with Support Services Critical to Power Empowerment

For the implementing organisations that strive to bring changes through women's engagement in the energy supply chain and decision-making and subsequent empowerment, addressing the entrenched social conditions and gender relations may not be an immediate possibility, as they work with constrained timelines and perhaps limited budgets. Electricity, as an enabling factor for women's empowerment, can possibly contribute to, rather than bringing about a paradigm change in gender relations. Just like any other sector, the electricity sector is also influenced by associated social sector policies and programmes, as far as gender dimensions are concerned. Thus, changes in gender relations would probably also require socio-cultural changes at a deeper level, over a period. Wider legislations such as land and inheritance rights (e.g. the Hindu Succession Act, 1956 in India allows equal inheritance rights for women and men among the Hindu religious community) and awareness for their proper implementation and opportunities for gainful employment will positively affect women's degree of empowerment through electrification. This, however, does not mean the noble goal of increasing electricity access, ownership of the electricity systems to women and involvement of women in supply chain should be abandoned but rather work in tandem with supporting policies and social networks to bring the desired changes in gender relations.

Disaggregated Data for Better Policy and Programme Formulation

Although the need of acquiring gender disaggregated data to inform policy and practice has been argued for a long time, data generation in the Indian subcontinent most often continues to use the household as the unit of analysis. This may be camouflaging the varied effects of electricity access on different gender groups, but it is also important to recognise the fact that the household or family is a strong institutional unit in the Indian subcontinent. Thus, clarity on the level and type of disaggregation and its collection (such as percentage of women-led enterprises and working women in a place, women in decision-making roles in a committee and percentage of such committees, percentage of women on company boards—large, medium and small, among others) can help to develop policies and programmes for gendered benefits. The larger challenge, however, is the precision and time in providing the appropriate data when engaging with policymakers that can allow them to better integrate gendered aspects into policymaking and programme formulation, going beyond acknowledging the traditional care giving roles of women. It will be worthwhile for researchers and practitioners to focus on collecting more data and empirical evidence on empowerment of women and share best practices with different stakeholders, especially with respect to organisation of electricity supply and women's empowerment in terms of the four conditional factors discussed in this chapter.

In sum, it is crucial to develop affirmative gendered policies and programmes for wider women's participation and decision-making in the process of organisation of supply and equally important is to enhance the procedure and processes that ensure better outcomes of the affirmative policies on ground. It is also vital to create the desired changes in the prevalent social norms and traditions to support women's participation. These factors will aid in meaningful participation and contribution of women in the energy transition process and leverage the opportunity to reshape the energy systems and move towards a demand-based delivery.

References

ADB. (2015). *Balancing the burden: Desk review of women's time poverty and infrastructure in Asia and the Pacific*. Retrieved from https://www.adb.org/sites/default/files/publication/177465/sdcc-balancing-burden.pdf

Arthur, C. (2010). *Women solar entrepreneurs transforming Bangladesh.* Retrieved from renewableenergyworld.com: https://www.renewableenergyworld.com/baseload/women-solar-entrepreneurs-transforming-bangladesh/#gref

Barefoot College. (2016). *The Barefoot Model for Global Sustainability White Paper.* Retrieved from https://www.barefootcollege.org/wp-content/themes/barefoot-college/download/barefoot-white-paper.pdf

Bijker, W., & Law, J. (1994). General introduction. In W. E. Bijker & J. Law (Eds.), *Shaping technology/building society: Studies in sociotechnical change* (pp. 1–14). Massachusetts Institute of Technology Press.

Blackden, M., & Wodon, Q. (2006). *Gender, time use, and poverty: Introduction.* MPRA Paper, University Library of Munich, Germany.

Bourdieu, P. (1977). *Outline of a theory of practice.* Cambridge University Press.

Cecelski, E. (2000). *The role of women in sustainable energy development.* NREL Technical Monitor; NREL/SR-550-26889.

Cecelski, E. (2006). *From the millennium development goals towards a gender sensitive energy policy research and practice: Empirical evidence and case studies.* Synthesis Report. Collaborative Research Group on Gender and Energy (CRGGE): ENERGIA/DfID.

Chandana, S. (2017). *Energy and gender in rural areas: Solar home systems and women in Bihar, India* (MPhil diss.). Centre for Sustainable Development, University of Cambridge.

Clancy, J., Winther, T., Matinga, M., & Oparaocha, S. (2012). *Gender equity in access to and benefits from modern energy and improved energy technologies: World development report background paper.* ETC/ENERGIA in Association Nord/Sør-konsulentene.

Dinkelman, T. (2011). The effects of rural electrification on employment: New evidence from South Africa. *American Economic Review, 101*(7), 3078–3108.

Dutta, S. (2018). *Supporting last-mile women energy entrepreneurs: What works and what does not.* ENERGIA the International Network on Gender and Sustainable Energy.

Dutta, S., Kooijman, A., & Cecelski, E. (2017). *Energy access and gender. Getting the right balance.* The World Bank.

Frontier Markets. (n.d.). Our solutions. Retrieved 2022, from https://www.frontiermkts.com/our-solutions

Glemarec, Y., Bayat-Renoux, F., & Oliver, W. (2016). Removing barriers to women entrepreneurs' engagement in decentralized sustainable energy solutions for the poor. *AIMS Energy, 4*(1), 136–172.

Govindan, M., & Murali, R. (2020). Gender inclusion in the energy policies: What are we missing? *Energy Review, 2*(7), 2–3. Retrieved from https://drive.google.com/file/d/1JCXrXxeFCxWgOHag-WLBULs8OLOTXiJL/view

Govindan, M., Murali, R., & Dholakia, D. (2020a). *Women in energy: Breaking stereotypes and inspiring change.* The Climate Technology Centre and Network

(CTCN). Retrieved from https://www.developmentaid.org/#!/news-stream/post/58820/new-ctcn-publication-women-in-energy

Govindan, M., Palit, D., Murali, R., & Sankar, D. (2020b). Gender in electricity policymaking in India, Nepal and Kenya. In G. Bombaerts, K. Jenkins, Y. A. Sanusi, & E. W. Guoyu (Eds.), *Energy justice across borders* (pp. 111–135). Springer Nature Switzerland AG.

Govt. of Maharashtra. (2001). *Resolution No.82/2001/MSA-2000/PK 415/K-2 dt.25.05.2001.* Retrieved from https://www.mahadiscom.in/wp-content/uploads/2020/03/Advt.-No.-04_2020.pdf

Grogan, L., & Sadanand, A. (2013). Rural electrification and employment in poor countries: Evidence from Nicaragua. *World Development, 43*, 252–265.

Hoffmann, V., Rao, B., Datta, U., Sanyal, P., & Surendra, V. (2017). *Poverty and empowerment impacts of the Bihar Rural Livelihoods Project, 3ie Grantee Final Report.* International Initiative for Impact Evaluation (3ie).

Hughes, T. (1983). *1983. Networks of power. Electrification in Western society, 1880–1930.* Johns Hopkins University Press.

ILO. (2010). *Women in infrastructure works: Boosting gender equality and rural development!* Retrieved from http://www.ilo.org/employment/Whatwedo/Publications/WCMS_150834/lang%2D%2Den/index.htm

IRENA. (2019). *Renewable energy: A gender perspective.* IRENA.

IUCN. (2017). *Energising equality: The importance of integrating gender equality principles in national energy policies and frameworks.* IUCN Global Gender Office.

IUCN. (2018). *Energizing equality: Sub-Saharan Africa's integration of gender equality principles in national energy policies and frameworks.* IUCN Global Gender Office.

Kabeer, N. (1999). Resources, agency, achievements: Reflections on the measurement of women's empowerment. *Development and Change, 30*, 435–464. https://doi.org/10.1111/1467-7660.00125

Kelley, D., Baumer, J., Brush, C., Greene, P., Mahdavi, M., Majbouri, M., ... Heavlow, R. (2016). *Women's entrepreneurship 2016/2017 report.* Global Entrepreneurship Monitor, pp. 1–91.

Kohlin, G., Pattanayak, S. K., & Wilfong, C. (2011a). *Energy, gender and development: What are the linkages? Where is the evidence?* Policy Research Working Paper 5800/Background Paper to the 2012 World Development Report. World Bank.

Kohlin, G., Sills, O., Pattanayak, S., & Wilfong, C. (2011b). *Energy, gender and development: What are the linkages? Where is the evidence?* Social Development Working Paper. No. 125. World Bank.

Labanca, N., & Bertoldi, P. (2018). Beyond energy efficiency and individual behaviours: Policy insights from social practice theories. *Energy Policy, 115*, 494–502. https://doi.org/10.1016/j.enpol.2018.01.027

Oparaocha, S., & Dutta, S. (2011). Current opinion in environmental sustainability. *Gender and Energy for Sustainable Development, 3*(4), 265–271.

Palit, D. (2013). Solar energy programs for rural electrification: Experiences and lessons from South Asia. *Energy for Sustainable Development, 17*, 270–279.

Palit, D., & Govindan, M. (2016, December 27). Why we need to measure how electricity empowers women. *Daily O*. Retrieved from https://www.dailyo.in/voices/electricity-women-empowerment-teri-united-nations-energy-policy/story/1/14745.html

Palit, D., Malhotra, S., Pandey, M. K., & Bankoti, N. (2015). Solar lighting for rural households: A case of innovative model in Bihar, India. In M. Schafer (Ed.), *Micro perspectives for decentralized energy supply*. International Conference.

Pande, P. (2014). Equality: As a social principle described in Vedic tradition. *Veda Vidya, 24* (July–December). Retrieved August 10, 2021, from https://msrvvp.ac.in/ved-vidya/24/19_Eng._Priyanka_pande_24.pdf

PIDG. (2018). *Mainstreaming gender equality to improve infrastructure development impact*. Retrieved from https://www.pidg.org/wp-content/uploads/2019/01/PIDG-KIT-gender-report-final_published.pdf

Ramani, P. (2015). Women empowerment in India from prehistoric to present. *International Journal of Research and Scientific Innovation, 3*(1), 197–201.

Rewald, R. (2017). *Energy and women and girls: Analyzing the needs, uses, and impacts of energy on women and girls in the developing world*. . https://www.oxfamamerica.org/explore/research-publications/energy-women-girls:Oxfam Research Backgrounder Series.

SEforALL. (2017). *People-centered accelerator*. Retrieved from https://www.seforall.org/sites/default/files/People-Centered_Accelerator.pdf

SEforAll. (2018). *Levers of change: How global trends impact gender equality and social inclusion in access to sustainable energy*. Sustainable Energy for All (SEforAll).

Shankar, A., Elam, A., & Glinksi, A. (2018). Building the evidence base for women's entrepreneurship in the energy. Retrieved from https://www.energia.org/cm2/wp-content/uploads/2018/07/RA-7-Womens-Entrepreneurship-Jun-27webv3.pdf

Singh, D. (2020, March 8). *International Women's Day 2020: MSEDCL sees rising no. of women technicians in state*. Retrieved from https://www.freepressjournal.in/mumbai/international-womens-day-2020-msedcl-sees-rising-no-of-women-technicians-in-state

Smith, K., & Dutta, K. (2011). Cooking with gas. *Energy for Sustainable Development, 15*, 115–116.

Sovacool, B. K., Clarke, S., Johnson, K., Crafton, M., Eidsness, J., & Zoppo, D. (2013). The energy-enterprise-gender nexus: Lessons from the multifunctional platform (MFP) in Mali. *Renewable Energy, 50*, 115–125.

Sovacool, B. K., Hess, D. J., Amir, S., Geels, F. W., Hirsh, R., Medina, L. R., et al. (2020). Sociotechnical agendas: Reviewing future directions for energy and climate research. *Energy Research & Social Science, 70*, 101617. https://doi.org/10.1016/j.erss.2020.101617

Standal, K., & Winther, T. (2016). Empowerment through energy? Impact of electricity on care work practices and gender relations. *Forum for Development Studies, 43*(1), 1–19.

UNFCCC. (n.d.). Solar Sister|Uganda, Rwanda, South Sudan. Retrieved from https://unfccc.int/climate-action/momentum-for-change/lighthouse-activities/solar-sister

UNHCR. (2022, July 12). UNHCR Rwanda. Retrieved from Nyabiheke Refugee Camp Profile: https://reliefweb.int/report/rwanda/unhcr-rwanda-nyabiheke-refugee-camp-profile-31-march-2021

Wilhite, H. (2008). New thinking on the agentive relationship between end-use technologies and energy-using practices. *Energy Efficiency, 1*, 121–130. https://doi.org/10.1007/s12053-008-9006-x

Wilhite, H. (2017). *Gender implications of energy use and energy access*. EEG State-of-Knowledge Paper Series. Oxford Policy Management.

Winther, T. (2008). *The impact of electricity: Development, desires and dilemmas* (New edition 1st ed.). Berghahn Books. JSTOR, www.jstor.org/stable/j.ctt9qd98v

Winther, T., Matinga, M. N., Ulsrud, K., & Standal, K. (2017). Women's empowerment through electricity access: Scoping study and proposal for a framework of analysis. *Journal of Development Effectiveness, 9*(3), 389–417.

Winther, T., Ulsrud, K., Matinga, M., Govindan, M., Gill, B., Saini, A., et al. (2020). In the light of what we cannot see: Exploring the interconnections between gender and electricity access. *Energy Research & Social Science, 60*, 101334.

CHAPTER 4

A Feminist Policy Analysis of the Gender and Climate Change Nexus in the Colombian Coal and Energy Sector

Ana Maria Mahecha Groot, Natalia Daza Niño, Silvia Rojas Castro, Kathrin Meyer, María Laura Rojas, and Kerstin Mohr

Introduction

As the sixth-largest exporter of coal globally, Colombia covers a significant share of the growing energy demand. It is estimated that the South American country owns more than 6.2 trillion tons of hard coal and thus disposes of the largest coal reserves in Latin America (Decoalonize, 2021). Colombia is under increasing pressure to restructure the national coal export sector because of the coal phase-out plans of importing countries

A. M. Mahecha Groot (✉)
Mundo Común, Bogotá, Colombia
e-mail: anamahecha@mundocomun.org

N. Daza Niño
Business and Human Rights Resource Centre, Bogotá, Colombia
e-mail: n.daza@uniandes.edu.co

N. Rocha Lawton, C. Forson (eds.), *Women and the Energy Sector*,
https://doi.org/10.1007/978-3-031-43091-6_4

and international climate agreements. Moreover, coal mining has repeatedly come under international criticism due to global reporting on human rights violations and massive environmental damage (Strambo & González Espinosa, 2020). Within this context, historically marginalised populations, including women, Indigenous and Black communities—mainly in the regions of La Guajira and Cesar, where the largest mines in the country are located—have suffered particularly from the negative socio-environmental impacts of the local coal mines. Simultaneously, if not planned and managed properly, plans on energy transition resulting from climate mitigation efforts are also expected to have negative impacts on these groups (Bainton & Holcombe, 2018; Kramarz et al., 2021).

Although various publications have pointed to some of the gender-differentiated impacts of Colombia's coal sector (Tobón, 2015; Blanco Vizarreta & Dongo Román, 2019), there is still a research gap on the incorporation and response of Colombian climate and coal sub-sector policies to these concerns particularly and more broadly to gender equality in the context of an increasing concern on energy transitions. Hence, a first step to fill this gap is to identify how the categories of gender and women are included and to which degree the claims and perspectives of women are incorporated in existing policies. We understand gender as a multidimensional structure of embodied social relations. Under a patriarchal gender structure, social relations produce a differentiated allocation of rights, privileges and responsibilities between women and men, with the first being who usually experience the highest burden (Connell, 2012).

S. R. Castro
European Center for Constitutional and Human Rights, Berlin, Germany
e-mail: rojas-castro@ecchr.eu

K. Meyer
German Development Cooperation, San José, Costa Rica
e-mail: kathrin.meyer@giz.de

M. L. Rojas
Transforma, Bogotá, Colombia
e-mail: marialaura.rojas@transforma.global

K. Mohr
Polis 180, Berlin, Germany
e-mail: kerstin.mohr@polis180.org

Gender is also a social and cultural construction based on—as well as simultaneously defining of—attributes, norms, roles, and attitudes that are considered appropriate for different people, e.g. men and women. As such, gender is not immutable or universal and can vary significantly among sociocultural contexts, especially outside Western cultures (Mohr et al., 2020, p. 2).

Such an analysis is highly relevant to describe possible gender inequalities that exclude women and their needs and views, which are necessary to undertake climate and coal policies that promote gender equality and, in doing so, are more efficient for everyone.

Drawing on feminist scholars (Arora-Jonsson, 2011; Bacchi, 1999; Ulloa, 2016), this study analyses existing policies of the Colombian climate and coal sector to examine how gender plays a role within them. Relying on a policy analysis of the climate and coal sectors in Colombia, the chapter will provide answers to the following questions: (a) how gender is perceived and incorporated in Colombia's climate and coal policies and (b) how current policies can be strengthened to promote gender equality and respond more efficiently to the climate crisis. In doing so, we will propose and elaborate on the concept of a "gender-responsive climate policy" (GRCP) (Mohr et al., 2020).

The policy analysis is based on the policy paper "Gender-Responsive Climate Policy: A Case Study of the Colombian Coal Sector", published in October 2020 (Mohr et al., 2020). Since its publication, the analysis was further expanded to shed light on gender representation within the Colombian coal sector and incorporate recent policy developments in the country.

In the sections to come, we first present a general context of Colombia, briefly describing the gender inequality and social conflicts in mining contexts and discussing why the latter matters for the Colombian climate policy. Section "Conceptual Approach" presents the study's conceptual approach, highlighting a feminist perspective. We acknowledge the socially constructed and dynamic nature of gender relations and recognise the importance that energy and coal policies should acknowledge gender differences, be more aware of gendered social relations and recognise women as agents of change who have a voice which is key to policymaking. Section "Methodology and Data" introduces the threefold research methods that we used for the policy analysis based on (1) critical discourse analysis, (2) corpus-informed document analysis, and (3) corpus-based generation of quantitative data. In Section "Results and Discussion of the Policy

Analysis", we present the policy analysis results and discussion, investigating whether a gender-responsive climate policy is already in place. As the last section of conclusions, Section "Conclusions" summarises our results and presents the recommendations derived from our findings.

The Colombian Case

Colombia, Coal, and Climate Change

Colombia is Latin America's third most populated country, with more than 50 million inhabitants. The country has experienced rapid economic growth over the past ten years, with an estimated gross domestic product (GDP) of USD 334 billion in 2018 (World Bank, 2022). Its natural resource endowments are its main economic drivers, including petroleum and coal products, coffee, flowers, and gold, with profits resulting from mining activities between 2010 and 2017 holding up to 30% of Colombia's annual GDP (Rincón Navas, 2018). Despite this, due to the lack of proper resource governance, state capture, and corruption, Colombia remains one of the most unequal countries in the world (World Bank, 2019).

As such, while energy production and extractive activities—and their profits—have been expected to contribute to improving the well-being of the communities in which these take place, these have in many cases worsened the living conditions of local populations. For instance, in regions such as Cesar and Magdalena, it can be evidenced how as extractive companies come to hold operations in particular communities, these regions face increased poverty, environmental degradation, and low participation in the job market which these companies are expected to promote (Rodríguez Garavito et al., 2017).

Women have also been particularly affected by extractive activities in the country. Due to their close relation to livelihoods, women are the first in experiencing the negative impacts of environmental damage in their food security and health (Tobón, 2015). Moreover, this dependency on livelihoods, added up to the extractive companies' men-focused job market, has resulted in a reduction in women's economic autonomy, as paid activities for them are limited. In some cases, regions in which these companies operate saw increases in sexual exploitation and trafficking as the only paid activities these women could access (Tobón, 2015).

Furthermore, while not standing as its main cause, natural resources have played a key role in fuelling Colombia's 52-year internal armed conflict. Resource-rich regions have dealt with increased pressures between different actors, ranging from transnational companies to the state. For instance, between 2002 and 2012, 80% of human rights violations registered in the country took place in regions in which natural resource extraction activities took place, including those related to energy production such as carbon mines and fossil fuel extraction plants (CINEP, 2012). While not all violence in these regions has been related to legal extraction activities, private extractive companies have been held responsible for hiring paramilitary groups to protect their businesses from both armed groups and civil society opposing their activities (Rodríguez Garavito et al., 2017). These tensions affect disproportionately women, who have faced sexual violence and displacement in these contexts of the armed and social conflicts (UN Women, 2016).

Colombia is likely to be among the countries highly affected by the impacts of the climate crisis (IDEAM-UNAL, 2018; IPCC, 2021), and the changes are expected to be endowed differently across marginalised social groups, including women. For instance, it has been reported that Black and Indigenous women have been disproportionately displaced by climate-induced tensions (Cárdenas et al., 2021). Furthermore, global environmental changes do not stand alone as the only impacts the country will face. Climate mitigation efforts including the country's commitment to reduce by 51% of its greenhouse emissions by 2030, as well as those resulting from the energy transition measures from Colombia's coal importers, will bring significant socio-economic challenges given the role coal exports have played for the country (Glynn et al., 2020; MADS, 2020; WRI, 2021). All these occur while simultaneously facing a peacebuilding process since 2016, the year in which a peace agreement was signed between the government and the *Fuerzas Armadas Revolucionarias de Colombia (FARC)*.

Climate and Coal Policies and their Challenges

Efforts on designing and implementing effective climate policies—including those related to a just transition (i.e. the phase-out of fossil fuels, especially coal, and the transition to renewable energies towards a climate-neutral economy, putting people first and ensuring that no one is left behind)—are thus deemed necessary to sustaining peace and facing the socio-economic challenges that this phenomenon will bring to the country. Up until today, Colombia's climate policy, mirroring international trends, has slowly shifted from an initial emphasis on mitigation to adaptation. At the international and national levels, an evolution in climate policy is evident; in the 1990s and early 2000s, there was a concentration of efforts to reduce greenhouse gases (GHG) emissions (mitigation) to combat climate change. After the 2000s, the adaptation agenda appeared as crucial as the mitigation agenda in climate policies due to the increasing manifestations of the climate crisis.

During the period from 1994 to 2010, the country's predominant approach was on mitigation. A second period corresponding to 2010–2014 focused on adaptation and disaster risk management. One of the primary triggers for this orientation change was the impacts of the "*La Niña*" phenomenon that hit the country at the end of 2010 and brought economic and social challenges (Rodríguez, 2020). The third period, corresponding to the years 2014–2018, was marked by the need and obligation of Colombia to adhere to and comply with the new international commitments: the Paris Agreement and the Sustainable Development Goals. Within this period, Colombia issued the "Third National Communication on Climate Change" (2015), the National Policy on Climate Change (2017), and Law 1931 of 2018. Thus, from 2018, the government reinforced the implementation stage of the National Climate Policy and the Climate Change Law. It is noteworthy to mention that the law also regulated the so-called Comprehensive Territorial Climate Change Management Plans (*PIGCCT* or *Planes Integrales de Gestión del Cambio Climático Territorial* in Spanish) and the Comprehensive Sectoral Climate Change Management Plans (*PIGCCS* or *Planes Integrales de Gestión del Cambio Climático Sectoriales* in Spanish). These instruments were created to identify, evaluate, prioritise, and adopt measures related to climate change according to Colombian territorial distribution and sectors of the economy under the scope of each ministry.

The development of these policy instruments represents a significant step forward for Colombia to fight the climate crisis. However, the real challenge lies in their implementation. As such, it is necessary to strengthen and articulate actions from state and non-governmental entities, civil society, and the entire population in general, to achieve compliance with the commitments made by the country while making sure no one is left behind. For this reason, climate policy needs to become gender-responsive. Colombia, in its updated Nationally Determined Contributions (NDC) and in its 2050 Long-Term Strategy to fulfil the commitments of the Paris Agreement, has pledged to include gender perspectives. However, as our analysis shows, there is still a long way to go in making Colombia's climate policy gender-responsive.

As far as this chapter is concerned, we look at the climate-gender-coal nexus through the coal sector as a case study. The reasons for choosing the Colombian coal sector as a case study are threefold and lie in (1) the urgent need here for an effective climate policy, (2) the high relevance of gender aspects, and (3) the current window of opportunity for the creation of a gender-responsive climate policy in this sector. At this moment, there is no such climate policy exclusive to the coal sector. The coal subsector is combined with the hydrocarbons sub-sector under the mining and energy sector umbrella. Despite the latter, it is pivotal to mention that the mining energy sector was one of the sectors prioritised in the action plan of the National Climate Change Policy. Thus, the Ministry of Mines and Energy, according to the guidelines defined in the National Climate Change Policy (MADS, 2017) and the obligations established in Law 1931 of 2018, was the first ministry in Colombia to adopt a *PIGCCS*. Through the *PIGCCS*, the mining and energy sector seeks to fulfil the mitigation and adaptation goals established in Colombia's NDC. The mitigation component aims to reduce emissions from the mining and energy sector by 2030 to 11.20 Mt CO2eq. Also, the mining and energy sector is expected to reach lower emissions in 2030 than those recorded in 2015, reaching its maximum peak in 2023, four years earlier than established in the NDC (Minenergía et al., 2021). Regarding the adaptation component, the specific goals for the energy mining sector by 2025 are as follows:

- Three planning instruments on hydrocarbon, coal mining, and electric power, with climate change guidelines to ensure comprehensive operating conditions under new scenarios of operational and environmental demands.
- An updated climate risk analysis methodology and a strategy to update it periodically at the national and corporate levels.
- An ecosystem-based adaptation project for the electricity sector that will help companies in the sector ensure compliance with their strategic objectives.

Although a sectoral climate policy is already in place, it is crucial to analyse its commitment to a successful transition of the coal sub-sector leading to social justice and equity with a strong gender perspective. How prepared are the mining regions for the transition? How are climate policies ensuring that the populations of these regions can be better off after the transition? How is this transition being differentiated with a gender perspective?

As mentioned before, we focus on La Guajira and Cesar, which are Colombia's two main coal-mining regions, producing around 90% of the country's coal (UPME, 2019). The Cerrejón mine, which is in La Guajira, is one of the largest open-pit coal mining areas in the world. The area covered more than 69,000 hectares in 2020 (Ulloa, 2020, p. 3) and has been steadily expanding since the end of the 1990s. In contrast, there is a larger number of mines in Cesar, none of them as big as El Cerrejón, such as El Corozo, La Loma-Pribbenow, El Descanso, El Hatillo, and Cerro Largo, among others (Herrera, 2010; País Minero, 2021).

Despite the massive exploitation of coal, both regions have not benefited from mining. In fact, the socio-environmental costs outweigh the economic value produced by coal mining, affecting mostly Black and Indigenous communities. According to regional reports, operating energy companies did not improve the situation of communities in the past but aggravated the local living conditions through water pollution, land displacement, and increasing environmental risks (Corral et al., 2021, p. 30).

For instance, in La Guajira—where around 45% of the population self-identifies as Indigenous (DANE, 2018)—groups affected by El Cerrejón include the Wiwa, Kogi, as well as the Wayúu. Approximately 55% of Wayúu living in La Guajira are affected by poverty and 25.7% by extreme poverty (Vidal Parra, 2019, p. 50; Mejía Curiel & Ramos Ruiz, 2014, p. 41). Since the El Cerrejón mine was developed, more than 70,000 Indigenous people have been displaced from their territory (Duro Montealegre, 2018, p. 29; Gerstenberg & González, 2019, p. 3). Similarly, Black and Indigenous communities in Cesar have struggled with the violation of their fundamental rights (Cardoso, 2015, p. 72).

In this context, Colombia is now facing the questions of how to handle its coal resources in the future and what an energy transition would look like. These are no minor questions, given the socio-economic consequences that the coal phase-out can have in Colombia and especially in the regions whose economies heavily depend on coal extraction. Also, regarding the commitment of Colombia to incorporate the gender perspective into the national climate policy, there are important ambitions that the country should put into practice, particularly in the coal sector, to ensure that any emerging climate policy is gender-responsive from the outset and promotes a just transition. However, gender-responsive climate policies, particularly for the coal sector, remain under-researched. Hence, in this case study we look at the inclusion of gender perspectives in current climate policies.

Conceptual Approach

During the last ten years, the field of gender, sustainability, and development has been fond of interest in the nexus between the current climate crisis and gender. Drawing on existing literature on the differentiated environmental impacts and contributions of men and women, mainly from ecofeminism and feminist political ecology, scholars have elaborated a large body of developments on these phenomena (Ojeda, 2012). While research on gender-responsive climate policy, both in the sense of persuading the need for gender to be assessed in climate policy and critiques on its implementation, has been explored, these studies have been mainly limited to general rather than sectoral explorations, with a few exceptions drawn in relation to, e.g. the agriculture and the mobility sector.

At the same time, recent critiques of the literature on energy transitions pointed out how academic concerns on this issue focused on the development of clean energies and green jobs while ignoring the global nature of transitions and the challenges it represents at all stages of the energy life cycle (Healy & Barry, 2017). In this sense, it is key to keep in mind, as Healy and Barry (2017) say, that just transition and distributional impacts and the role of labour in low-carbon transitions should be addressed more systematically in energy justice analyses. The latter would allow explaining impacts in the context of fossil fuel phase-out processes, including mine closures, which has been an under-researched subject, as well as the socio-economic and gendered impacts of carbon-related extractive activities (Bainton & Holcombe, 2018; Sesele et al., 2021). Facing this scenario, we consider addressing the inquiry on how climate and energy policies in countries in which these transition processes are likely to occur are dealing with gender—and how they could and should—be deemed necessary.

As asserted in the introductory section, our research was approached from a feminist perspective. In doing so, this was developed from feminist research ethic, in which both the process of gathering data and the findings resulting from it aim to understand gendered oppression and provide strategies towards gender equality (Hesse-Biber & Leavy, 2006). As such, we depart from recent feminist critiques regarding gender mainstreaming in environmental and climate policy, which have expressed their concerns on the reproduction of gender stereotypes of women as victims or connected and caring of nature and the approach to mainstreaming as a gender number game in which fulfilling gender quotas has become equated with substantial representation (Lau et al., 2021; Westholm & Arora-Jonsson, 2018). Thus, our evaluation of gender considerations at a policy level will not be limited to identifying efforts on gender, but it will analyse the discourses through which gender is presented, as well as it will analyse levels of responsiveness.

These last criteria will be based on the following scale: Gender-negative and gender-blind describe the fact scenarios in which gender inequalities are not recognised. By ignoring gender differences, a perpetuation of existing gender discrimination can continue, or new forms of discrimination can occur. In contrast, gender-aware and gender-sensitive concepts recognise gender differences but do not provide impulses to overcome them. Approaches that pursue the recognition and transformation of gender inequalities are described as gender-responsive and gender-transformative. While gender-transformative approaches strive to break down traditional gender roles and thereby transform prevailing inequalities, gender-responsive approaches react with multifaceted approaches to respond to unequal structures (GIZ, 2018).

Finally, departing from a feminist standpoint on energy systems, we understand the relation between gender and energy is not limited to acknowledging the gendered impacts and contributions of energy systems but rather reading from feminist lenses the gendered power structures in which it lies (Bell et al., 2020).

Methodology and Data

For the analysis, we used a methodology of triangulation (McNeill & Chapman, 2005) which consisted of three research methods: (1) critical discourse analysis (Van Dijk, 2008), (2) corpus-informed document analysis (Tognini-Bonelli, 2001), and (3) corpus-based generation of quantitative data (Fest, 2015). In using these methods, we elaborated an in-depth analysis on (a) how the gender-climate change nexus was conceptualised in gender, climate, and energy sector policies and (b) the extent to which these policies were gender-responsive/climate-responsive.

To do so, we elaborated a specialised corpus (Baker, 2006) consisting of 32 documents (see ANNEX I) in which the specialisation criteria consisted of official government documents on policies and laws regarding (a) climate change, (b) energy and coal sector, and (c) gender policies. The specific documents which were selected for each corpus were based on the documents identified through interviews with gender and climate policy experts as the most relevant in the field. These include public policy roadmaps, documents of Colombia's National Council for Economic and Social Policy (CONPES), national development policy plans, government regulatory frameworks, law enactments, and presidential decrees. All documents included in the corpus were those corresponding to the time range between 1993 and 2020, since in Colombia relevant regulations on the environment (Law 99/1993), gender (2013), and energy (Law 143/1994) were formulated in this time frame. To analyse trends within policy subjects, we separated the main corpus into three sub-corpuses according to the three themes of interest: gender policy, climate policy, and mining and energy policy.

To run the analysis, each corpus was loaded separately into the corpus linguistic software LancsBox.v.4.v (Brezina et al., 2018). The software provided information regarding the frequency, collocation, keyness, and context of the gender-related/climate-related words identified based on the initial literature review on gender and climate change, which served as a basis for the corpus-based document analysis. In being triangulated with the critical discourse analysis, this analysis was able to identify not only the degrees to which policies incorporated gender perspectives and the tendencies this incorporation followed but also subjacent ideological structures (2006) on representations of women—more broadly gender—and the extent to which these offered opportunities to disrupt hegemonic discourses on gender and promote social transformation.

Finally, we generated corpus-based quantitative data to obtain information such as the percentage of documents in which women and gender appear as a working area as well as the comparative frequency in which it takes place compared to other social areas (race, ethnicity, class). These data were then used to complement and strengthen the documents' analysis on the findings of each corpus. Furthermore, the data were analysed using information based on the findings of an initial literature review about gender and climate change/gender and energy/gender and just transitions and gender and coal mining, as well as a set of interviews carried out with female environmental defenders and female-led environmental organisations in La Guajira and Cesar.

Results and Discussion of the Policy Analysis

Social stressors, including gender, shape people's vulnerability towards climate change. At the same time, they manifest in people's resilience and adaptive capacities, as well as the possibility to engage meaningfully in conversations on climate action and policy. As such, designing policies that provide scenarios that facilitate participation, respond to the needs that result from these stressors, and enhance transformation towards ending inequality is deemed necessary in a changing climate. When making a case for gender, policies in the context of the current climate crisis which aim for gender responsiveness should address the different ways women and men cope with the impacts of climate change, take climate action, and push for the transformation of social structures which means moving beyond gender stereotypes in order to involve women in climate policy and provide adequate mechanisms and budgets for climate change-related gender considerations to be effectively taken into account and implemented (Lau et al., 2021). The following section will thus analyse the extent to which these considerations come into place in Colombia's existing climate, gender, and energy policies.

Climate Change Policy and Gender

In the climate change policy corpus (CPCC), the incorporation of a gender perspective was evaluated through the iteration and use of *gender-associated words.* The results of the analysis suggest that efforts to incorporate a gender perspective in climate policy documents are to a large extent absent. Out of 23 documents, 10 made use of gender-associated words. However, the iteration of these words was not always related to the integration of a gender perspective. In fact, the use of these words in certain contexts can evidence how climate policies, even if these do not intend to incorporate a gender perspective, can reaffirm and reproduce hegemonic discourses on gender, being language central to gender oppression (Lazar, 2005). For example, in at least four documents of the corpus, the iteration of the word "*hombre*" ("man") was used as a replacement for the word "human being" as in "mankind", reproducing the patriarchal logic in which the norm of humanity is male, androcentric (Cameron, 1992). Furthermore, in documents such as the *PIGCCT* Cesar, the only reference to women was through the word "*madres*" (mothers) reproducing the idea that the recognition of women is relegated to its reproductive role (Fraser, 2013).

Only 25% of the CCPC (6 out of 23 policy documents) denoted some degree of incorporation of a gender perspective. From these, at least four general tendencies can be drawn. First, gender issues continue to be equated to women issues. Evidence of this is the fact that no policy document referred to diverse gender identities nor addressed men and masculinities as a key component to a gender-responsive policy.

Second, for the most part, policy documents failed to address gender from the lenses of *intersectionality*. Considering Mara Viveros' (2016) definition of intersectionality, what is expected in policymaking is for categories of difference (race, gender, class, ability, etc.) not to be approached solely by themselves but pointing out and addressing the specificity of experiences, oppressions, and needs the very intersection of these produces. In fact, only one policy document, the National Strategy for the Reduction of Emissions from Deforestation and Forest Degradation (*ENREDD+ or Estrategia Nacional para la Reducción de Emisiones por Deforestación y Degradación de Bosques in Spanish*), recognises to some extent (five iterations) the specificities of the positionality of Black women and rural women. However, there is no consideration of disability and indigeneity regarding incorporating gender perspective. Third, because these policies tend to oversimplify the incorporation of a gender perspective, the subjects it touches upon tend to be extremely limited, resulting in subjects such as gender and energy transitions being wholly ignored.

Finally, most documents assert—especially at the level of diagnosis and planning—these will consider gender issues (either explicitly through "gender mainstreaming", a "gender perspective", or including women/gender as a variable) without an in-depth explanation on how this will be developed throughout the policy or by limiting it to only one section of the policy. These cases thus evidence the reproduction of a broader tendency in "gender mainstreaming" in which it is seen as a technocratic element that can be "added and stirred" into policy documents, far from promoting real opportunities for gender equality in climate policy (Westholm & Arora-Jonsson, 2018).

The extent to which each of these policies incorporates a gender perspective draws interesting insights worth noting. On the one hand, two policies evidenced a basic or gender-aware incorporation of a gender perspective. The ecosystem-based climate change adaptation guidebook for Colombia (AbE or *Guía de adaptación al cambio climático basada en ecosistemas en Colombia* in Spanish), a policy document about ecosystem-based adaptation, evidences considerations in one section on how these actions could affect or strengthen the capacities of communities, with an emphasis on vulnerable populations which included women, children, and the elderly. Identifying women as vulnerable in relation to the differentiated impacts of adaptation strategies might be useful to the incorporation of a gender perspective, being presented by itself, without further explanation or providing other dimensions of women in relation to adaptation strategies and risks deflecting attention from gendered power structures (Arora-Jonsson, 2011). Nonetheless, in this case, it reproduces the idea that women's vulnerability is due to an "inherit nature" rather than the product of social relations. The second policy document, the Climate Change Considerations for Land Use Planning (CCOT or *Consideraciones de Cambio Climático para el Ordenamiento Territorial* in Spanish), which introduces considerations on climate change for spatial planning and territorial development, elaborates about the incorporation of a "gender perspective" as a suggested element to be added in when undertaking the considerations presented. To do so, the policy also asserts that the introduction of this perspective might benefit of including sex-disaggregated data as well as validation of this data with local communities. However, the incorporation is one that appears as a suggested item, not as a mandatory one, while at the same time lacking a properly developed explanation on what this perspective consists of. This, once again, points to the reduction of gender perspective as a mere technocratic "add-on".

On the other hand, the four policy documents left were classified as gender sensitive, two of which fall under the "highly gender-sensitive" category. The first document classified as gender sensitive is the National Plan for Disaster Risk Management (PNGRD or *Plan Nacional de la Gestión del Riesgo de Desastres* in Spanish), the main policy document on risk and disaster management. From the beginning, this policy sets as a commitment incorporating a gender perspective in all actions executed. Despite lacking an in-depth explanation on what this perspective consists of, there are a few elements which evidence how this will be implemented. First, the policy calls for training on gender awareness to be delivered to all public officials working under the National System for Disaster Risk Management, a key element when it comes to introducing a gender perspective in policies. Furthermore, under its strategic objective #5 related to education, the policy establishes educational programmes with a gender-oriented perspective, which will be introduced at a local level through an interactive and participatory process. This explanation is limited to one specific policy point, despite its commitment to incorporate a gender perspective in all its actions.

The second document that was classified as gender sensitive was *PIGCCT* Guajira—the territorial climate change plan for La Guajira. Even though this document does not call explicitly for the incorporation of a gender perspective in the plan as a whole, it does incorporate elements which respond to such logic. First, the plan provides a partial diagnosis related to the gender-climate change nexus, providing insights on existing gender-disaggregated data related to (1) the differential impact of climate change in food security, especially for female-headed households given existing statistics in Colombia concerning this population; (2) how climate change might worsen respiratory diseases suffered by women as a result of the use of traditional stoves and fuels in Colombia; and (3) the low levels of gender-specific investment assigned by the department of La Guajira according to 2017 data for adaptive capacity building.

Furthermore, the policy calls for further disaggregated data (including gender-specific data) on existing health conditions which climate change might exacerbate to build policies with special considerations on the adaptive capacities of the health system. Furthermore, the plan commits to the promotion and implementation of an agenda of environmental education with a gender perspective recognising existing local experiences and knowledge of women on water usage and storage for the implementation of the intersectoral environmental education agenda. Finally, the plan identifies the need for higher investment to build gender-related adaptive capacities. Although these elements point to a limited number of subjects related to the gender-climate change nexus in which the document touches on, it does evidence an effort to provide a diagnosis of the situation in this regard and some plans (not related at all to the diagnosis) to elaborate on. Moreover, it is one out of three documents which recognise, at least to some extent, in this case, related to water management, the knowledge of women and how this might contribute to the climate change agenda.

The last two documents—the ENREDD+ and the 2020 updated NDC—fall under a "gender-sensitive" classification, given the inclusion of aspects which lie between the gender-sensitive and gender-responsive degrees of integration. The first one, the ENREDD+, refers to the national plan for reducing emissions from deforestation and forest degradation, under the UNFCCC REDD+ framework. To start off, the plan introduces gender equality as one of the central principles of the strategy. According to the document, this is done acknowledging the central role it plays in addressing climate change and vice versa. Furthermore, ENREDD+ incorporates existing gender-disaggregated data to elaborate on a partial diagnosis related to the gender-climate change nexus. This includes data on (1) the effects that using traditional cooking and heating fuel and stoves has in respiratory diseases in Colombia, an activity that affects women disproportionately and may worsen with climate change, and (2) how access to resources varies between men and women and how it might have differential impacts regarding equality of participation in the conservation of ecosystems. While limited, it does provide information on the context and the interplay of gender and resources, both directly affected by the policy. Moreover, the policy incorporated the voices of women to the policymaking process, who accounted for 28% of the total participants who participated in it. According to the document, this helped specifically in providing insights on gender roles, perceived risks, and forest management.

At the level of the implementation, ENREDD+ draws on specific plans for the inclusion of a gender perspective. For example, the plan establishes specific platforms such as the *Plataforma de Mujeres Negras* (Black Women Platform) recognising both the gendered and racial impacts of climate change as well as the relation these women have with climate change and forest management. In addition, the plan establishes specific participation and capacity-building quotas for women through the implementation process and indicators through which the progress made in these areas can be tracked.

However, the policy lacks elements to be fully gender-responsive. First, while guaranteeing women's participation is necessary, it is not sufficient by itself. If attention is not paid to pre-existing inequalities (which means ensuring these are broadly addressed and not only in the context of the strategy), in most cases, meaningful participation of women will hardly take place. This is a common critique to REDD+-related programmes (Larson et al., 2015). Moreover, emphasis on participation is insufficient if women are not seen as key actors in the development of policies. In the document, while Black and Indigenous organisations are granted the responsibility over specific objectives and plans (as well as participation spaces), no women's organisation is granted this role. Once again, proving that participation is not sufficient by itself as an indicator of gender-responsiveness.

Finally, the 2020 updated NDC is the last highly gender-sensitive policy in the CPCC corpus. From the beginning, this document asserted a gender perspective was to be incorporated as a "transversal and integrative" element. It is worth noting that, compared to the NDC submitted by the Colombian government in 2015 which did not include a gender perspective, the document indeed presented a higher ambition in relation to this aspect. The document has a clear explanation of what this perspective consists of, recognising both the differentiated impacts of climate change on women and their women's role and capabilities in relation to environmental issues. Furthermore, the policy calls for the inclusion of climate change-specific considerations in the existing process of establishing institutional coordination and updating the national gender policy following the Lima Work Programme—which was established in 2014 to "advance gender balance and integrate gender considerations into the work of Parties and the secretariat in implementing the Convention and the Paris Agreement so as to achieve gender responsive climate policy and action" (UNFCCC, n.d.). All of this is necessary when addressing a complex nexus such as the gender-climate change nexus. Moreover, women's groups were invited to participate in the process of policy building, along with other interest groups. However, the interests of these groups are not explicitly presented in the document or any website, making it impossible to know if the concerns of the women participants were indeed included in the document.

Despite being introduced as a transversal element, the gender perspective appears related to specific goals and objectives, with no explanation. For example, when it comes to the objectives on climate change management, the gender perspective appears only related to the objective of "Education, capacity building and sensibilization" while ignored in objectives related to "Information, science, technology and innovation". A similar logic is evidenced in the sectoral measures related to the reduction of greenhouse emissions, in which gender equality (under SDG No. 5) is considered in measures regarding energy efficiency but not in measures regarding domestic water management, which makes no sense considering women's role in water management in these settings. While establishing far more details than other documents, as well as welcoming other documents, the way in which the incorporation of a gender perspective is evidenced in specific measures and objectives fails to go beyond the already criticised “add-and-stir” approach.

Mining Energy Sector Policy and Gender

In the framework of the public policy of the mining energy sector, a total of six key sector documents were analysed. These documents were chosen to include guidelines for both responses to the climate crisis and the gender approach. However, only two out of the six documents, the National Policy for the Formalization of Mining in Colombia (PNFMC or *Política Nacional para la Formalización de la Minería en Colombia* in Spanish) and the Gender Guidelines for the mining and energy sector (LGSME or *Lineamientos de Género para el sector minero energético* in Spanish), presented some degree of gender-responsiveness through mentions of gender-associated words.

It is worth pointing out that the nature of these documents at the level of its main policy target is radically different, and hence they should be regarded in different ways. While the LGSME is a sector-specific gender policy, the PNFCM is a policy in which the main goal is not gender equality but the formalisation of mining in the country. This can be seen in the fact that out of the 134 times that gender-associated terms were mentioned, 131 of them can be found in the *LGSME*, while only 2 can be found in the PNFMC.

It is worth noting that the LGSME opens the door for a further discussion on gender at the level of the mining and energy sector and other climate change-related sectors. Opening up this conversation around gender should be pointed out as a critical step on the pathway to develop more gender-responsive policies. Two central narratives on gender were identified in the LGSME. The first one addresses gender inequality from an integrationist viewpoint, showing the lack of integration of women as both the root cause and the solution to gender inequality within the sector. Specifically, to achieve gender equality, the policy deems as crucial elements the lack of participation by women in mining projects, the job market, and the built capacities of the sector. However, the document ignores the inequalities produced in the areas where the mining sector operates, such as environmental damage on specific resources which are vital to women such as water or sexual trafficking activity in which women are the most affected (Tobón, 2015). Accordingly, the solutions provided by the policy are limited to the integration of women and ignore both their role in transformative processes that position them as leaders and agents of change in their communities and their knowledge and experience in resource management.

Closely related to this first discourse that ignores the agency of women, the second narrative in the LGSME shows women as vulnerable victims of gender-based violence within their communities. This falls into the representation of women solely as victims rather than as active subjects capable of transforming both their lives and communities.

In regard to the PNGFM, the narratives incorporate a gender-sensitive approach. Although gender is not one of the main priorities of the plan, nor is it highly developed within the plan, the policy recognises the demands on gender equality, gender-associated violence resulting from mining activities such as sex trafficking, and the need for these to be addressed. Nonetheless, when undertaking these efforts, the policy does not perceive women as active agents of the process, but rather as subjects to be acted upon either by the state or by the companies of the formal mining sector.

Moreover, it is worth noting that gender narratives rarely intersect with those of climate change. Indeed, this intersection was made evident only on two occasions within the *LGSME*, which recognised the risks that the environmental crisis might pose to women, as well as the role these might play in adaptation to climate change (p. 7, 2020). In the case of the *PNGFM*, although climate change is addressed in two opportunities, none of them establish a link between climate change and gender issues.

Finally, as in the climate policy corpus, ideas on "gender" are limited to addressing women issues as no mentions of diverse sexual orientations and gender identities occur.

How Do Gender Policies Engage with the Climate Crisis and the Energy/Coal Sector?

In the gender policy corpus (GPC) the concern for climate change, evidenced through the iteration of the words "*cambio climático*" (climate change), "*variabilidad climática*" (climate variability), and "*crisis climática*" (climate crisis), added up for a total frequency of eight occurrences. Nonetheless, this concern is not reflected in all documents nor did it in equal depth through the ones in which it appeared. As such, the subject lacked explicit recognition in one of the most important national gender mainstreaming policies: *CONPES* Document 161 on gender equality for women.

First off, climate change as a concern only took place in the Guidelines for the National Public Policy on Gender Equality for Women (*LPPNG or Lineamientos de Política Pública Nacional de Equidad de Género para las Mujeres* in Spanish) and the Guidelines for the Territorialisation of the National Public Policy on Gender Equity (*OTPPNG or Orientaciones para la Territorialización de la Política Pública Nacional de Equidad de Género* in Spanish). Moreover, out of eight occurrences, the LPPNG accounted for seven of them, while the *OTTPNG* only accounted for one. When addressed, climate change appears under the concern of the risks and vulnerabilities posed to women because of the extreme weather events that result from climate variability. In this regard, it could be asserted that the discourse through which climate change-gender nexus is established in gender policies is one in which women appear solely as victims of climate change. In fact, the demands the *OTTPN* makes to policymakers regarding the gender-climate change nexus are primarily based on its negative impact on women associated to "natural disasters".

In the case of the *LPPNG*, this discursive trope remains equal. Climate change appears either to assert the need to create protocols which prevent, attend, and safeguard women (specifically rural women) from the effects of climate change; to address the need to establish a link between natural disasters, climate change, and women as a means to diminish factors of risk and vulnerability for women; to shed lights on the fact that rural women, in being small scale producers, will be the most harshly affected by climate change; to evidence how seasonal emergencies in the past have had disproportionate negative effects on women; and how climate variability poses a threat to women and their families.

This discourse of "women as victims" with a special emphasis on "natural disasters" proves problematic to a truly gender-transformative approach to the climate crisis. As such, the nexus is built on the imagery of women as victims of a wild, uncontrollable nature on the need for a paternalistic state to intervene (Brown, 1995; Ojeda, 2012). Hence, the policy fails to recognise women as active political subjects themselves who design actions and solutions to the climate crisis, while it reproduces the stereotype of women as victims in need of a paternalistic saviour. In fact, in the policies, political spaces with regard to climate change are open for women to express their "needs and demands" as victims, but no political participation is open for women to provide solutions to the crisis (Ojeda, 2012, p. 41).

While the policy does recognise the differentiated impacts of one of the multiple dimensions of climate change on women, the narrative it contains does not allow for a transformative approach to the nexus to take place. It fails to recognise, on one hand, the need to assess the relation between the state, women, and nature and, on the other hand, women's political role related to risk management and the building of solutions to the climate crisis. The effort of the LPPNG, should be noted, does open the door for the nexus to be developed, including a call for specific policy guidelines on the nexus to be developed (Ojeda, 2012).

Finally, reducing the climate crisis to the "natural disaster" dimension results in a depoliticization of the phenomenon as is understood solely as the enacting of the wild forces of nature. From a feminist point of view, an approach which ignores the power relations (including gendered power relations) which sustain the climate crisis stands as highly problematic (Allen et al., 2019; True & Parisi, 2013). At the policy level, this leaves no room for related discussions on political decisions, for example, on the need of just transitions that contribute to clean, sustainable, and, most importantly, fairer energy systems, which can prove useful to women's social and economic autonomy.

Conclusions

This chapter showed how the recognition of the climate-gender nexus has evolved during the last ten years in Colombia's policies on climate change, gender, and the mining and energy sector. The policy analysis revealed, on the one hand, that climate and energy policies are currently not gender-responsive and incorporate stereotypical understandings of gender roles. On the other hand, it showed that gender policies have not sufficiently incorporated concerns about the climate crisis. Thus, the question remains of how Colombia should strengthen political pathways and policy interventions that can enable conditions in the country to transform social injustices and inequalities and engender the energy transition.

In this chapter, we have attempted to show the urgency of formulating and implementing a gender-responsive climate policy based on intersectional approaches. Hence, in the coal sector, a GRCP can foster gender equality—which requires not only including women in the industry's job market but acknowledging and attending the specific harms the industry has done to women and their communities—and in turn, increased gender equality can lead to better climate policies and the achievement of a just transition.

Particularly, the policy analysis revealed that climate and energy policies in Colombia are currently not gender-responsive, in most cases with no references to women and gender at all, and when doing so, they incorporate stereotypical understandings of gender. The analysis further showed that gender policies have not sufficiently incorporated concerns about the climate crisis. In the cases in which it was included, it tended to depart from the idea of women as "helpless victims" of nature, once again reproducing gender stereotypes. Thus, despite the political attention to finding adaptive solutions to the impacts of climate change and finding pathways for just transitions that benefit all, there is still a policy gap in relation to gender in relation to climate change and the low-carbon energy transition.

Climate change and energy policy should be ambitious not only in relation to GHG emission reduction targets; it also has a great potential to lead the process of articulating climate change, coal, and gender agendas. According to the interviews carried out with female environmental defenders and female-led environmental organisations in the regions of La Guajira and Cesar (Mohr et al., 2020), energy transition presents itself as an opportunity for their demands to be listened to. Women are not only victims of environmental harm from both mines and climate alterations but have been at the forefront of resistance and organising despite the stigmatisation and marginalisation they have been forced to face. Women should hence be acknowledged as active agents who contribute to building innovative solutions to the climate crisis. Furthermore, due to the exclusion they and their communities have faced, the policies implemented until today lack knowledge of the needs of their communities. The energy transition can be thus seen as an opportunity to transition to a just and sustainable society in which the demands of these communities are heard. In contrast to these interviews, the policy analysis in this chapter showed that none of these concerns has been incorporated by policymakers in a serious manner. Until that does not occur, gender equality will not be achieved, and climate change will not be overcome.

The research for this publication was part of the project "Colombia, Gender and Climate: Women as Actors for Transformation". The project was carried out by Polis180 with collaboration from Transforma and received funding from GIZ. The views expressed in this chapter are those of the authors and do not necessarily reflect the official positions of Polis180, Transforma, or GIZ.

The gender-associated words used were such as "género" (gender), "géneros" (genders), "mujer" (woman), "mujeres" (women), "sexo" (sex), "sexos" (sexes), "hombre" (man), "hombres" (men), "madres" (mothers) and "madre" (mother) "LGBT", "LGTBQ", "LGTBI", "LGTB+", "orientaciones sexuales diversas" (diverse sexual orientations), and "orientaciones de género diversas" (diverse gender identities).

References

Allen, E., Lyons, H., & Stephens, J. C. (2019). Women's leadership in renewable transformation, energy justice and energy democracy: Redistributing power. *Energy Research & Social Science, 57*, 101233.

Arora-Jonsson, S. (2011). Virtue and vulnerability: Discourses on women, gender and climate change. *Global Environmental Change, 21*(2), 744–751.

Bacchi, C. L. (1999). *Women, policy and politics: The construction of policy problems.* Sage.

Bainton, N., & Holcombe, S. (2018). A critical review of the social aspects of mine closure. *Resources Policy, 59*, 468–478.

Baker, P. (2006). *Using corpora in discourse analysis.* Continuum.

Bell, S. E., Daggett, C., & Labuski, C. (2020). Toward feminist energy systems: Why adding women and solar panels is not enough. *Energy Research & Social Science, 68*, 101557.

Blanco Vizarreta, C., & Dongo Román, M. (2019). *Género e industrias extractivas en América Latina.* DAR.

Brezina, V., Timperley, M., & McEnery, A. (2018). # LancsBox v. 4. x.

Brown, W. (1995). *States of injury: Power and freedom in late modernity.* Princeton University Press.

Cameron, D. (1992). *Feminism and linguistic theory.* Springer.

Cárdenas, J., Downing, C., & Vélez, J. (2021). Climate-driven Recruitment and Other Conflict Dynamics in Colombia. http://collections.unu.edu/eserv/UNU:8329/MEACFindings8.pdf

Cardoso, A. (2015). Behind the life cycle of coal: Socio-environmental liabilities of coal mining in Cesar, Colombia. *Ecological Economics, 120*, 71–82.

CINEP. (2012). *Minería, conflictos sociales y violación a derechos humanos en Colombia.* Bogotá.

Connell, R. (2012). Gender, health and theory: conceptualizing the issue, in local and world perspective. *Social Science & Medicine, 74*(11), 1675–1683.

Corral, F., Santamaría, R., Mejía, A., Sánchez, Y., Cardoso, A., & Malz, N. (2021). *Hechos, realidades y perspectivas de la minería de carbón en el Cesar y La Guajira.* Colectivo de Abogados José Alvear Restrepo.

DANE. (2018). Geovisor CNPV 2018 – Grupo Étnico, Censo Nac. Poblac. Vivienda CNPV 2018. Retrieved November 10, 2021, from http://geoportal.dane.gov.co/

Decoalonize. (2021). Still Burning: Coal, colonialism and resistance. https://still-burning.net/book/

Duro Montealegre, R. M. (2018). *Territorio y re-existencia: alternativas al neo-extractivismo hegemónico: caso mujeres-lideresas Wayúu. La criminalización de las resistencias como última estrategia desarrollista de despojo en América Latina.* Fundación CICCUS.

Fest, J. (2015). Corpora in the Social Sciences-How corpus-based appraches can support qualitative interview analyses. LFE. Revista de Lenguas para Fines Específicos.

Fraser, N. (2013). *Fortunes of feminism: From state-managed capitalism to neoliberal crisis.* Verso Books.

Gerstenberg, F., & González, P. (2019). La minería de carbón en Colombia y la situación económica de las mujeres rurales: la comunidad El Hatillo (Cesar, Colombia). *Ambiente y Desarrollo*, Jg. 23. https://doi.org/10.11144/Javeriana.ayd23-45.mccs.

GIZ. (2018). Gender-responsive project management. A practical Guide.

Glynn, P. J., Błachowicz, A., & Nicholls, M. (2020). *Incorporating just transition strategies in developing countries nationally determined contributions* (pp. 1–21). Springer International Publishing.

Healy, N., & Barry, J. (2017). Politicizing energy justice and energy system transitions: Fossil fuel divestment and a "just transition". *Energy Policy, 108*, 451–459.

Herrera, H. (2010). Explotación de carbón en el Cesar. Retrieved December 2021, from https://www.semillas.org.co/es/revista/explotaci

Hesse-Biber, S. N., & Leavy, P. (2006). *Emergent methods in social research.* Sage.

IDEAM-UNAL. (2018). Variabilidad Climática y Cambio Climático en Colombia. http://documentacion.ideam.gov.co/openbiblio/bvirtual/023778/variabilidad.pdf

IPCC. (2021). Summary for policymakers. In *Climate change 2021: The physical science basis. Contribution of Working Group I to the Sixth Assessment Report of the Intergovernmental Panel on Climate Change* [MassonDelmotte, V., P. Zhai, A. Pirani, S.L. Connors, C. Péan, S. Berger, N. Caud, Y. Chen, L. Goldfarb, M.I. Gomis, M. Huang, K. Leitzell, E. Lonnoy, J.B.R. Matthews, T.K. Maycock, T. Waterfield, O. Yelekçi, R. Yu, and B. Zhou (eds.)]. Cambridge University Press. In Press.

Kramarz, T., Park, S., & Johnson, C. (2021). Governing the dark side of renewable energy: A typology of global displacements. *Energy Research & Social Science, 74*, 101902.

Larson, A. M., Dokken, T., Duchelle, A. E., Atmadja, S., Resosudarmo, I. A. P., Cronkleton, P., ... & Selaya, G. (2015). The role of women in early REDD+

implementation: lessons for future engagement. *International Forestry Review, 17*(1), 43–65.
Lau, J. D., Kleiber, D., Lawless, S., & Cohen, P. J. (2021). Gender equality in climate policy and practice hindered by assumptions. *Nature Climate Change, 11*(3), 186–192.
Lazar, M. M. (2005). Politicizing gender in discourse: Feminist critical discourse analysis as political perspective and praxis. In *Feminist critical discourse analysis* (pp. 1–28). Palgrave Macmillan.
MADS. (2017). Política Nacional de Cambio Climático. Bogotá, Ministerio de Ambiente y Desarrollo Sostenible.
MADS. (2020). Actualización de la Contribución Determinada a Nivel Nacional de Colombia (NDC). Retrieved October 16, 2021, from https://www4.unfccc.int/sites/ndcstaging/PublishedDocuments/Colombia%20First/NDC%20actualizada%20de%20Colombia.pdf
McNeill, P., & Chapman, S. (2005). *Research methods* (3rd ed.). Routledge.
Mejía Curiel, E. B., & Ramos Ruiz, J. L. (2014). El capital social en los resguardos indígenas wayuu del municipio de Riohacha (La Guajira – Colombia). *Revista Colombiana de Ciencias Sociales, 5*(1), 39–54.
Minenergía, GIZ, CIAT. (2021). Plan Integral de Gestión del Cambio Climático del Sector Minero Energético. Retrieved October 14, 2021, from https://pigccme.minenergia.gov.co/public/web/
Mohr, K., Rojas Castro, S., Meyer, K., Mahecha Groot, A. M., Daza Niño, N., & Rojas, M. L. (2020). Gender responsive climate policy. A case study of the Colombian coal sector. *Polis, 180.*
Ojeda, D. (2012). Género, naturaleza y política: Los estudios sobre género y medio ambiente. *HALAC, I*, 55–73.
País Minero. (2021). Integran dos contratos de Drummond para operación de Mina El Corozo en Cesar. Retrieved July 14, 2021, from https://www.paisminero.com/mineria/carbon-colombiano/23402-integran-dos-contratos-de-drummond-para-operacion-de-mina-el-corozo-en-cesar
Rincón Navas, C. M. (2018). La tributación nacional en las transferencias y regalías territoriales de Colombia: una búsqueda de equidad fiscal entre la nación y las entidades territoriales (2000–2017). *Investigación y Desarrollo, 26*(2), 125–158.
Rodríguez, G. A. (2020). *Retos para enfrentar el cambio climático en Colombia.* Universidad del Rosario.
Rodríguez Garavito, C. A., Rodríguez Franco, D., & Durán Crane, H. (2017). *La paz ambiental: retos y propuestas para el posacuerdo.* Dejusticia.
Sesele, K., Marais, L., & van Rooyen, D. (2021). Women and mine closure: A case study of policy in South Africa. *Resources Policy, 72*, 102059.

Strambo, C., & González Espinosa, A. C. (2020). Extraction and development: Fossil fuel production narratives and counternarratives in Colombia. *Climate Policy, 20*(8), 931–948.31.

Tobón, G. (2015). *Mujeres, conflictos socioambientales y la resolución 1325 de las Naciones Unidas*. Red Nacional de Mujeres, Bogotá.

Tognini-Bonelli, E. (2001). *Corpus linguistics at work* (Vol. 6). John Benjamins Publishing.

True, J., & Parisi, L. (2013). Gender mainstreaming strategies in international governance. In *Feminist strategies in international governance* (pp. 59–78). Routledge.

Ulloa, A. (2016). Feminismos territoriales en América Latina: defensas de la vida frente a los extractivismos. *Nómadas, 45*, 123–139.

Ulloa, A. (2020). The rights of the Wayúu people and water in the context of mining in La Guajira, Colombia: Demands of relational water justice. *Human Geography, 13*(1), 6–15.

UN Women. (2016). *El género y el papel de las mujeres en el proceso de paz de Colombia*. ONU Mujeres.

UNFCC, The Enhanced Lima Work Programme on Gender. (n.d.). Retrieved July, 2022, from https://unfccc.int/topics/gender/workstreams/the-enhanced-lima-work-programme-on-gender#:~:text=In%202014%20the%20COP%20established,responsive%20climate%20policy%20and%20action

UPME. (2019). Carbón. Retrieved February 2022, from https://www1.upme.gov.co/simco/Cifras-Sectoriales/Paginas/carbon.aspx

Van Dijk, T. A. (2008). *Discourse and power*. Macmillan International Higher Education.

Vidal Parra, S. (2019). The water rights-based legal mobilization of the Wayúu against the Cercado Dam: An effective avenue for court-centered lawfare from below? *Antípoda. Revista de Antropología y Arqueología, 34*, 45–68.

Viveros, M. (2016). La interseccionalidad: una aproximación situada a la dominación. *Debate Feminista, 52*, 1–17.

Westholm, L., & Arora-Jonsson, S. (2018). What room for politics and change in global climate governance? Addressing gender in co-benefits and safeguards. *Environmental Politics, 27*(5), 917–938.

World Bank. (2022). Colombia Data. Retrieved January, 2022, from https://datos.bancomundial.org/pais/colombia

World Bank, World Development Indicators. DataBank. (2019). Retrieved January 2022, from https://databank.worldbank.org/reports.aspx?source=2&series=SI.POV.GINI&country=COL

WRI. (2021). Colombia's early plans for green jobs and just transition. Retrieved January 2022, from https://www.wri.org/just-transitions/colombia

PART II

Women and Gender Inequality in Energy Transition

CHAPTER 5

Empowering Women in a Climate-Changing World Through Climate-Resilient Energy Access

Katrin Lammers, Camille Belmin, Zakia Soomauroo, and Martha M. Hoffmann

Introduction

The climate crisis is the most fundamental threat confronting humanity today. According to the Intergovernmental Panel on Climate Change (IPCC), severe, widespread, and irreversible impacts are expected by the end of the century if no additional mitigation efforts are undertaken (IPCC, 2014b). However, while implications can be observed globally,

K. Lammers
Reiner Lemoine Institut gGmbH, Berlin, Germany
e-mail: katrin.lammers@rl-institut.de

C. Belmin
Social Metabolism & Impacts, Potsdam Institute for Climate Impact research (PIK), Potsdam, Germany

International Institute for Applied Systems Analysis, Wittgenstein Center (IIASA, OeAW, University of Vienna), Wien, Austria

N. Rocha Lawton, C. Forson (eds.), *Women and the Energy Sector*,
https://doi.org/10.1007/978-3-031-43091-6_5

climate change does not affect communities and individuals uniformly and equitably. In particular, there is abundant evidence that women, especially poor women, tend to be disproportionately affected by the multiple impacts of climate change. This is notably true for those living in rural areas, who tend to be most reliant on natural resources for their livelihoods. Furthermore, this gendered vulnerability tends to re-enforce pre-existing inequalities between men and women (Eastin, 2018). As such, climate change could undermine existing efforts to promote gender equality in many parts of the world, and it is therefore crucial to understand the factors supporting women in building resilience to climate change.

Affordable and clean energy access for all, defined as the Sustainable Development Goal (SDG) 7 by the United Nations, is a fundamental necessity for human well-being and resilience (GEA, 2012; McCollum et al., 2018). It is also well documented that access to clean and affordable energy can contribute to women's empowerment and gender equality (SDG 5), as it improves health, reduces time poverty, enables access to information and education, and creates economic opportunities (Rewald, 2017; OECD & IEA, 2016; Das et al., 2020; Rotondi et al., 2020). Yet the specific role that energy access plays in reducing gender inequality in the context of climate change has been poorly investigated. Currently, the bulk of the literature focuses on the effect of energy access on gender

---

Z. Soomauroo (✉)
Reiner Lemoine Institut gGmbH, Berlin, Germany

Reiner Lemoine Foundation, Berlin, Germany

Chair of Sustainability Economics of Human Settlements, Technische Universität Berlin, Berlin, Germany

Mercator Research Institute on Global Commons and Climate Change, Berlin, Germany
e-mail: zakia.soomauroo@rl-institut.de

M. M. Hoffmann
Reiner Lemoine Institut gGmbH, Berlin, Germany

Graduate School EnergySystemTransition of the Reiner Lemoine Foundation, Berlin, Germany

Department of Energy and Environmental Management, University of Flensburg, Flensburg, Germany
e-mail: martha.hoffmann@rl-institut.de

equality (Pearl-Martinez, 2020; Nathwani & Kammen, 2019), on the impact of climate change on gender equality (Eastin, 2018; Aipira et al., 2017; Jerneck, 2018), or on the role that energy access plays in climate change adaptation. However, to our knowledge, there is a dearth of research regarding the gender-energy-climate nexus.

Scope of the Chapter

Here, we attempt to address this research gap by providing a reflection on the triangular dynamics between climate change impacts, gender equality, and energy access. To do so, we reviewed and extracted relevant insights from the three strands of literature, namely, on the effects of (i) climate change on gender equality, (ii) energy access on gender equality, and (iii) climate change on energy systems. This allowed us to identify the specific benefits of energy access for women's empowerment in the context of climate change, as illustrated in Fig. 5.1. We also summarised the measures

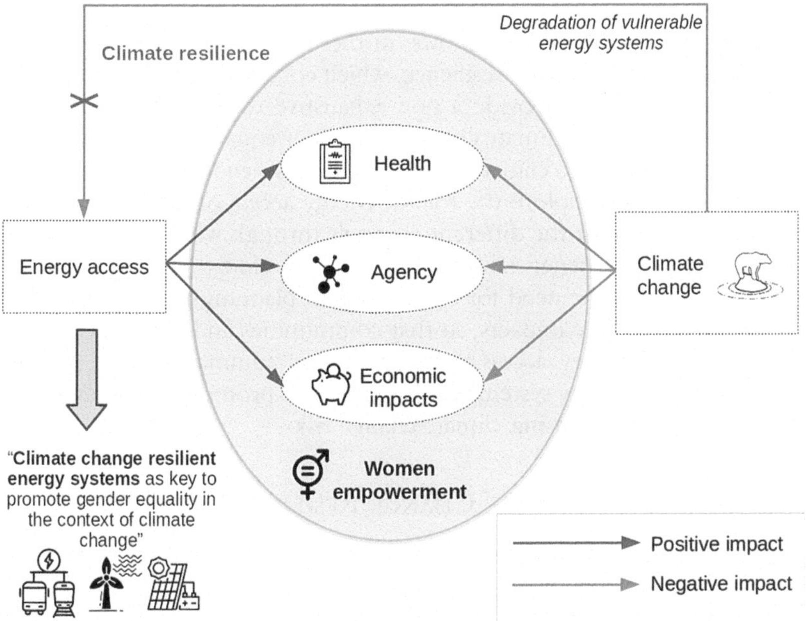

Fig. 5.1 Linkages between climate change impacts, women's empowerment, and energy access

needed to ensure that energy systems are resilient to the impacts of climate change so that women can reap the benefits of energy access. We primarily focused on rural areas in the so-called Global South, whose communities often lack access to clean and affordable energy, face higher climate risks, and have lower resilience.

On the Use of the Word 'Gender'

Throughout this chapter, we adopt a binary definition of gender, meaning that we only take into consideration two genders: male and female. The reason for this simplification is that the bulk of the existing literature on climate change and gender does not account for non-binary or trans people (van Daalen et al., 2020). However, we do acknowledge the complex reality of gender diversity and that climate change may disproportionately affect the experiences of non-binary, trans, and non-heterosexual people (van Daalen et al., 2020).

Outline of the Chapter

The chapter is structured as follows: in the upcoming section, we introduce the concept of climate resilience, which constitutes the framework of this chapter. We then provide a non-exhaustive review of the literature covering the effects of climate change on gender equality and demonstrating that vulnerability to climate change is a highly gendered phenomenon. Following that, we explore the role of energy access on building climate resilience and present the different channels through which energy access can be particularly beneficial for women facing climate change impacts. Next, we evaluate the need for energy system planning that accounts for future climate change impacts, so that communities and women can reap the benefits of energy access. Our conclusion summarises that climate change-resilient energy system planning is key to promote gender equality in a world facing increasing climate change risks.

Climate Change Resilience

The ability to preserve one's existence, well-being, and livelihood in the face of climate change is called climate resilience. The concept of climate resilience is focal not only to assess the impacts of climate change but also to define suitable adaptation measures. In the following, a general introduction to the concept is presented.

General Concept of Resilience

Resilience in general defines how individuals, communities, or societies continue to thrive and develop under shocks and stresses (Walker et al., 2004). Vulnerability, a commonly used term in gender and climate studies, is considered the opposite of resilience. Vulnerability accounts for the degree to which a person, process, or system is susceptible to, or unable to cope with, the adverse effects of shocks and stresses (DFID, 2011).

While talking about resilience, it is important to first define the context ('resilience of what?') and the cause of potential disturbances ('resilience to what?') as highlighted in Fig. 5.2 (DFID, 2011). For example, the context of resilience can be a specific community, a group of people (e.g. women), or an energy system, whereas the disturbances can be oil spills or storms.

Every system or process has a certain capacity to deal with these disturbances depending on their (i) exposure, (ii) sensitivity, and (iii) adaptive capacity. Exposure is the degree to which an entity (e.g. a whole community), a group (women), or an infrastructure is exposed to a disturbance (e.g. climate change). Sensitivity is the likeliness to experience adverse consequences while being exposed to hazards (Brooks et al., 2014), for example, extreme weather events. Adaptive capacity refers to the ability to cope and recover. This relates to factors that allow communities to anticipate and effectively plan for stresses and disasters, as well as to learn from experience of previous hazards and to act on the lessons learned (Brooks et al., 2014).

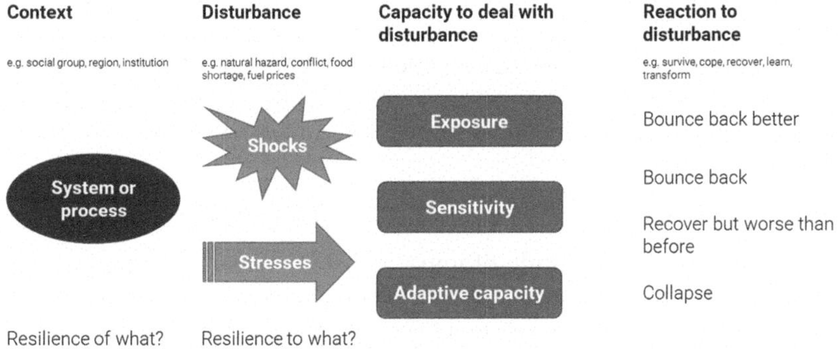

Fig. 5.2 Resilience framework, based on DFID (2011)

The three factors—exposure, sensitivity, and adaptive capacity—then influence how the system or process reacts to and is influenced by the disturbance (DFID, 2011). To relate this approach to the commonly used terms 'vulnerability' and 'impact', one can summarise that (i) exposure and sensitivity indicate the potential impact and (ii) the potential impact and adaptive capacity together are indicators for vulnerability.

Application to Climate Change Impacts on Women and Energy Systems

In this chapter, we concentrate on climate resilience as we focus on disturbances caused by the impacts of climate change. The context to which we apply climate resilience is either women or energy systems. Climate resilience describes the strength to prepare for, withstand, and recover from shocks and stresses caused by the impacts of climate change (Centre for Climate and Energy Solutions, 2021). Climate change-induced shocks are sudden events like storms, landslides, and extreme weather events that have a relatively short duration but can severely impact communities (DFID, 2011). In contrast, climate change-related stresses are long-term trends that increase vulnerabilities, such as rising temperatures, degradation of natural resources, or decreased agricultural productivity (DFID, 2011). It is important to note that the exposure to these climate change-induced shocks and stresses is often difficult to influence as it is mainly determined by geographical position, and relocation is often a last resort solution.

We will elaborate on the specific vulnerability of women in the face of climate change-induced disasters in the following section. Various measures have the potential to strengthen a community's resilience towards the impacts of climate change (IPCC, 2014a). This includes the development of proper disaster risk management and building adaptive capacity. Access to energy can play a pivotal role, and therefore we discuss the implications of energy access in later sections of this book chapter.

Gendered Vulnerability to Climate Change

Understanding the multitude of impacts that climate change has specifically on women is important to understand the role of energy access in supporting women facing climate change. In the following, we provide a non-exhaustive review of these impacts in three main domains: economic impacts, health, and agency. Throughout this section, we adopt the resilience and vulnerability framework presented in the previous section. The insights are summarised in Table 5.1.

Table 5.1 Overview over climate change impacts on women

Economic	*Agency and social wellbeing*	*Health*
Lower socio-economic status	Exclusion and inequality in decision making and resource accessibility	Higher risk of mental health issues
Lack of asset ownership (e.g. property)	Gendered cultural expectations	Less provision of reproductive and maternal health
Less survival skills	Limited autonomy	Higher rate of deaths from heat waves
Patriarchal nature of relief efforts management	Discriminatory access to food resources in disaster risk management	Higher risk of death and injury in disasters
Higher burden of care work	Precarious relationship to migration (male-out migration, own migration opportunities)	Higher exposure to food insecurity due to climatic stresses
Lower level of formal education	Deplorable makeshift refugee camps conditions	Lack of gender-disaggregated health data
Inequitable household dynamics	Inequitable socio-economic group dynamics (caste system)	Gender-based violence

On Vulnerability and Diversity

In this chapter, we will often refer to women as vulnerable, but we would like to stress that there is nothing intrinsically vulnerable or passive about women. Women are actually major agents of change in the climate crisis (Listo, 2018). Women can provide knowledge and experiences useful in a climate-changing world (Commission on the Status of Women, 2008). Women also tend to hold central responsibilities in the household and at the community level, which place them in a good position to advocate for such strategies. Instead, the disproportionate risks faced by women are the results of deep-rooted gender roles and perceptions imposed on women by patriarchal structures, rather than physiological or biological differences.

Women as Agents of Change: Global
'Women often have a strong body of knowledge and expertise that can be used in climate change mitigation, disaster reduction and adaptation strategies. Furthermore, women's responsibilities in households and communities, as stewards of natural and household resources, positions them well to contribute to livelihood strategies adapted to changing environmental realities.' (Commission on the Status of Women, 2008)

Regardless of our focus on women in this chapter, we also acknowledge that there are cases in which men face higher risks due to climate change. For example, men suffer a higher risk of suicide following extreme weather catastrophes or physical strain due to working outside under difficult conditions (Cianconi et al., 2020). That said, men generally have greater agency due to higher economic, social, and political power.

Last, although the empirical evidence described in this chapter shows that women are a particularly vulnerable group in the face of climate change, it is important to highlight that additional social structures such as ethnicity, race, religion, and caste as well as demographics such as age, education, and socio-economic background further intersect with gender identities. These aspects infer differentiated vulnerabilities and resilience within the heterogeneous group of women and society as a whole (Lau et al., 2021). In the boxes below, we provide two examples that underline the intersectional nature of climate change impacts.

Impact of Temperature Change: South Africa

In a study looking at the impact of temperature changes on income by different types of households in South Africa, Flatø et al. (2017) found that female-headed households were more impacted than male-headed households. They also found households headed by widows, never-married women, and 'left-behind' migrant households were the most affected types of households.

Impact of Hurricane Katrina: USA

When Hurricane Katrina struck the USA in 2005, those the most affected were African Americans, a historically vulnerable socio-economic group in the USA. Specifically, African American single mothers faced additional difficulties because they relied more heavily than married women on the community network, which was then eroded due to the resulting displacement. (Litt, 2008)

Economic Impacts

Improving women's access to financial and material resources is key to securing women's livelihoods and thereby strengthening their adaptive capacity (Rao et al., 2020). In this subsection, we describe how climatic shifts often limit livelihood options for women and decrease financial security and stability.

Ownership

Women are likely to have less control over assets like land or capital. This makes women more vulnerable to stresses and shocks, as 'assets, especially land, are critical because ownership can provide physical protection, a way to mitigate and manage crises, and adapt livelihood strategies to changing environmental conditions' (Deere & Doss, 2006). Globally, women own on average about 10–12% of the world's agricultural land (Raney, 2011), revealing a huge gender gap. In situations of climatic shocks and stresses, the lack of agricultural ownership can aggravate disparities in crop production and limit women's adaptive capacity (Eastin, 2018).

Income, Care Work, and Women's Dependency

'Rising familial burdens [...], declining food and water security, coupled with a relative inability to employ productive assets to cope with climatic shocks, should decrease female income generation capabilities, mobility, opportunities to build human capital, and access to formal credit markets. The result can reinforce gender disparities in the household division of labor and heighten female reliance on male income, which increases the opportunity costs of divorce and reduces women's independence.' (Eastin, 2018)

Care Work

Women globally provide more care work than men do. This care work is most commonly financially unpaid and reduces women's ability to earn a similar income to men, undermining their financial independence. Especially in rural areas, survival tasks such as securing water, food, and wood for cooking and heating often fall under women's responsibility, adding a further burden (Enarson et al., 2000). For example, firewood collection takes an average of 1.4 hours a day and even up to 5 hours per day in some countries, such as Sierra Leone (IEA, 2017). Climate change can exacerbate a scarcity of resources like firewood or water. This may increase the time needed to collect these resources and aggravate women's time poverty, as longer distances need to be travelled.

Multiple Caretaking Burdens of Women: Kenya
Women living around Mount Kenya in Kenya and Mount Kilimanjaro in Tanzania are heavily affected by drought and water shortages because they also bear the burden of caretaking. As a result, they devote significant time to both locating new water sources and delivering it to individuals under their care (Enarson et al., 2000).

Higher disaster risks, reduced livelihoods, or crop failure is likely to result in higher migration rates of men seeking opportunities in urban areas. The women left behind have to take on the additional burden of household chores. Limited financial autonomy, longer distances to crucial resources, and less time available increase women's sensitivity and exposure and decrease their adaptive capacity.

Family Commitments Lowering Income of Women: Pakistan
In Pakistan, where cotton crops are being destroyed by irregular monsoon rain and flooding, men are migrating for work, whereas women are unable to relocate due to family commitments, resulting in lower income for themselves (Rao et al., 2019).

Health

Climate change also affects the physical and mental well-being of people. In the following, we explore how climate stresses and shocks specifically affect women's health.

Health and Climate Stresses

A review paper by the Global Gender and Climate Alliance (2016) found that a majority of their examined studies revealed that women are more likely to suffer from climatic changes. The shortage of critical resources during climate change-induced droughts and crises can result in the discrimination of women. Due to gender hierarchies, girls are more likely to receive less food during periods of food shortages, making them more susceptible to malnutrition and diseases (World Health Organization and others, 2014).

Health and Climate Shocks
Women have a higher risk for death or injury from extreme weather events (Global Gender and Climate Alliance, 2016). When natural disasters strike, pre-existing discriminatory practices are exacerbated, and their detrimental health impact on women and girls is intensified. There is also a direct link to a higher risk of partner violence and food insecurity (Flatø et al., 2017).

Higher Number of Female Deaths from Heatwaves: Global
A string of studies found that women are more likely to die in heatwaves than men (Azhar et al., 2014; Fouillet et al., 2006; Rey et al., 2007). This is also observed in nine European cities, demonstrating that women in higher-income countries are also more vulnerable to heatwaves than men (D'Ippoliti et al., 2010; Toulemon & Barbieri, 2008). There may be some physiological reasons for an increased risk among elderly women, whereas social factors can also be important in determining the risk of negative impacts of heatwaves.

After large-scale natural disasters, primary healthcare system breakdowns can lead to an increased number of miscarriages as well as maternal and infant mortality (van Daalen et al., 2020). In addition, after disasters, pregnant women often experience heightened health risks, and there is a general trend of reduced access to reproductive and maternal care services.

Female Death Rate After the 2004 Indian Ocean Tsunami: Bangladesh
Women accounted for up to 80% of those who died during the 2004 Indian Ocean tsunami, and death rates among women were almost four times higher than those among men in the 1991 Bangladesh cyclone (DFID, 2011). Compared to children and men, more women died due to societal norms, rather than 'by chance'. (UN Gender Perspectives, 2008)

Agency

Agency is the 'ability to define one's goals and act upon them' (Kabeer, 1999). Yet women are restricted by certain structures, such as social obligations and lack of entitlements, inhibiting women's agency. This increases women's sensitivity to climate change impacts and limits their adaptive capacity (Jackson, 2003).

Education and Information

Girls are more likely than boys to quit formal education prematurely in periods of crisis, driven by traditional gender roles and societal expectations (Chigwanda, 2016). This causes them to be less informed about climate change and less likely to be involved in political decision-making processes that will affect climate change adaptation and resilience (Kwauk et al., 2019). Due to this, the perception of women has less weight in relief effort management planning, leading to less financial capacity and lower reparations (Neumayer & Plümper, 2007).

Agency in Times of Climate Stresses

On a structural level, relief efforts after disasters are almost exclusively managed and controlled by men, systematically excluding women, their needs, competencies, and experiences from contributing to these efforts (Enarson et al., 2000). As a result of migration, precarious and unsafe working conditions in destination areas could further negatively affect the personal development of both young men and women (Rao, 2019).

Diversification of Activities to Generate Income: Namibia

In Namibia, women in the village of Epyehsona started to diversify their activities in times of drought, for example, processing Marula nuts, doing basketry, or selling chickens. This was identified as an adapting strategy to climate variability (Angula, 2010).

Girls and women are less prone to be taught essential survival skills (Kwauk et al., 2019), such as climbing trees, which could prove useful during a flood. In addition, cultural expectations also add to the detriment of women's autonomy during times of emergency. For example, during the 1991 flood in Bangladesh, it was reported that wearing the traditional sari hindered women's ability to escape to a safe place (Juran & Trivedi, 2015).

In the aftermath of disasters, women are more likely than men to be displaced, to be sexually assaulted, to be victims of violence, and to face other human rights violations (Heslin et al., 2019). Cultural norms may also prevent women from moving freely during times of disaster.

Role of Energy Access on Climate Resilience and Gender Equality

Energy access can be defined as having reliable access to both a minimum level of electricity at the household and community level and clean and efficient cooking fuels (Perera et al., 2015). Access to affordable and clean energy, the SDG 7, is a central pillar of human development and well-being (GEA, 2012; McCollum et al., 2018). It is also necessary to achieve other SDGs like ending poverty, advancing health, improving education, water access and sanitation, addressing climate change impacts, and achieving gender equality (United Nations, 2020). This section shows how energy access, among manifold requirements, is also essential to build up climate resilience within communities and reduce gendered vulnerability to climate change. Again, we follow the three main dimensions of (i) economic impacts, (ii) health, and (iii) agency.

Economic Impacts

Energy access enables more diverse livelihood opportunities for communities in general, and more specifically for women, reducing their burden of care work and supporting their financial autonomy. This increases communities' autonomy and resilience to climate change and empowers women.

Communities' Resilience Through Diversification of Livelihood and Increased Productivity

Energy access is essential in all three economic sectors and its underlying activities (see Fig. 5.3), some of which can strongly support the resilience and adaptive capacity in the face of climate change: information and communications technology (ICT), media, agriculture, health, education, and consumer goods.

Electricity access in agrarian communities can increase resilience to climate variability and increase agriculture efficiency. Agricultural activities

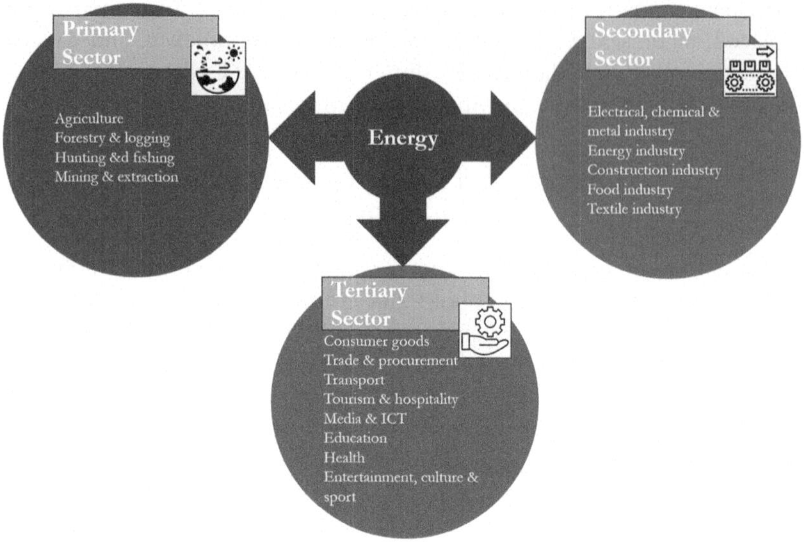

Fig. 5.3 Economic activities with strong energy input links, based on Stuart (2017)

heavily depend on water availability, but water supplies are threatened by fluctuations in precipitation patterns, with an increasing trend towards recurring extreme events (floods or droughts) caused by climate change. Increasingly unreliable water sources are therefore a challenge for farmers. Through the application of storage, pumps, and efficient irrigation systems, energy access can increase productivity and robustness in the agricultural sector while enabling food preservation, for example, through cooling technologies. This increases overall food security (Best, 2014; Stuart, 2017).

Improved Agricultural Practice Increasing Resilience: Senegal
A case study in rural Senegal demonstrated that electrification projects increased the resilience of communities by improving agricultural practices. The majority of agricultural activities are heavily dependent on underground water supplies due to low precipitation levels over the last few decades. The application of pumps and irrigation systems enabled them to cope with adverse effects of climate change (Perera et al., 2015).

Water purification technologies, especially desalination technologies, will play an increasing role in providing access to clean water in many countries, particularly among coastal communities (Ferroukhi et al., 2015). Given the increased risk of water scarcity, the application of water purification systems enabled by energy access adds to the increased resilience of communities.

Exacerbated Drinking Water Scarcity Without Electricity: Sahel Region
Due to the widespread lack of energy access in the Sahel region, it is estimated that around 68 million people have to find ways to transport water from as far as 10 kilometres every day. In addition to that, water tables are found 100 metres in the ground at most. Without energy access, people cannot tap into those groundwater resources and face great difficulty coping with prolonged drought conditions (Perera et al., 2015; Ferroukhi et al., 2015).

Last, some case studies have shown that electrification can open up new livelihood opportunities by introducing electricity-powered applications. For example, in a case study of a rural community in Senegal analysed by Perera et al. (2015), access to electricity created new market segments and livelihood opportunities, including tailoring, hairdressing, furniture making, welding, new restaurants, and others. The local production and provision of goods combined with wider livelihood options supports the overall independence of communities, creates diverse economic activities, and thus increases their climate resilience (UNDP, 2020). However, conclusions on the effect of electrification on the diversification of livelihoods should be avoided, as limited causal evidence exists on this subject (Pelz, 2020).

Women's Financial Autonomy
Electrification can increase women's financial autonomy and bargaining power and therefore contributes to reducing their vulnerability to climate change. By reducing the time spent on household chores, electrification can increase women's employment in the job market (Dinkelman, 2011; Grogan & Sadanand, 2013). In addition, access to electricity increases the income of female entrepreneurs, although male entrepreneurs are more likely to benefit from the promotion of productive uses of electricity (Pueyo et al., 2020).

Women's lack of asset ownership, particularly land, makes them more vulnerable to climate stresses and crop failures (Deere & Doss, 2006; Eastin, 2018). Access to clean and affordable energy can contribute to greater women's employment, earnings, and ultimately higher control and ownership of assets and reduce this gendered vulnerability (Das et al., 2020).

Electrification can also help women adapt to climate change. In communities where crop and livestock farming is the primary source of livelihood, pursuing non-agricultural activities has been noted as an adjustment strategy in times of crop failures. For example, Kuvare et al. (2008) noted that women are more likely to sew and sell traditional drinks or pottery items. Access to electricity at the household level or community level can help women engage in these non-agricultural activities to increase their revenues by mechanising work and helping them pursue financial autonomy.

Reduced Reliance on Natural Resources

Energy access can enhance women's resilience to climate change by reducing the household's reliance on natural resources, which may become more scarce, for example, with the increased frequency of droughts in areas affected by desertification. Traditionally, women are responsible for managing the resources necessary to feed their families. In rural areas, in particular, women often take care of subsistence crops and small livestock, collect water, and gather fuelwood. Therefore, women are likely to be more affected by the scarcity of natural resources.

Scarcity in fuelwood and water may also increase women's and girls' exposure to harassment or sexual assault, as they need to travel a longer distance to collect natural resources. This can be especially crucial in areas of conflict. For example, during the war in Darfur, many rapes with predominantly female victims were reported by Médecins Sans Frontières (MSF) International (2005), and 88% of these women were pursuing their daily activities like fetching water by the river bed. Access to pumped water through electrification and access to clean and sustainable fuels make resources available directly in the household and help to reduce women's exposure to violence in times of conflict.

Finally, wood scarcity can also affect women's reproductive choices and can lead parents to have more children who might help them to collect wood (Aggarwal et al., 2001). This effect could increase in the future due to higher risks of droughts in many parts of the world. Access to clean cooking fuels can reduce reliance on firewood, prevent women from walking longer distances and being exposed to violence, and improve their ability to make reproductive choices independent of necessity (Belmin et al., 2021).

Health

There is nexus between energy access and health in the context of climate change impacts.

Improved Health Services
Rural health centres are able to extend their services if they have access to a reliable electricity supply. This allows the operation of a broader range of medical equipment and the storage of vaccines and medications through cooling options (freezers and fridges) (Perera et al., 2015), both allowing for more treatment options. In addition, with access to electricity, rural communities gain independence from urban infrastructures (e.g. hospitals, health, and vaccination centres) and are enabled to provide first aid and accelerate the recovery process in case of an emergency like a natural disaster (Gippner et al., 2012). However, access should be reliable for communities to truly benefit from energy supply in health services (IEA, 2014).

Higher Number of Health Workers in Electrified Areas: Nepal
In rural areas of Nepal with access to electricity, there is an average of 11 health workers per 10,000 people, compared to two health workers per 10,000 people in non-electrified areas (Perera et al., 2015).

Improved Health Conditions
Indoor air pollution caused by traditional cooking technologies results in a heavy burden of respiratory diseases and annually 2.8 million premature deaths (IEA, 2017). This burden once again falls predominantly on women and children, as they spend more time indoors on cooking and household chores. The health condition of people—and women in particular—can therefore be improved with access to clean cooking energy. This can enhance their physical resilience to the adverse health effects of climate change and put them in a better position in the face of climate change impacts.

Specifically, access to clean cooking fuels can improve women's health during pregnancy (Wang et al., 2019), enabling them to endure disaster-related stress better during pregnancy, delivery or postpartum, and thus reduce the number of maternal deaths related to climate change-induced disasters.

Agency

Access to clean and affordable energy comes with several co-benefits that increase people's adaptive capacities, particularly women. We focus on three fundamental mechanisms: access to information and communications technology (ICT), access to education, and decreased burden of care work, which applies more specifically to women. Closely connected to these aspects, we give a short overview of changes in social norms evoked by energy access that may lead to more gender equality.

ICT for Communities

Electrification is a prerequisite for adaptation measures that require ICT (see Fig. 5.3). If the electricity supply is guaranteed, communities are empowered to access relevant information, such as weather forecasts, and communicate with relevant institutions in case of an emergency. This allows for better preparation of incoming disturbances. In addition, the implementation and reliable operation of early warning systems are facilitated once a community has access to electricity (Perera et al., 2015).

Additionally, electricity allows greater access to information and knowledge by using ICT (television, radio, phones) for the general public and improves personal circumstances, with access to services and social safety nets as co-benefits (Perera et al., 2015).

Impact of ICT Access for Farmer: Nepal

Farmers in Nepal can benefit from access to ICT as they can access early warning systems about floods that could affect their crops and livestock while at the same time monitoring current market prices to optimise their profit (S. Giri, 2011).

Improved Quality of Education and Learning

Education is a central requirement for people's adaptive capacities and disaster preparedness (Hoffmann & Blecha, 2020; Hoffmann & Muttarak, 2017). Access to clean and affordable energy is beneficial for climate adaptation as it facilitates access to high-quality education, which has proven necessary for people's adaptive capacities and preparedness for disasters (Hoffmann & Blecha, 2020; Hoffmann & Muttarak, 2017). Energy access enables education through various channels. First, access to clean and affordable energy enables children to attend school as it reduces the burden of household work, which falls predominantly on women and children (Das et al., 2020). Second, reliable electricity access creates higher-quality educational tools, methods, and materials, attracting more skilled teachers to rural communities (Gippner et al., 2012; Perera et al., 2015). Third, it also enables access to a more diverse set of sources of information, which enhances the adaptive capacity building of communities and, in turn, increases climate change awareness and preparedness (Stuart, 2017). Last, children benefit from extended potential learning periods through electric lighting and more time availability, resulting in better learning outcomes (Gippner et al., 2012).

> ***Increased Schooling Time Through Energy Access: Nepal***
>
> In rural Nepal, boys from electrified households spend on average 0.4 more years in school than boys from non-electrified households. The study revealed important gender differences, as girls from electrified household spent only 0.2 more years in school than girls from non-electrified households. (Perera et al., 2015)

Decreased Care Work for Women

As we previously introduced, women tend to bear the burden of household and care work. This creates highly gendered time poverty, which tends to be more prevalent in rural communities without access to energy. This time poverty can be reduced by access to clean and affordable energy. For example, access to improved cooking technologies leads to less time spent on fuel collection. However, the way women use their freed-up time is highly context dependent (Mahat, 2006).

Women's Empowerment and Change in Social Norms
Generally, access to clean and affordable energy can help women empower and overcome patriarchal structures that increase their vulnerability to climate change and can even cost their lives during disasters. Even though social factors like intra-household power dynamics or cultural factors can act as barriers to the positive effects of clean and affordable energy on women's empowerment (Rosenberg et al., 2020; Standal & Winther, 2016), evidence exists that access to clean and affordable energy enhances girls' and women's access to education, improves their control over assets, increases asset ownership, and increases their confidence (Das et al., 2020). In addition, electric appliance ownership like T.V. (La Ferrara et al., 2012) or mobile phones (Rotondi et al., 2020) can improve their access to information on reproductive health and expose them to alternative norms. In general, empowerment and education allow women to face climate-related risks with more resilience.

Energy access can serve as a long-term enhancer of women's status and empowerment and at the same time help prevent the exacerbation of gender inequalities in the case of disasters or economic and climatic shocks and stresses.

Climate-Resilient Energy Systems

For both communities and women to benefit from energy access and improve their resilience to climate change, it is fundamental that energy systems are inherently resilient to the effects of climate change. Therefore, energy planning that takes climate change into account is essential. If climate change suddenly (shocks) or gradually (stresses) leads to disruptions in energy supply, communities cannot benefit from the positive effects of energy access. Climate change shocks and stresses can affect both large and interconnected energy systems or small decentral standalone systems. Since this chapter focuses on the vulnerabilities of women living in rural areas and takes many examples of the Global South, we decide to focus on smaller decentral (off-grid) energy systems, as they are the most prevalent type of energy infrastructure in rural areas of the Global South. In this section, we take a closer look at how reliable access to energy can be achieved in the face of climate change.

Climate Resilience or the Blind Spot of Energy System Planning

Currently, electricity access and climate resilience are barely linked in current adaptation and electrification strategy planning. A review on the topic compiled by Perera et al. (2015) concludes that 'only a few' studies linked energy access with climate change impacts and the integration of climate resilience considerations. Ebinger and Vergara (2011) also emphasise the importance of integrated risk-based planning processes in the energy sector to address occurring climate change impacts and build resilience. Another review conducted by Schaeffer et al. (2012) highlights that '(...) climate impacts research is fundamental in developing tools to assist energy planners and policymakers in avoiding surprises and overcoming potential energy systems' bottlenecks (...)' (Schaeffer et al., 2012). Looking at the findings of the previous sections, it becomes evident that efforts to develop and strengthen resilient energy system planning are needed to sustain the positive impacts of electrification on women's climate resilience.

> ***Climate-Resilient Infrastructure Planning Is Reasonable***
> The UN estimates in their Climate Action Pathway report of 2020 that without climate change mitigation and adaptation actions, the global cost increase of climate-related disasters response adds up to a total of USD 2.7 trillion over the next 20 years. The cost of making infrastructure resilient is only about 3% of this amount. (Global Climate Action, 2020)

Climate Change Impacts on (Off-Grid) Energy Systems

To implement resilient energy system planning, it is first necessary to understand which climate change-induced threats impact energy systems. Focusing on rural areas and energy access provision, Stuart (2017) identified four main climate risks affecting the components, efficiency, and demand of (off-grid) energy systems: (i) temperature increase, (ii) fluctuating precipitation leading to droughts, torrential rains, and floods, (iii) extreme weather events such as storms and cyclones, and (iv) sea-level rise.

Increased temperature reduces system efficiency and increases cooling needs leading to higher energy demands (Lammers et al., 2020; Schaeffer et al., 2012; Stuart, 2017). Most impacts on energy systems caused by sea-level rise, precipitation fluctuation (e.g. causing floods), and extreme

Fig. 5.4 Climate change impacts on technical equipment. Solar panels destroyed by hail and storm (*left*), house destroyed by a storm (*middle*), electricity lines destroyed (*right*), Bulon Don island, Thailand. (Taken by Katrin Lammers, 2017)

weather events primarily affect the physical state of technical components (damage of equipment) (Lammers et al., 2020; Schaeffer et al., 2012; Stuart, 2017); see Fig. 5.4 for examples from the field. These damages lead to disruptions in energy supply and may cause significant setbacks for affected communities and particularly women.

Measures to Increase Climate Resilience

Adaptation measures to mitigate the previously mentioned four climate change risks can be divided into engineering and sociopolitical measures (Asian Development Bank, 2013). Engineering measures directly aim at increasing the resilience of the technical assets to the impacts of climate change, while sociopolitical actions examine the energy system as embedded into its environment and include, for example, the surrounding community, geographical characteristics of the sites, decision-making processes, and land use planning.

Based on a study conducted by Lammers et al. (2020), we compile potential technical adaptation measures in Table 5.2. Either individual or combined, these measures can address and improve energy system resilience. They should be selected depending on the hazard occurring in the specific area of installation.

In general, a diverse energy mix and modular and distributed systems are of great value as they enhance energy security by eliminating reliance on a single source or system. This increases the ability of the energy system to withstand shocks and stresses and thus leads to higher system reliability. This will sustain the contribution of reliable and climate-resilient energy supply to greater climate resilience in rural areas. In the following, a selection of measures to enhance the resilience of energy systems at the technical and sociopolitical levels is described.

Table 5.2 Overview of technical adaptation measures to mitigate climate risks of (off-grid) energy systems based on Lammers et al. (2020)

Measure	*Explanation*
Diversification of energy mix and high renewable energy shares	Low dependency on fossil fuels Reduced likeliness of supply shortcuts and full system failure Local energy generation according to the most optimal resource available
Modular and distributed systems	Reduced need for transmission lines being exposed to e.g. storms or landslides Reduced likeliness to supply shortcuts and full system failure (if one part of the modular system fails, another part is able to operate independently) Interconnecting several mini-grids to larger supply systems to enhance energy security
Energy container solutions	More flexibility to rearrange the system's position Allows for reactive actions in case of changing hazard zones Allows to easily raise the energy system above the ground in case of frequent flooding or land slides
Concrete-sided buildings to house sensitive energy system equipment (instead of metal)	More resistance to wind and salt corrosion Stays cooler than metal housing
Underground cabling	Higher resistance to wind and, to a certain extent, flood surges Cooling effect on cables for increased efficiency
Cooling for substations, transformers, inverters, and storage technology	Increased efficiency and lifetime of components

Energy System Design

With the global temperature rise, it will be beneficial to install cables underground to take advantage of the cooling effect of the ground and to place sensitive components of the energy system, such as batteries and inverters, in a cool environment (e.g. proper concrete building with air circulation or even air conditioning). These measures also help protect energy systems' components in coastal areas from ageing and efficiency loss due to salt corrosion. To prevent the risk of flooding, installing energy supply systems in areas of high elevation or far from the shorelines, rivers, and lakes is recommended. If this is not applicable (e.g. due to scarce land resources), raising the energy system equipment above the ground (e.g. on poles) is a promising option to reduce damages caused by floods. To minimise the impact of storms on electrical infrastructure, it is also beneficial to invest in underground cabling to reduce their exposure to storms. Deeply grounded mounting structures of solar power stations can mitigate damages caused by strong winds. Containerised solutions make it easier to move energy infrastructure to other areas if, for example, sea-level rise, frequent flooding, or other hazards jeopardise its operation.

Operation and Maintenance, Management and Standardisation

There are also operational and social adaptation measures to combat the risk of climate change impacts on energy systems: more robust operational and maintenance procedures and integrated and holistic land use planning and management (Asian Development Bank, 2013). The development of emergency plans and routines and an improvement of forecasting and modelling approaches can help detect upcoming bottlenecks, enabling proper emergency supply and security (Asian Development Bank, 2013; Stuart, 2017). Furthermore, to support mainstreaming of resilient energy planning, specific design codes, standards, and requirements for climate-resilient energy system planning can be integrated (Asian Development Bank, 2013).

Participatory Energy System Planning

Instead of top-down approaches, participatory energy planning can empower affected communities and thus increase their resilience (Rahman & Ghosh, 2016). The aim of participatory planning is that communities can directly get involved in the planning and implementation of the measures that will affect them, leading to a higher public acceptance. They thus develop the resilience to manage the risks of natural disasters while getting support from organisations with predictive capabilities (Rahman & Ghosh, 2016). Different tools to include the public in political planning and decision-making in participatory processes exist: frequent meetings and workshops between community representatives and politicians, public hearings, training, surveys, and joint field trips as well as offering the option of community ownership and management are some of them

(Stober et al., 2021). Ecosystem-based adaptation (EbA) approaches address climatic hazards and adaptation to long-term climatic change and are often related to participatory planning together with affected communities (UNDRR, 2021). They are another vital tool to increase resilience. EbA deals with landscape restoration (including afforestation and restoring land with grasses, shrubs, or trees) to reduce erosion and landslides that may affect communities and their energy systems (UNDRR, 2021). Community- and ecosystem-based solutions to improve climate resilience, which entail combining traditional knowledge and modern science, tend to be more sustainable and accepted by communities (Kabisch et al., 2017). Often, the involvement of women in participatory, community-based, and ecosystem-based measures and planning processes increases the chances of implementation and acceptance (Duflo, 2012; Corner, 1997; Rahman & Ghosh, 2016).

The Involvement of Women in Energy Projects Is Key

As with all transitions, if only specific stakeholders are involved in the planning process, also electrification projects may result in unfavourable impacts, such as an exacerbation of existing inequalities (Ahlborg, 2017). By taking a gender perspective on energy planning, Lieu et al. (2020) showed that women's viewpoints are dominantly absent in existing energy transitions across different geographical spaces and should explicitly be included ex-ante. The focus must be on creating, funding, and supporting an ecosystem of institutions and organisations that integrate women into the heart of energy planning. In her report on the role of women in sustainable energy development, Cecelski (2000) finds that women are the mainstream users and often producers of energy. Without women's involvement in renewable energy projects, these projects carry a particular risk of being inappropriate and failing (Cecelski, 2000). Another important aspect is women's empowerment through their involvement in energy planning and provision (Standal & Winther, 2016). If women take direct responsibility, their reputation and role within the communities are often strengthened (Standal & Winther, 2016).

These examples underline the importance of involving women in energy decision-making and planning. However, to be successful, energy projects have to be applied within a local and cultural context, considering cultural aspects in general and addressing women's needs specifically (Cecelski, 2000).

Women Empowerment: Afghanistan
The status of women in Bamiyan (Afghanistan) increased within their communities after being trained as solar engineers for local electrification projects. They became 'natural' members of the women village council. (Standal & Winther, 2016)

Integrated Planning for a Just Transition

For energy systems to withstand current and future climatic changes, they must consider these changes in technical designs and integrate social structures and adaptation needs into the planning procedure. Energy systems planned without considering climate change are prone to fail partially or entirely. Consequently, the benefits of energy access on rural populations, particularly women, will cease, increasing their vulnerability. Given that both climate change impacts and the electrification challenge are crucial issues to consider while discussing gender equality, it is evident that a just and sustainable transition requires integrated planning.

Conclusion

Climate change leads to large-scale shifts in the world's economic, societal, and natural systems. These global changes are rapidly redesigning the realities and livelihoods of humanity, which is simultaneously the main victim and main driver of such changes (IPCC, 2014b). The negative impacts of these global changes require systems, societies, and individuals to be capable of quickly adapting to those changes favouring those with the highest resilience (Vyas et al., 2009).

In this chapter, we explored the three-way relationship between climate change impacts, energy access, and gender inequalities. Because women are particularly vulnerable to the effects of climate change, their empowerment depends on improving their overall climate resilience. By reviewing the literature, we demonstrated that energy access is a central brick allowing adaptation and lowering vulnerability. Reliable access to energy is a prerequisite for better access to education, health services, and information. It can create job opportunities, increase agricultural productivity, empower communities, and increase their independence from urban areas.

In addition, access to energy is particularly beneficial to women in the context of climate change in many ways. Access to energy (i) reduces the burden of household chores, which predominantly falls on women, (ii) improves women's health, and (iii) can support financial independence. These three factors contribute to building women's climate resilience and reducing gender inequalities exacerbated by climate change.

Electrification projects that are resilient to the adverse effects of climate change are therefore vital in promoting these positive impacts. However, in order to achieve these long-term benefits, it is crucial to consider energy systems' vulnerability when planning and implementing them. Therefore, we present a set of measures that can enhance energy systems' resilience, ranging from technical to sociopolitical solutions. Technical measures to increase the resistance of the energy systems' components towards the impacts of climate change are as important as creating an ecosystem of organisations trying to integrate climate aspects and women into energy planning.

We believe a holistic approach is essential to create and sustain positive correlations between energy access and building adaptive capacities in order to empower gender equality. The focus should not only be on short-term least-cost options but include climate resilience criteria and involve those affected. We encourage decision-makers, planners, and project developers to incorporate climate resilience and gender aspects into energy planning to ensure that energy infrastructure can guarantee the resilience of communities in the long run and help empower women in the face of future climate change impacts.

Contribution of Authors All authors equally designed, conceptualised, and wrote the chapter. The group determined the topic and scope of this chapter. Martha M. Hoffmann organised and moderated the writing process of the book chapter. She reviewed and commented on each section and worked on their integration. Zakia Soomauroo performed research on gendered vulnerability of climate change. Camille Belmin investigated the impact of energy access on women. Katrin Lammers introduced the concept of (climate change) resilience and how energy access contributes to climate change resilience. She also presented the concept of planning for climate-resilient energy systems and gave a brief overview of measures to achieve this.

Acknowledgements The authors are grateful for the financial support of the Heinrich Böll Foundation, the Reiner Lemoine Foundation, and TUB PAS scholarship. The authors kindly thank Ulli Lich for her helpful comments and feedback on the manuscript. And most importantly, let us acknowledge all the powerful women surrounding us and acknowledge the power in ourselves—the power that only we are ... able to rouse.

> "I raise up my voice—not so that I can shout, but so that those without a voice can be heard. We cannot all succeed when half of us are held back."—Malala Yousafzai

References

Aggarwal, R., Netanyahu, S., & Romano, C. (2001). Access to natural resources and the fertility decision of women: The case of South Africa. *Environment and Development Economics, 6*(2), 209–236. https://doi.org/10.1017/S1355770X01000122

Ahlborg, H. (2017). Towards a conceptualisation of power in energy transitions. *Environmental Innovation and Societal Transitions, 25*, 122–141. https://doi.org/10.1016/j.eist.2017.01.004

Aipira, C., Kidd, A., & Morioka, K. (2017). Climate change adaptation in pacific countries: Fostering resilience through gender equality. In *Climate change adaptation in pacific countries* (pp. 225–239). Springer.

Angula, M. N. (2010). *Gender and climate change: Namibia case study*. (Tech. Rep.). Heinrich Böll Stiftung.

Asian Development Bank. (2013). *Guidelines for climate proofing investment in the energy sector*. https://www.adb.org/sites/default/files/institutional-document/33896/files/guidelines-climate-proofing-investment-energy-sector.pdf

Azhar, G. S., Mavalankar, D., Nori-Sarma, A., Rajiva, A., Dutta, P., Jaiswal, A., Sheffield, P., Knowlton, K., & Hess, J. J. (2014). Heat-related mortality in India: excess all-cause mortality associated with the 2010 Ahmedabad heat wave. *PLoS One, 9*(3), e91831.

Belmin, C., Hoffmann, R., Pichler, P.-P., & Weisz, H. (2021). Fertility transition powered by women's access to electricity and modern cooking fuels. *Nature Sustainability*, 1–9. https://doi.org/10.1038/s41893-021-00830-3

Best, S. (2014). Growing power: Exploring energy needs in small-holder agriculture. *International Institute for Environment and Development (IIED), Discussion Paper April 2014*. https://pubs.iied.org/sites/default/files/pdfs/migrate/16562IIED.pdf

Brooks, N., Aure, E., & Whiteside, M. (2014). Final report: Assessing the impact of (ICF) programmes on household and community resilience to climate variability and climate change (Tech. Rep.). Department for International Development (DFID). https://doi.org/10.12774/eod_cr.june2014.brooksetal

Cecelski, E. (2000). The role of women in sustainable energy development. *National Renewable Energy Laboratory (NREL), Energy, Environment & Development, NREL/SR-550-26889.* https://www.nrel.gov/docs/fy00osti/26889.pdf

Center for Climate and Energy Solutions (C2ES). (2021). *Climate resilience portal.* Online Portal. https://www.c2es.org/content/climate-resilience-overview/

Chigwanda, E. (2016). *A framework for building resilience to climate change through girls' education programming.* Center for Universal Education at Brookings.

Cianconi, P., Betrò, S., & Janiri, L. (2020). The impact of climate change on mental health: A systematic descriptive review. *Frontiers in psychiatry, 11*, 74.

Commission on the Status of Women. (2008). *Gender perspectives on climate change.* Retrieved from http://www.un.org/womenwatch/daw/csw/csw52/issuespapers/Gen-der%20and%20climate%20change%20paper%20final.pdf

Corner, L. (1997). Women's participation in decision-making and leadership—A global perspective. In *Conference on Women in Decision-Making in Cooperatives—Asian Women in Co-operative Development Forum (ACWF).* https://iknowpolitics.org/sites/default/files/decision-making26leadership_global.pdf

D'Ippoliti, D., Michelozzi, P., Marino, C., de'Donato, F., Menne, B., Katsouyanni, K., Kirchmayer, U., Analitis, A., Medina-Ramón, M., Paldy, A., Atkinson, R., Kovats, S., Bisanti, l., Schneider, A., Lefranc, A., Iniguez, C., & Perucci, C. (2010). The impact of heat waves on mortality in 9 European cities: results from the EuroHEAT project. *Environmental Health, 9*(1), 1–9.

Das, I., Klug, T., Krishnapriya, P., Plutshack, V., Saparapa, R., Scott, S., Sills, E., Jeuland, M., Kara, N., & Pattanayak, S. (2020). *A virtuous cycle? Reviewing the evidence on women's empowerment and energy access, frameworks, metrics and methods.* Energy Access Project.

Deere, C. D., & Doss, C. R. (2006). The gender asset gap: What do we know and why does it matter? *Feminist Economics, 12*(1–2), 1–50.

DFID. (2011). *Defining disaster resilience: A DFID approach paper* (Tech. Rep.). Department for International Development (DFID) London UK. https://assets.publishing.service.gov.uk/government/uploads/system/uploads/attachment_data/file/186874/definingdisaster-resilience-approach-paper.pdf

Dinkelman, T. (2011). The effects of rural electrification on employment: New evidence from South Africa. *American Economic Review, 101*(7), 3078–3108. https://doi.org/10.1257/aer.101.7.3078

Duflo, E. (2012). Women empowerment and economic development. *Journal of Economic Literature, 50*(4), 1051–1079. https://doi.org/10.1257/jel.50.4.1051

Eastin, J. (2018). Climate change and gender equality in developing states. *World Development, 107*, 289–305. https://doi.org/10.1016/j.worlddev.2018.02.021

Ebinger, J., & Vergara, W. (2011). *Climate impacts on energy systems*. World Bank. https://doi.org/10.1596/978-0-8213-8697

Enarson, E. P., et al. (2000). *Gender and natural disasters*. ILO Geneva.

Ferroukhi, R., Nagpal, D., Lopez-Peña, A., Hodges, T., Mohtar, R. H., Daher, B., Mohtar, S., & Keulertz, M. (2015). *Renewable energy in the water, energy & food nexus* (Tech. Rep.). IRENA.

Flatø, M., Muttarak, R., & Pelser, A. (2017). Women, weather, and woes: The triangular dynamics of female-headed households, economic vulnerability, and climate variability in South Africa. *World Development, 90*, 41–62.

Fouillet, A., Rey, G., Laurent, F., Pavillon, G., Bellec, S., Guihenneuc-Jouyaux, C., Clavel, J., Jougla, E., & Hémon, D. (2006). Excess mortality related to the August 2003 heat wave in France. *International Archives of Occupational and Environmental Health, 80*(1), 16–24. https://doi.org/10.1007/s00420-006-0089-4

GEA. (2012). *Global energy assessment*. Cambridge University Press. https://ideas.repec.org/b/cup/cbooks/9780521182935.html

Gippner, O., Dhakal, S., & Sovacool, B. K. (2012). Microhydro electrification and climate change adaptation in Nepal: Socioeconomic lessons from the Rural Energy Development Program (REDP). *Mitigation and Adaptation Strategies for Global Change, 18*(4), 407–427. https://doi.org/10.1007/s11027-012-9367-5

Global Climate Action. (2020). *Climate action pathway, climate resilience, executive summary* (Tech. Rep.). United Nation FCCC. Retrieved from https://unfccc.int/sites/default/files/resource/ExecSumm_resilience_0.pdf

Global Gender and Climate Alliance. (2016). *Gender and climate change: A closer look at existing evidence* (Tech. Rep.). https://wedo.org/gender-and-climate-change-a-closer-look-at-existing-evidence-ggca/

Grogan, L., & Sadanand, A. (2013). Rural electrification and employment in poor countries: Evidence from Nicaragua. *World Development, 43*, 252–265. https://econpapers.repec.org/article/eeewdevel/v_3a43_3ay_3a2013_3ai_3ac_3ap_3a252-265.htm

Heslin, A., Deckard, N. D., Oakes, R., & Montero-Colbert, A. (2019). Displacement and resettlement: Understanding the role of climate change in contemporary migration. In *Loss and Damage from Climate Change* (pp. 237–258). Springer.

Hoffmann, R., & Blecha, D. (2020). Education and disaster vulnerability in Southeast Asia: Evidence and policy implications. *Community, Environment and Disaster Risk Management, 12*(4), 1401. https://doi.org/10.3390/su12041401

Hoffmann, R., & Muttarak, R. (2017). Learn from the past, prepare for the future: Impacts of education and experience on disaster preparedness in the Philippines and Thailand. *World Development, 96*, 32–51. https://doi.org/10.1016/j.worlddev.2017.02.016

IEA. (2014). World Energy Outlook 2014.

IEA. (2017). *Energy Access Outlook 2017.* https://www.iea.org/reports/energy-access-outlook-2017providing-energy-access-for-all-by-2030

IPCC. (2014a). Climate Change 2014: Impacts, adaptation, and vulnerability. Part B: Regional aspects. Contribution of Working Group II to the fifth assessment report of the intergovernmental panel of climate change. In Barros, V. R. et al. (Eds.), (pp. 1327–1370). Cambridge University Press. https://www.ipcc.ch/site/assets/uploads/2018/02/WGIIAR5-PartB_FINAL.pdf

IPCC. (2014b). Climate Change 2014: Mitigation of climate change. Contribution of Working Group III to the fifth assessment report of the intergovernmental panel on climate change (Tech. Rep.). http://www.ipcc.ch/report/ar5/wg3/

Jackson, C. (2003). Gender analysis of land: Beyond land rights for women? *Journal of Agrarian Change, 3*(4), 453–480.

Jerneck, A. (2018). What about gender in climate change? *Twelve feminist lessons from development. Sustainability, 10*(3), 627.

Juran, L., & Trivedi, J. (2015). Women, gender norms, and natural disasters in Bangladesh. *Geographical Review, 105*(4), 601–611.

Kabeer, N. (1999). Resources, agency, achievements: Reflections on the measurement of women's empowerment. *Development and Change, 30*(3), 435–464.

Kabisch, N., Korn, H., Stadler, J., & Bonn, A. (Eds.). (2017). *Nature-based solutions to climate change adaptation in urban areas.* Springer International Publishing. https://doi.org/10.1007/978-3-319-56091-5

Kuvare, U., Maharero, T., & Kamupingene, G. (2008). Research on farming systems change to enable adaptation to climate change (Tech. Rep.).

Kwauk, C., Cooke, J., Hara, E., & Pegram, J. (2019). Girls' education in climate strategies. Opportunities for improved policy and enhanced action in nationally determined contributions (Tech. Rep.).

La Ferrara, E., Chong, A., & Duryea, S. (2012). Soap operas and fertility: Evidence from Brazil. *American Economic Journal: Applied Economics, 4*(4), 1–31. https://doi.org/10.1257/app.4.4.1

Lammers, K., Bertheau, P., & Blechinger, P. (2020). Exploring requirements for sustainable energy supply planning with regard to climate resilience of Southeast Asian islands. *Energy Policy, 146*, 111770. https://doi.org/10.1016/j.enpol.2020.111770

Lau, J. D., Kleiber, D., Lawless, S., & Cohen, P. J. (2021). Gender equality in climate policy and practice hindered by assumptions. *Nature Climate Change, 11*(3), 186–192. https://doi.org/10.1038/s41558-021-00999-7

Lieu, J., Sorman, A. H., Johnson, O. W., Virla, L. D., & Resurrección, B. P. (2020). Three sides to every story: Gender perspectives in energy transition pathways in Canada, Kenya and Spain. *Energy Research & Social Science, 68*, 101550. https://doi.org/10.1016/j.erss.2020.101550

Listo, R. (2018). Gender myths in energy poverty literature: A critical discourse analysis. *Energy Research & Social Science, 38*, 9–18.

Litt, J. (2008). Getting out or staying put: An African American women's network in evacuation from Katrina. *NWSA Journal, 20*(3), 32–48.

Mahat, I. (2006). Gender and rural energy technologies: Empowerment perspective. A case study of Nepal. *Canadian Journal of Development Studies / Revue canadienne d'études du développement, 27*(4), 531–550. https://doi.org/10.1080/02255189.2006.9669172

McCollum, D. L., Echeverri, L. G., Busch, S., Pachauri, S., Parkinson, S., Rogelj, J., Krey, V., Minx, J. C., Nilsson, M., & Riahi, K. (2018). Connecting the sustainable development goals by their energy inter-linkages. *Environmental Research Letters, 13*(3), 033006. https://doi.org/10.1088/1748-9326/aaafe3

Médecins Sans Frontières (MSF) International. (2005). *The crushing burden of rape: Sexual violence in darfur*. https://www.msf.org/crushing-burden-rape-sexual-violence-darfur

Nathwani, J., & Kammen, D. M. (2019). Affordable energy for humanity: A global movement to support universal clean energy access. *Proceedings of the IEEE, 107*(9), 1780–1789.

Neumayer, E., & Plümper, T. (2007). The gendered nature of natural disasters: The impact of catastrophic events on the gender gap in life expectancy, 1981–2002. *Annals of the Association of American Geographers, 97*(3), 551–566.

OECD, & IEA. (2016). Energy and air pollution: World Energy Outlook special report 2016.

Pearl-Martinez, R. (2020). Global trends impacting gender equality in energy access. *IDS Bulletin, 51*(1), 7–26.

Pelz, S. (2020). Impacts of grid electricity access on rural non-farm entrepreneurship and employment in Ethiopia and Nigeria. http://pure.iiasa.ac.at/id/eprint/16938/

Perera, N., Boyd, E., Wilkins, G., & Itty, R. P. (2015). *Literature review on energy access and adaptation to climate change* (Tech. Rep.). Evidence on Demand. https://doi.org/10.12774/eod_cr.may2015.pereran

Pueyo, A., Carreras, M., & Ngoo, G. (2020). Exploring the linkages between energy, gender, and enterprise: Evidence from Tanzania. *World Development, 128*, 104840. https://doi.org/10.1016/j.worlddev.2019.104840

Rahman, M., & Ghosh, S. (2016). Increasing resilience by the participatory planning approach. In *Construction research congress 2016*. American Society of Civil Engineers. https://doi.org/10.1061/9780784479827.154

Raney, T. (2011). The state of food and agriculture. *Women in Agriculture: Closing the Gap for Development, 2.*

Rao, N. (2019). From abandonment to autonomy: Gendered strategies for coping with climate change, Isiolo County, Kenya. *Geoforum, 102*, 27–37. https://doi.org/10.1016/j.geoforum.2019.03.017

Rao, N., Mishra, A., Prakash, A., Singh, C., Qaisrani, A., Poonacha, P., Vincent, K., & Bedelian, C. (2019). A qualitative comparative analysis of women's agency and adaptive capacity in climate change hotspots in Asia and Africa. *Nature Climate Change, 9*(12), 964–971. https://doi.org/10.1038/s41558-019-0638-y

Rao, N., Singh, C., Solomon, D., Camfield, L., Sidiki, R., Angula, M., Poonacha, P., Sidibé, A., & Lawson, E. T. (2020). Managing risk, changing aspirations and household dynamics: Implications for well-being and adaptation in semi-arid Africa and India. *World Development, 125*, 104667.

Rewald, R. (2017). Energy and women and girls: Analysing the needs, uses, and impacts of energy on women and girls in the developing world. *Oxfam Research Backgrounder Series.*

Rey, G., Jougla, E., Fouillet, A., Pavillon, G., Bessemoulin, P., Frayssinet, P., Clavel, J., & Hémon, D. (2007). The impact of major heat waves on all-cause and cause-specific mortality in France from 1971 to 2003. *International Archives of Occupational and Environmental Health, 80*(7), 615–626.

Rosenberg, M., Armanios, D. E., Aklin, M., & Jaramillo, P. (2020). Evidence of gender inequality in energy use from a mixed-methods study in India. *Nature Sustainability, 3*(2), 110–118. https://doi.org/10.1038/s41893-019-0447-3

Rotondi, V., Kashyap, R., Pesando, L. M., Spinelli, S., & Billari, F. C. (2020). Leveraging mobile phones to attain sustainable development. *Proceedings of the National Academy of Sciences, 117*(24), 13413–13420. https://doi.org/10.1073/pnas.1909326117

Schaeffer, R., Szklo, A. S., de Lucena, A. F. P., Borba, B. S. M. C., Nogueira, L. P. P., Fleming, F. P., Troccoli, A., Harrison, M., & Boulahya, M. S. (2012). Energy sector vulnerability to climate change: A review. *Energy, 38*(1), 1–12. https://doi.org/10.1016/j.energy.2011.11.056

S. Giri, Y. M. (2011). *Using mobile phones to reduce the adversities of climate change in rural Nepal (Case Study), ICTs and agricultural adaptation to climate change*. Practical Action Nepal Office.

Standal, K., & Winther, T. (2016). Empowerment through energy? Impact of electricity on care work practices and gender relations. *Forum for Development Studies, 43*(1), 27–45. https://doi.org/10.1080/08039410.2015.1134642

Stober, D., Suškevičs, M., Eiter, S., Müller, S., Martinát, S., & Buchecker, M. (2021). What is the quality of participatory renewable energy planning in Europe? A comparative analysis of innovative practices in 25 projects. *Energy Research & Social Science, 71*, 101804. https://doi.org/10.1016/j.erss.2020.101804

Stuart, C. (2017). *Energy and climate change adaptation in developing countries* (Tech. Rep.). E.U. Energy Initiative Partnership Dialogue Facility.

Toulemon, L., & Barbieri, M. (2008). The mortality impact of the August 2003 heat wave in France: Investigating the 'harvesting' effect and other long-term consequences. *Population Studies, 62*(1), 39–53.

U.N. Gender Perspectives. (2008). Integrating disaster and risk reduction into climate change adaptation. *United Nations International Strategy for Disaster Reduction, Geneva, Switzerland*, 1–87.

UNDP. (2020). *Issues brief: Resilient livelihoods & value chain development.* UNDP. https://www.undp.org/publications/undp-issue-brief-resilient-livelihoods-value-chains

UNDRR. (2021). *Nature-based solutions for disaster risk reduction—Words into action* (Tech. Rep.). https://www.undrr.org/words-action-nature-based-solutions-disaster-risk-reduction

United Nations. (2020). *Affordable and clean energy: Why it matters.* https://www.un.org/sustainabledevelopment/wp-content/uploads/2016/08/7_Why-It-Matters-2020.pdf

van Daalen, K., Jung, L., Dhatt, R., & Phelan, A. L. (2020). Climate change and gender-based health disparities. *The Lancet Planetary Health, 4*(2), e44–e45.

Vyas, Y., Knill, P., Foy, T., Harrold, M., Steele, P., Tanner, T., Hirsch, D., Oostermann, M., Rooimans, J., Debois, M., Lamin, M., Liptow, H., Mausolf, E., Verheyen, R., Agrawala, S., Caspary, G., Paris, R., Kashyap, A., Sharma, A., Mathur, A., Sharma, M., & Sperling, F. (2009). *Poverty and climate change. Reducing the vulnerability of the poor through adaptation* (Tech. Rep.). World Bank.

Walker, B., Holling, C. S., Carpenter, S. R., & Kinzig, A. (2004). Resilience, adaptability and transformability in social-ecological systems. *Ecology and Society*. https://www.ecologyandsociety.org/vol9/iss2/art5/

Wang, Y., Chen, X., & Ren, S. (2019). Clean energy adoption and maternal health: Evidence from China. *Energy Economics, 84*, 104517. https://doi.org/10.1016/j.eneco.2019.104517

World Health Organization and Others. (2014). *Gender, climate change and health*. World Health Organization.

CHAPTER 6

The Energy Transition and Gender Considerations of the Workforce Transformation: A Critical Review of the Dutch Case

Marielle Feenstra and Maria Kottari

Introduction: Socio-economic Challenges Ahead of Energy Transition and Climate Action

The severity of the climate effects emerging across the globe in conjunction with the scientific data corroborating their future intensification (IPCC, 2021) brings the quest to reduce carbon emissions to the epicentre of policymaking and policy advocacy agendas of multiple actors. The transition to a low-carbon, renewable energy system away from fossil-fuel-based power generation and consumption is the most prominent measure brought forward to combat climate change.

M. Feenstra (✉)
Delft University of Technology, Delft, The Netherlands
e-mail: marielle@75inq.com

M. Kottari
Amsterdam, The Netherlands

N. Rocha Lawton, C. Forson (eds.), *Women and the Energy Sector*,
https://doi.org/10.1007/978-3-031-43091-6_6

Given the crucial role of the energy system in the broader socio-economic processes and developments, the shift to low-carbon energy production and consumption comes along with innovation imperatives and opportunities. Those innovations are not merely technological but also social and, as such, inherently political. A decarbonised future necessitates the social acceptability of climate action and energy transition. In contrast, policymakers must face some parallel emerging socio-economic trends, such as digitalisation and new business models. In addition, the creation of different social imaginaries regarding technology, especially the energy-related technologies, challenges the different functions of human and non-human actors (Kottari & Cornelis, 2022). The energy sector faces the task of initiating a transition in which energy supply and security become less self-evident and less conventional than what society experienced in recent decades. The system in which supply was paramount is changing to a demand-oriented energy market responding to the needs of the energy consumers (Feenstra, 2021).

Many scholars (Standal & Feenstra, 2021; Hanke et al., 2021) have conducted studies about the socio-economic benefits of a transition to a renewable energy-based system. However, these benefits are not restricted to mitigating climate change or achieving significant carbon emissions reduction. Instead, those foreseen benefits are principally associated with a possible transformation of energy governance. They entail creating a more accessible, decentralised, inclusive, and, ultimately, democratic energy system and the radical changes they can induce in the workforce and labour market.

Compared to an energy system based on fossil fuel power generation, renewable energy can provide comparative advantages. First, the emerging renewable energy technologies (including solar, wind, geothermal, and biogas) can be deployed in different scales: small, local, and district-level and from other actors and with a variety of investors. Those conditions will facilitate further the market entrance of renewable energy communities (Caramizaru & Uihlein, 2020) at least in the EU context and will invert the 'traditional' roles of power generation and distribution, infrastructures' ownership, management, and profit (Hanke et al., 2021; Räty & Carlsson-Kanyama, 2010). Additionally, the perpetuity and abundance of renewable energy, enabled by technology, can facilitate and enhance more equitable access to energy sources. Furthermore, renewable energy can assist in alleviating issues such as household-level energy poverty (Kottari & Cornelis, 2022). To become a key and effective just transition enabler, renewable energy management has to consider issues of social justice (Pablo García et al., 2020) and therefore encompass elements of equity, inclusion, and exclusion that are present in different geographical and political contexts. Within this conceptual framework, gender considerations, particularly the socially constructed gender roles, become very relevant to resonate with existing inequalities in society to access and power over energy sources (Clancy et al., 2020). Given the influence that gender dynamics have within any sociopolitical context and decision-making process, climate change action and the process of energy transition (Fathallah & Pyakurel, 2020) should be assessed through gender lenses.

The ageing population in the Global North causes an ever-increasing shrinkage of the workforce. Combined with the challenge of accelerating the energy transition, the energy sector is feeling the urgency to utilise as much labour potential as possible. This challenge is also relevant for other sectors involved in the energy transition, such as construction. In this chapter, however, we mainly understand the energy sector as the entire energy-related industry in the Netherlands (plus mineral extraction, water companies, waste management, and energy supply).

Women have traditionally been under-represented in the energy-related industry, while the challenges within the energy transition require a diversity of insights and skills. IRENA (2018) has calculated that, worldwide, the number of jobs in the renewable energy sector will increase from 10.3 million in 2017 to almost 29 million in 2050. In other words, the energy sector is growing with the task of the energy transition, bringing the untapped labour potential into the spotlight to meet the rising demand for well-skilled workers. However, studies on labour are blind to the effects of the transition on the quality of jobs, the evolution of working hours, and gender (Pablo García et al., 2020). These shifts in labour demands will subsequently shift investments and resources from the conventional energy sector to ultimately achieve a low-carbon and resource-efficient economic system. The challenge for labour markets and human capital agendas globally is to maximise the benefits for workers and help assure a fair sharing of unavoidable adjustment costs, such as direct and indirect job losses, employees skills shortages, and bottlenecks, for instance, the rise of new technologies and digitalisation disrupting the traditional employment patterns, job quality, and gender considerations (IEA, 2017; ILO, 2018, 2019; World Bank Group, 2019).

Following these introductory remarks, the chapter will further focus on exemplifying the gendered aspect of the energy transition debate and respective policy initiatives (Section "Research Methodology") and providing some additional insights about labour market transformation (Section "Engendering the Energy Transition"). Then, placing the whole narrative into the specific Dutch perspective (Section "Energy Transition and Workforce Transformation: Gender Considerations"), the chapter will deliver a concise outline of the gender considerations of the Dutch energy sector (Section "Energy Transition and Workforce Transformation: The Dutch Case"). Finally, the chapter will conclude with specific policy recommendations (Section "Concluding Remarks").

Research Methodology

This chapter is based on the research and work done by the Dutch NGO 75inQ. [1] 75inQ is an NGO that has the mission to accelerate the energy transition (SDG7) by improving gender equality (SDG5). 75inQ wants to contribute to the aim that by 2030 the stakeholders of clean and affordable energy systems are reflecting the diversity of society, offering equal opportunities for all actors involved in the energy system. The energy transition to meet the world's needs for electricity, heating, cooling, and transport in a sustainable way is one of the greatest challenges facing humanity in the twenty-first century, both in terms of meeting the needs of the present and in terms of effects on future generations. Improving energy access for all and making energy cleaner are key to achieving most of the United Nations 2030 Sustainable Development Goals, which cover issues ranging from climate action to gender equality. Focusing on universal access to energy, increased energy efficiency, and the increased use of renewable energy through new economic and job opportunities is crucial to creating more sustainable and inclusive communities and resilience to environmental issues like climate change. 75inQ leads towards equal representation in the political and economic decision-making process. Gender diversity adds value to the impact and efficiency of organisations that manage to include at least 35% of women in decision-making positions.

A quick scan performed by 75inQ in coordination with the Dutch Platform Talent for Technology [2] (PTvT) commissioned by the Dutch

[1] Both authors are part of the 75inQ research team. MF is co-founder and scientific director of 75inQ (www.75inq.com).

[2] Platform Talent for Technology (PTvT https://ptvt.nl/) is a public-private collaboration to promote STEM education, training, and work to stimulate young people to choose a career in technology.

Energy Top Sector provides a first picture of the role of girls and women [3] in the energy transition sector in the Netherlands. The Top Sector Energy has been implementing a human capital agenda (HCA-TSE) since 2012 financed by the Dutch national government. [4] The aim of this agenda is to contribute to solutions for the quantitative and qualitative bottlenecks in the labour market. Sufficient availability of well-qualified people on the labour market is, after all, an important precondition for solving today's climate and energy issues and therefore also an important theme within the Climate Agreement of the Dutch government.

The methodology on which this chapter is based is a combination of literature review, desk review of policy documents (from the European Commission and the Dutch National Government), and statistical data from the Dutch Central Bureau of Statistics on labour market participation. Not all data are gender disaggregated, which is a limitation for this research and for research on the gender-energy nexus in general (Feenstra, 2021). Where data was lacking or provided a fragmented picture of the situation, observations of the authors and insight knowledge on gender and energy transition in the Dutch labour market are added. These observations are clearly identified in this chapter.

Engendering the Energy Transition

At first, in this chapter, gender is considered a social construct of 'a system of socially defined roles, privileges, attributes and relationships between men and women which are learned and not biologically determined' (Khamati-Njenga & Clancy, 2003). Gender relations are dynamic depending on time, place, and context. Hence the gender dimensions of energy and climate policy differ across social, cultural, economic, and political contexts. An intersectional approach to gender relations pinpoints further variation across various social factors, such as age, class, ethnicity, income level, social status, marital status, and sexual identity. Such an intersectional approach reveals a potential range of needs and capacities to meet those needs, which may require targeted forms of action rather than generic policy instruments (Feenstra, 2021).

[3] In this chapter we distinguish girls (still in school and educational training programs) and women (age 18 and older).

[4] Top Sector Energy: https://www.topsectorenergie.nl/en

Gender dynamics in the energy transition process have attracted growing research interest in academia and other stakeholders such as public institutions and private companies (Clancy et al., 2020). A rather concrete agenda is lining up around, as also discussed in the previous section, two, although distinct, still interconnected subject matters. One is related to the new emerging energy supply and demand dynamics and the other is to the growing need for skills and diversity in the workforce in the renewable energy industry. Applying a gender perspective to the energy transition reveals whether gender influences the uptake of a more sustainable energy system and if the energy transition equitable impacts women and men (Feenstra, 2021). Likewise, including gender considerations in the workforce transformation affirms the substantial co-benefits of decision-making diversity in the low-carbon energy sector (IRENA, 2018) and the ever-expanding talent pool and skills variation the multidisciplinary industry of renewable energy requires.

There is a considerable absence of credible empirical data (Fathallah & Pyakurel, 2020) that can systematically analyse gender issues and energy systems and delve into the gender-related energy consumption patterns and energy-saving behaviours. As such, we observe a gender disparity and 'blindness' in the design and implementation of respective energy policies. Gender remains a problematic concept to embed in energy policies and energy transition given the fact that the sector is male dominated in terms of employees, decision-makers, and producers (Clancy & Feenstra, 2019), (IEA Energy and Gender), (IRENA, 2018). The following arguments resonate with the relevance of the gender-energy nexus.

First, women's gendered role goes beyond household energy use in agriculture, food processing, service, and manufacturing. Second, households are not a homogenous entity; instead, intrahousehold allocations of time and resources differ between genders. The manifestation of energy poverty affecting mainly women also confirms that (Feenstra, 2021; Kottari, & Cornelis, 2022). Third, women are energy users and participants of the energy supply chain. Fourth, women have diverse roles in designing, adapting, and using new energy technologies, either acting as consumers or as energy professionals. All those arguments are valid in the current energy transition debate, and the design and implementation of sustainable energy policies should address them (Feenstra, 2021).

Energy Transition and Workforce Transformation: Gender Considerations

Although definitions of green employment vary among scholars and organisations (Czako, 2020; ILO, 2018, 2019; IRENA, 2018; UNEP/ILO/IOE/ITUC, 2008), some recurring themes exist across the different reports. In general terms, a green job is related to producing and distributing clean energy and environmental goods and services in any economic sector and enterprise (UNEP/ILO/IOE/ITUC, 2008). The production and distribution of clean energy include energy production from renewable energy sources and technologies, efficient power and raw materials use, CO_2 emissions cut-down, and climate change adaptation. In addition, the United Nations Environmental Program (UNEP) stresses that green jobs must be decent, offering the workers adequate wages and safe working conditions while protecting their labour rights.

The question of how green policies affect labour markets and economic competitiveness is relatively new and under-researched. However, it is the component of defining green jobs because they have been created exclusively due to green growth and climate change policies. As the green economy develops, demand will increase for lower-skilled labour to operationalise and implement the policies and projects.

The facts prove that a low-carbon economy will be a significant source of job growth in the future of work (ILO, 2018). The shift towards energy production from renewable sources and energy efficiency can result in the net creation of some 24 million new jobs by 2030, counterbalancing the estimated loss of around 6 million jobs by 2030 (ILO, 2018). IRENA (2018) showcases that the renewable energy sector (including hydropower) employed 10.3 million people, directly and indirectly, in 2017, representing an increase of 5.3% compared to the numbers reported in 2016.

Although economic growth and employment creation go, in essence, hand in hand, the positive effects of environmental sustainability in the job market are not straightforward. Therefore, an adequate, coherent framework of policies and measures is a prerequisite for ensuring decent working conditions within the green employment sector. Furthermore, those policies need to allow the employees to transition smoothly towards new employment sectors and safeguard those that would unavoidably lose their positions due to lower activity in industries that contribute to environmental degradation, such as the energy-intensive industry and mining.

The representation of women across the whole spectrum of energy-related industries is low. When entering and advancing in the conventional and renewable energy sectors, they face various barriers. Those barriers are mostly related to long-term perceptions around the gender roles in many industrial activities (Clancy & Feenstra, 2019), and as such women must overcome difficulties entering the energy industries but mainly to retain leadership positions, engineering roles, get promotions, and training. The under-representation of women in technical roles in the energy sector is primarily because science, technology, engineering, and mathematics (STEM) education is a highly male-dominated domain. Women STEM graduates have fewer chances to gain a position in their field and level of education (IRENA, 2019). The renewable energy sector is a relatively new employment sector, and as such, it provides an opportunity for past failures regarding the workforce gender gap to be avoided (Kottari, 2020) (Fig. 6.1).

Locating women's positions in the energy sector is not easy. Data is available to a limited extent, insufficiently accessible, lacks specification in the industry, at the job level, and is generally not broken down between male, female, or other. Based on existing and accessible data, the energy labour market remains a grey area because of the overrepresentation of men and the large share of the fossil industry in the current energy landscape. However, the data poverty and their lack of specification validate the areas for improvement. The call for more specific data to paint a more nuanced picture as an incentive for policy development towards a more inclusive energy sector and to monitor the implementation of

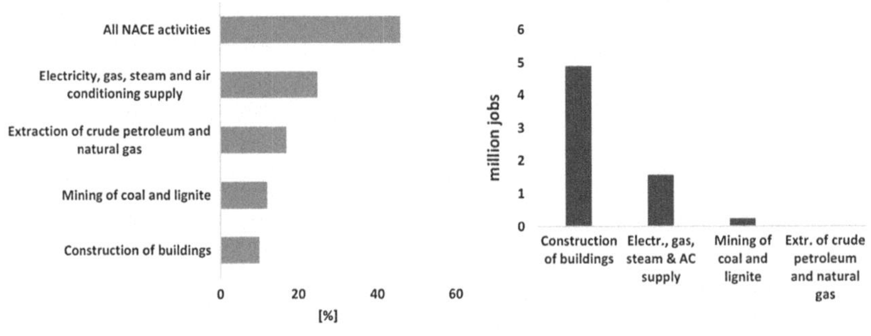

Fig. 6.1 Women's share in employment in selected economic activities and women's employment in the same sectors, EU-28, 2018 (Czako V., 2020)

interventions is clear. The lack of gender-disaggregated data provides limited insight into the real numbers (Teladia et al., 2017). The data is now fragmented over various reports and lacks a combination of qualitative and quantitative data so that figures and statistics interpret the observations and experiences of those involved. Data not only helps identify women's place in the energy transition, but it also has a robust monitoring function to measure progress in gender equality in the energy sector and promptly identify bottlenecks in the energy transition's labour market. Many statistics can be broken down by gender if specifically requested (Feenstra & Creusen, 2021).

Despite the persistent lack of gender diversity in the energy industry, the benefits of hiring and retaining women in the energy workforce are well documented (Stephens et al., 2019). Studies reveal that companies investing in the female workforce and including gender-diverse teams have more innovation and revenue growth than their competitors (Pearl-Martinez & Stephens, 2016; Stephens et al., 2019). Furthermore, increased gender parity in professional groups results in more effective and inclusive results in decision-making. In addition, women are more likely to support sustainability and environmental action in the workplace, and companies with more women on the board of directors are more likely to invest in renewable energy and strategies that reduce carbon emissions (IRENA, 2019).

Energy Transition and Workforce Transformation: The Dutch Case

The energy transition in the Netherlands brings certain elements that characterised the Dutch context. First of all, the Netherlands are part of the global commitments of climate change reduction and a loyal member of the European Union. Hence, the international conventions and agreements are shaping the legal context of the Netherlands. The necessity of the current energy transition from fossil fuels to sustainable energy sources to reduce greenhouse gas emissions to limit further global warming is scientifically proven (IPCC, 2014) and endorsed by the latest multilateral climate agreement, the Paris Agreement (UNFCCC, 2015). This ambition is also expressed in the Dutch National Climate Agreement (2019) that sets a target for CO_2 emissions reduction by 49% in 2030, increasing to 95% in 2050. The Climate Agreement entered into force on 28 June

2019 through the Climate Act, and the Dutch National Energy and Climate Plan (NECP) (Dutch NECP) is based on its content, in particular its objectives and measures towards reducing greenhouse gas (GHG) emissions by 49% by 2030. To meet the EU's energy and climate targets for 2030, EU countries need to establish a 10-year integrated national energy and climate plan (NECP) for the period from 2021 to 2030. After the EU Regulation on the Governance of the Energy Union and Climate Action (EU Regulation, 2018/1999) and the Clean Energy for All Europeans package (European Commission (2016)) adopted in 2018–2019, the European Green Deal (European Commission (2019)) and the new EU Climate Law (EU Regulation, 2021/1119) further shape EU and Dutch policy. All in all, these developments show that energy transition policy is in movement and shaped through different governance mechanisms. This significant ambition to reduce climate change, stipulates multiple stakeholders to work together and requires a strong labour market. Therefore, the labour market and human capital have been identified as one of the priorities within the Dutch Climate Agreement, emphasising sufficiently skilled workers for the transition to sustainable energy systems (Feenstra & Creusen, 2021).

Secondly, the political context in the Netherlands shapes the Dutch energy transition. The existing policy framework for the energy transition in the Netherlands is based on the 2017 Coalition Agreement of the installed government after the 2017 elections. The energy transition ambitions of the Dutch government are rooted in the Energy Agreement of 2013 and the Climate Agreement of 2019 which entered into force through the Climate Act on 28 June 2019. In turn, these are shaped by UN climate negotiations, EU climate law, and groundbreaking national climate litigation through the case of Urgenda v. the Netherlands (Bakker, 2021). The latter was taken by the civil society organisation Urgenda and a large group of concerned citizens to challenge the adequacy of Dutch climate policy. The courts legally obliged the Netherlands to mitigate GHG emissions at a faster pace than mandated by the EU by 2020, spurring on various additional last-minute GHG reduction programmes, including for the energy efficiency of homes. As a result of the Urgenda judgements, the Netherlands lobbied for higher decarbonisation targets within the EU, that is, at a level of reducing GHG emission by 55% by 2030, compared to 1990 emissions level. The Commission supported this objective in its proposal for the European Green Deal which was adopted by the EU Member States in December 2020.

Another contextual factor in the Netherlands is the collaboration with private actors in the energy system. The Netherlands has long-time experience in public-private partnerships in energy planning and project management. Within the EU, the decarbonisation and climate change policy as formulated in the EU Cohesion Policy 2014–2020 pursues a triple-helix innovation model which involves co-creation and cooperation between three stakeholder groups: public sector, academia, and business. In some regions, the model has become the quadruple helix when civil society is added as the fourth stakeholder group. This is an essential element of all programmes and projects funded by the EU. In achieving the decarbonisation ambition of the EU, many energy efficiency projects are receiving EU grants as a catalyst. Within the triple- and quadruple-helix cooperation, the role of the public sector is becoming less dominant. The corporate sector and NGOs are increasingly working together in energy efficiency projects, leaving government actors to a more facilitating and stimulating role.

In light of these contextual factors, the Dutch energy transition ambition can only be realised when setting priorities and shaping the strategies to be implemented in policies. The labour market and human capital have been identified as one of the priorities within the Dutch Climate Agreement (2019). The share of women in the labour market within the energy sector is very limited (22%) and encourages a target group approach to use untapped labour potential to accelerate the transition to a sustainable and fair energy system. Based on the data available at Statistics Netherlands, of the workforce in the Dutch energy sector, on average 22% of them is female. These data do not distinguish between the sustainable and fossil energy sectors, neither on job level nor educational level. The consequence of this is that of the average 22% female employees in the energy sector, the service-oriented and support professions within the fossil industry also count.

One of the strategies of the human capital agenda is to stimulate more women and girls to choose a career in the energy sector. This strategy is an important part of the solution to tackle the labour shortage in the sector, promote diversity in the energy sector, and organise the existing culture of the sector in an inclusive and sustainable way (Feenstra & Creusen, 2021). The added value of diversity in teams is that the quality of decision-making improves. Decision-making about the energy transition and the design and implementation of appropriate interventions require insight into the technical, economic, and social challenges of the energy transition. Below we indicate for each challenge what the role of women is and to what extent the under-representation of women in the energy sector hinders the engendering of the energy transition. These challenges can then inspire policymakers to tailor policies to resonate with the technical, economic, and social elements of the energy transition.

Technical Task: Untapped Labour Potential

The Dutch '*poldermodel*' is a stakeholder participation model, consulting major actors in the policy field when a new policy or agreement is designed. The energy sector is no exception to this political tradition. The Energy Agreement of 2013 was established by 40 organisations, including government, employers, trade unions, nature and environmental organisations, social organisations, and financial institutions. Implementing the Energy Agreement must result in an affordable and clean energy supply plus creating at least 15,000 full-time jobs in the clean technology market. At 23 February 2018, the stakeholder consultation started feeding into the Climate Agreement. The government created five sectoral climate round table discussions and three task forces in the fields of finance, innovation, and labour market and education. Citizens were invited to contribute to the Climate Agreement by submitting plans, ideas, suggestions, and questions online through the site internetconsultatie.nl. The Internet consultation was published on 24 August 2019 and open for reactions till 4 October 2019. There were 1602 reactions from citizens and organisations. A Climate Council delegation visited five locations during which citizens could participate, plus the Climate Council organised in each of the twelve provinces a consultation meeting to engage in dialogue.

When the Climate Agreement was drawn up, 150 parties worked together within 5 climate tables (electricity, industry, built environment, agriculture and land use, and mobility) (Dutch National Climate Agreement, 2019). Within the 141 professions identified by the climate tables, 76 professions can be identified as contributing to the energy transition. Of these, 69% of these professions can be regarded as technical professions. A first start in gaining more insight into women in professions that contribute to climate improvement and energy transition. Information can be obtained by requesting the Central Bureau of Statistics Netherlands to split the professions of the climate tables according to gender, job level, and energy professions. As long as this request has not yet been made, we are dependent on figures already available.

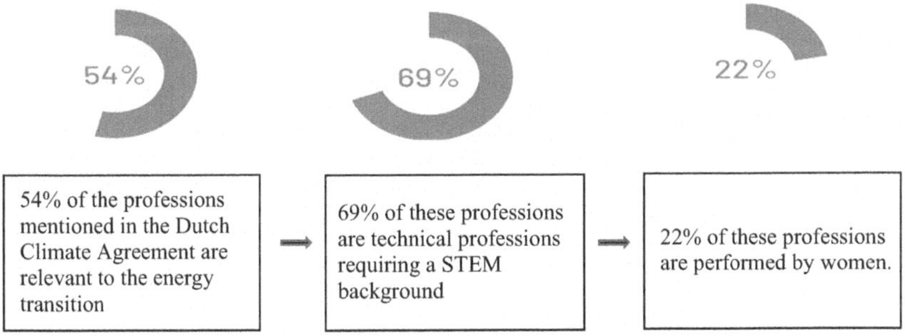

Fig. 6.2 Percentage of occupations relevant to the energy transition and female participation, Credits: 75inQ based on CBS figures (CBS, 2019)

When we compare these figures[5] with the available figures from the Central Bureau of Statistics Netherlands and put them next to the number of employees in the energy and industrial sector, we see that only 22% of these professions are performed by women (Fig. 6.2). From these figures it can be concluded that there are opportunities to increase the labour potential by focusing on more women and girls in technology. The energy transition as a technical task requires both the deployment of new and different disciplines of the labour force and the retention of the high-quality technical skills of the labour force that they have developed within the current energy system based on fossil energy sources.

Economic Challenge: Demand-Oriented Energy Market

The current energy transition is not only a technological transition from fossil to sustainable energy sources but also an economic transition. The energy sector is faced with the task of initiating a transition in which supply and security become less self-evident and less adequate than what society has been built on in recent decades. The system in which supply was paramount is changing to a demand-oriented energy market (Feenstra, 2021).

[5] Exact data on how many workers are currently employed in these professions, broken down into men and women and job level, are not yet available.

We are currently confronted with rising energy prices due to the changes in the gas supply. The security of the national supply of Dutch gas is no longer a given since there is a political agreement to reduce gas mining in the North of the Netherlands (the province of Groningen). The decline in gas production had limited financial consequences, given the relatively low gas price at the time. Policymakers are confident that the well-functioning Title Transfer Facility (TTF) gas trading system and a growing supply of liquefied gas (LNG) would permanently guarantee the security of gas supply at acceptable prices (Van den Beukel & Van Geuns, 2021). The extreme gas price increases experienced in 2021 exposes the harrowing by-product of this supply-driven demand; the affordability of our energy consumption is under pressure. A conservative estimate calculates that 660,000 households in the Netherlands are unable to heat their homes, and they live in energy poverty (Straver et al., 2021).

There is a strong difference in participation in the energy transition between men and women and population groups. Income inequalities, quality of housing, care for dependent relatives (children and/or informal care), personal energy dependence (e.g. diet, physical disability, work activities, mobility), and financial opportunities to invest in energy efficiency are causing a growing gap between those who can participate in the energy transition and those who are left out (Feenstra et al., 2021). Insight into the economic tasking of the energy transition requires different skills and knowledge than the technical task. Diversity in the energy sector, both in the backgrounds and in the training and skills of employees in the sector, contributes to an energy transition that reflects the diversity of needs of energy users. The economic priority of the energy transition is closely related to the social impact as described below.

Social Impact: A Just and Inclusive Energy Transition

The diversity of society is currently insufficiently reflected in the current energy sector, while the task of the energy transition requires broad commitment and social support for fundamental system change. With the growing inequality in access to affordable sustainable energy, the international agreement to realise 'sustainable energy for all' is coming under pressure. The Sustainable Development Goals are not only an international commitment but are also the driving force for local action plans and require embedding in national policy. In addition, the European

Commission has expressed its commitment to promoting 'Clean Energy for All Europeans' and has made reporting on member states' tackling energy poverty mandatory in the National Energy and Climate Plans (NECPs). The Dutch government has used the climate agreement as input for the Dutch NECP. However, the terms girls, women, gender, and diversity are mentioned only once in the human capital paragraph emphasising the need for a sufficient and well-qualified labour force. The paragraph 'Policy efforts on the labour market, training and employment in relation to the transition' does make recommendations for more inclusive access to the labour market (e.g., people with a distance to the labour market caused by disabilities are mentioned in this context), but there is no further strategy of a gender-specific approach. The attention to energy poverty in the Dutch NECP is limited but meets the minimum requirements of the European Commission. However, with the current energy prices, the social challenge in the energy transition is fully in the spotlight. It shows that in the current energy transition, all energy consumers must be considered.

The impact of energy-related behaviour is important for policymaking and demand-driven product development. Energy policymakers and managers need to be aware of the different patterns of domestic energy use between men and women. A study in four European countries found significant differences in energy expenditure between single women and men. In general, men appeared to expend more energy, mainly related to transport and outdoor activities (Räty & Carlsson-Kanyama, 2010). In family households, women spend more time on unpaid, energy-intensive domestic work such as cooking and washing. Dutch working women spend an average of 4.5 hours a day on unpaid housework (men 2.5 hours). The Netherlands is therefore in fourth place (behind Malta, Slovenia, and Cyprus) as an EU member state, where women spend the most hours of unpaid work in the household (EIGE, 2021). The European Institute for Gender Equality has found that women are more sustainable consumers than men, as they value eco-labelled products and green sourcing more highly. They also appeared to be more willing to change their energy-related behaviour in favour of sustainable options (EIGE, 2012). Research shows that a gender perspective in future intervention programs is essential, as energy conservation affects the timing and types of household tasks, resulting in a potential increased workload for women (Carlsson-Kanyama & Lindén, 2007).

Concluding Remarks

This chapter on women in the Dutch energy transition workforce provides a first picture of the role of women in the technical task, economic challenge, and social impact of the energy transition and the need for the energy sector to use the untapped labour potential of women to accelerate the energy transition. Decision-making about the energy transition and the design and implementation of appropriate interventions require insight into the technical, economic, and social challenges of the energy transition. Diversity in the energy sector contributes to more demand-oriented energy transition interventions that better reflect diversity in society.

The impact of energy-related behaviour is essential for policymaking and demand-driven product development. Energy policymakers and managers need to be aware of the different patterns of domestic energy use between men and women, bringing the social impact of energy transition in connection with the technology challenge. Räty and Carlsson-Kanyama found significant differences in energy expenditure between single women and men in their study in four European countries. In general, men appeared to spend more on energy, mainly related to transport and outdoor activities, whereas women spend more time on unpaid, energy-intensive domestic work such as cooking and washing in family households. Dutch working women spend an average of 4.5 hours a day on unpaid housework (men 2.5 hours). Therefore, the Netherlands is in fourth place (behind Malta, Slovenia, and Cyprus) as an EU member state where women spend the most hours of unpaid work in the household (EIGE, 2021). The European Institute for Gender Equality has found that women are more sustainable consumers than men, as they value eco-labelled products and green sourcing more highly. They also appeared to be more willing to change their energy-related behaviour in favour of sustainable options (16). Research shows that a gender perspective in future intervention programs is essential, as energy conservation affects the timing and types of household tasks, resulting in a potentially increased workload for women (Carlsson-Kanyama & Lindén, 2007).

The climate change challenge and the energy transition ambition in the Netherlands are prioritised by Dutch policy at the national government level supported by the corporate sector. However, the 22% share of women in the Dutch energy transition lags dramatically behind the ambition towards an inclusive energy sector that can reflect the diversity in society

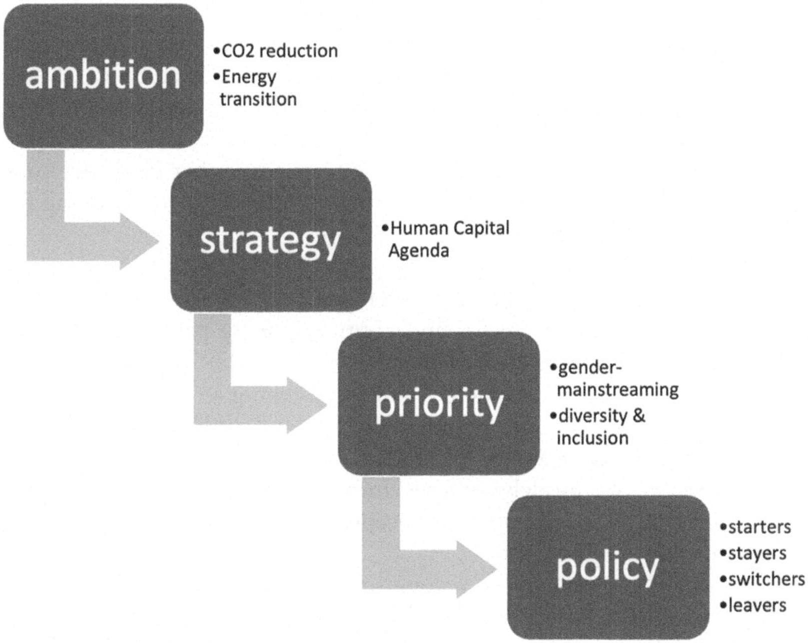

Fig. 6.3 Visual representation of gender-responsive energy policy roadmap. (Authors' own visualisation)

to develop policies and design products resonating with the diversity in needs of energy consumers. The human capital agenda of the energy transition requires extra attention and continuous monitoring to sustainably accelerate the energy transition for everyone. Furthermore, the following four observations are made about the labour market in the Dutch energy sector (Fig. 6.3):

- Starters: are mainly men due to an overrepresentation of men in STEM education.
- Stayers: women working in the energy sector are facing limitations to grow in their career and to experience limit mobility between jobs and positions. This is partly caused by limited support networks within their companies and the lack of possibilities to adjust their working hours to care responsibilities.

- Switchers: women switching from either fossil fuel companies to the renewable sector or women entering the energy sector coming from other professions and/or other sectors. These women have a different need to get supported with onboarding programs, training, and mentoring networks.
- Leavers: women leaving the energy sector to either work for non-energy-related professions or decide to temporarily stop working for caring giving responsibilities.

Each of the four groups need different support systems and policies to keep them in the energy sector and continue taking part in the human capital of the energy transition. Starters would benefit from targeted policy on breaking through prejudices and stereotyping around gender roles through active interventions within education and in the media to visualise the energy sector as a potential and attractive employer for women. In addition, visible role models within technical education are of great importance to dispel the persistent misconceptions about career opportunities for girls in the energy sector.

Stayers can be supported by recognising and breaking the existing glass ceilings within the sector. This starts with data collection, analysis, and reporting of the progression of women within the organisation. Ambitious goals should be set and measured with progress (and lack thereof) reported regularly. Engaged top management and shareholders should become acquainted with the underlying drivers of employees and patterns of inequality and approach diversity as a measurable success factor. Women in the energy sector would be supported by networks of women within and beyond the companies they are working for. These networks and the activities associated with them are being pulled out of the voluntary sphere and are part of women's leadership and development programs within the sectors.

Switchers need increasing recognition of non-technical skills, such as creative and critical thinking, communication, collaboration, and problem-solving skills, as part of the recruitment process and qualifications. Also, the energy sector should invest in and facilitate professional networks to promote the lateral entry and promotion of women from outside the energy sector to access the informal professional networks that contribute to successful sectoral mobility. Work-to-work trajectories with a focus on women can be set up through public-private partnerships to reach the group of women who do not have the professional network themselves to successfully transfer to one of the technical professions relevant to the energy transition.

Monitoring, indexing, and reporting outflows are important elements of a gender-inclusive personnel policy to enable life stage-related interventions and prevent losing human capital in the long run. However, employers in the energy sector should request gender-disaggregated data to gain insight into the actual numbers of women in the energy sector, to monitor the diversity and the labour mobility of women in the energy sector. The desired data to deepen the insight into the mobility of female professionals in the energy sector is based on the following indicators:

- Influx of girls in the energy sector (the starters)
- Lateral entry of women from another sector to the energy sector (the switchers)
- Throughput of women in the energy sector to various positions (the stayers)
- Outflow of women from the energy sector (the leavers)

Only then can a policy intervention be drafted, implemented, and monitored to meet the inclusion and gender-mainstreaming strategy of the human capital priority of the energy transition ambition of actors in the energy sector. The Dutch case study demonstrates the necessity to move beyond stating an ambition that resonate with the increasing gender inequalities in society. A diverse and inclusive human capital agenda is a necessity to meet the technological, economic, and societal challenges of the energy transition and climate change.

References

Bakker, C. (2021). Climate change litigation in the Netherlands: The Urgenda case and beyond. In *Climate change litigation: Global perspectives* (pp. 199–224). https://doi.org/10.1163/9789004447615_010

Caramizaru, E., & Uihlein, A. (2020). Energy communities: An overview of energy and social innovation, EUR 30083 EN, Publications Office of the European Union, . ISBN 978-92-76-10713-2, https://doi.org/10.2760/180576, JRC119433

Carlsson-Kanyama, A., & Lindén, A. L. (2007). Energy efficiency in residences—Challenges for women and men in the North. *Energy Policy, 35*(4), 2163–2172. https://doi.org/10.1016/j.enpol.2006.06.018

Centraal Bureau voor de Statistiek (CBS). (2019). Werkgelegenheid/Banen Nijverheid en Energie. https://www.cbs.nl/nl-nl/cijfers/detail/83451NED?q=hernieuwbare%20energie
Clancy, J. S., & Feenstra, M. H. (2019). Women, gender equality and the energy transition in the E.U., Study for the FEMM Committee of the E.U. Parliament, PE 608.867.
Clancy, J. S., et al. (2020). Engendering the energy transition: Setting the scene. In J. S. Clancy, G. Özerol, N. Mohlakoana, M. Feenstra, & L. S. Cueva (Eds.), *Engendering the energy transition*. Palgrave Macmillan. https://link.springer.com/chapter/10.1007/978-3-030-43513-4_1
Czako, V. (2020). Employment in the Energy Sector Status Report 2020. Joint Research Centre, European Commission, Publications Office of the European Union, ISBN: 978-92-76-18206-1, JRC120302
Dutch National Climate Agreement. (2019). https://www.klimaatakkoord.nl/binaries/klimaatakkoord/documenten/publicaties/2019/06/28/national-climate-agreement-the-netherlands/20190628+National+Climate+Agreement+The+Netherlands.pdf
Dutch National Energy and Climate Plan (NECP). (n.d.). https://ec.europa.eu/energy/sites/ener/files/documents/nl_final_necp_main_en.pdf
EU Regulation. (2018/1999). EU Regulation of the European parliament and of the council of 11 December 2018 on the governance of the energy union and climate action. https://eur-lex.europa.eu/legal-content/EN/TXT/PDF/?uri=CELEX:32018R1999&from=EN
EU Regulation. (2021/1119). EU Regulation of the European parliament and of the council of 30 June 2021 establishing the framework for achieving climate neutrality 'European Climate Law'. https://eur-lex.europa.eu/legal-content/EN/TXT/PDF/?uri=CELEX:32021R1119&from=EN
European Commission. (2016). Clean Energy for all Europeans (COM (2016)860 final. https://ec.europa.eu/energy/sites/ener/files/documents/com_860_final.pdf
European Commission. (2019). The European Green Deal (COM (2019) 640 final. https://eur-lex.europa.eu/resource.html?uri=cellar:b828d165-1c22-11ea-8c1f-01aa75ed71a1.0002.02/DOC_1&format=PDF
European Commission, National Energy and Climate Plans (NECPs). (n.d.). https://ec.europa.eu/energy/topics/energy-strategy/national-energy-climate-plans_en
European Institute for Gender Equality (EIGE). (2012). Gender equality and climate change report. http://www.eige.europa.eu/content/document/gender-equality-and-climate-change-main-findings
European Institute for Gender Equality (EIGE). (2021). Gender inequalities in care and consequences for the labour market. https://eige.europa.eu/sites/default/files/documents/20204948_mh0220888enn_pdf.pdf

Fathallah, J., & Pyakurel, P. (2020). Addressing gender in energy studies. *Energy Research and Social Science, 65*, 101461. https://doi.org/10.1016/j.erss.2020.101461

Feenstra, M., et al. (2021). Humanising the energy transition: Towards a national policy on energy poverty in the Netherlands. *In Frontiers in Sustainable Cities*. https://doi.org/10.3389/frsc.2021.645624

Feenstra, M. H. (2021). Gender just energy policy. *Engendering the Energy Transition in Europe*. https://doi.org/10.3990/1.9789036551960

Feenstra, M. H., & Creusen, A. (2021). Women in the energy transition. Representation of women in the energy sector: Opportunities and recommendations, 75inQ. *Expertise Center Gender & Energy*. https://75inq.com/services/library/rapportage-vrouwen-in-de-energietransitie/

Florian Hanke, et al. (2021). Do renewable energy communities deliver energy justice? Exploring insights from 71 European cases. *Energy Research and Social Science, 80*, 1022444. https://doi.org/10.1016/j.erss.2021.102244

International Energy Agency (IEA). (2017). Digitalization and energy. *Technical Report*. https://www.iea.org/reports/digitalisation-and-energy

International Energy Agency (IEA). (n.d.). Energy and gender. https://www.iea.org/topics/energyandgender/

International Labour Organisation (ILO). (2018). World Employment and Social Outlook 2018: Greening with Job. https://www.ilo.org/global/publications/books/WCMS_628654/lang%2D%2Den/index.htm

International Labour Organisation (ILO). (2019). Skills for a greener future: A global view. https://www.ilo.org/skills/pubs/WCMS_732214/lang%2D%2Den/index.htm

International Renewable Energy (IRENA). (2018). Renewable energy and jobs: Annual review 2018. https://www.irena.org/-/media/Files/IRENA/Agency/Publication/2018/May/IRENA_RE_Jobs_Annual_Review_2018.pdf

International Renewable Energy (IRENA). (2019). Renewable energy. A gender perspective. https://irena.org/-/media/Files/IRENA/Agency/Publication/2019/Jan/IRENA_Gender_perspective_2019.pdf

IPCC. (2014). Climate change 2014: Synthesis report. In R. K. Pachauri & L. A. Meyer (Eds.), *Contribution of Working Groups I, II and III to the fifth assessment report of the intergovernmental panel on climate change core writing team*. IPCC. https://www.ipcc.ch/report/ar5/syr/

IPCC. (2021). Climate change 2021: The physical science basis. In V. Masson-Delmotte, P. Zhai, A. Pirani, S. L. Connors, C. Péan, S. Berger, N. Caud, Y. Chen, L. Goldfarb, M. I. Gomis, M. Huang, K. Leitzell, E. Lonnoy, J. B. R. Matthews, T. K. Maycock, T. Waterfield, O. Yelekçi, R. Yu, & B. Zhou (Eds.), *Contribution of Working Group I to the sixth assessment report of the intergovernmental panel on climate change*. Cambridge University Press. https://www.ipcc.ch/report/sixth-assessment-report-working-group-i/

Khamati-Njenga, B., & Clancy, J. S. (2003). Concepts and issues in gender and energy. *ENERGIA*. http://www.africa-adapt.net/media/resources/80/Energia-gender-energy.pdf

Kottari, M. (2020). Fostering green jobs. Opportunities & Challenges for the European Union's (EU) Clean Energy Transition. In A. Rządkowska (Ed.), *Europe-energy-climate: The quest for the clean energy transition in the EU*. European Solar Network (ESP) Press.

Kottari, M., & Cornelis, M. (2022). Energy poverty (re)Invented? Concept and regulatory gaps in the EU amidst the decarbonisation process. In C. Rubio-Bellido & J. Solis-Guzman (Eds.), *Energy poverty alleviation*. Springer. https://doi.org/10.1007/978-3-030-91084-6_1

Labanca, N., et al. (2020). Transforming innovation for decarbonisation? Insights from combining complex systems and social practice perspectives. *Energy Research and Social Science, 65*, 101452. https://doi.org/10.1016/j.erss.2020.101452

Pablo García, et al. (2020). Just energy transitions to low carbon economies: A review of the concept and its effects on labour and income. *Energy Research and Social Science, 70*, 101664. https://doi.org/10.1016/j.erss.2020.101664

Pearl-Martinez, R., & Stephens, J. (2016). Toward a gender diverse workforce in the renewable energy transition. *Sustainability: Science, Practice and Policy, 12*(1). https://doi.org/10.1080/15487733.2016.11908149

Räty, R., & Carlsson-Kanyama, A. (2010). Energy consumption by gender in some European countries. *Energy Policy, 38*(1), 646–649. https://doi.org/10.1016/j.enpol.2009.08.010

Standal, K., & Feenstra, M. (2021). Engaging the public for citizens energy production in Norway: Energy narratives, opportunities and barriers for an inclusive energy transition. In F. Karimi & M. Rodi (Eds.), *Energy transition in the Baltic Sea Region*. Routledge. https://library.oapen.org/handle/20.500.12657/52575

Stephens, J. C., et al. (2019). Women's leadership in renewable transformation, energy justice and energy democracy: Redistributing power. *Energy Research and Social Science, 57*, 101233. https://doi.org/10.1016/j.erss.2019.101233

Straver, K. et al. (2021). Energy poverty and the energy transition. *Towards improved energy poverty monitoring, measuring and policy action*. Retrieved December 2020, from https://publications.tno.nl/publication/34637565/tst4jN/TNO-2020-energy.pdf

Teladia, A. et al. (2017). The evidence base for gender and inclusion in sustainable energy. *Sustainable energy for all*. Retrieved December 2021, from https://www.seforall.org/system/files/gather-content/P2DataandEvidenceWorkingPaper.pdf

United Nations Environment Programme (UNEP), International Labour Organization (ILO), International Organisation of Employers (IOE), & International Trade Union Confederation (ITUC). (2008). Green jobs: Towards decent work in a sustainable, low-carbon world. https://www.ilo.org/wcmsp5/groups/public/@ed_emp/@emp_ent/documents/publication/wcms_158733.pdf

United Nations Framework Convention on Climate Change (UNFCCC). (2015). The Paris Agreement. https://unfccc.int/sites/default/files/english_paris_agreement.pdf

Van den Beukel, J., & Van Geuns, L. (2021). De afnemende leveringszekerheid van aardgas in Nederland, HCCS. Retrieved August 2022, from https://hcss.nl/report/afnemende-leveringszekerheid-aardgas-nl/

World Bank Group. (2019). World Development Report 2019: The changing nature of work. https://www.worldbank.org/en/publication/wdr2019

CHAPTER 7

How to Build a Gender-Balanced Solar Sector Workforce in the Brazilian Energy Transition

Aline Cristiane Pan, Kathlen Schneider, Natália Helena Ribeiro Chaves, Larissa Boing, Patrícia Betti, Aline Kirsten Vidal de Oliveira, and Izete Zanesco

Context

The energy sector is generally male-dominated, and the importance of women's performance must be discussed in order to introduce positive impacts and gender equity in the sector. Specifically in renewable energy, the participation of women is only 32% worldwide (IRENA, 2019). The underrepresentation is even more pronounced in STEM areas (Science,

A. C. Pan
Universidade Federal do Rio Grande do Sul (UFRGS), Porto Alegre, Brazil
e-mail: aline.pan@ufrgs.br

K. Schneider (✉) • L. Boing • A. K. V. de Oliveira
Universidade Federal de Santa Catarina (UFSC), Florianópolis, Brazil

N. Rocha Lawton, C. Forson (eds.), *Women and the Energy Sector*,
https://doi.org/10.1007/978-3-031-43091-6_7

Technology, Engineering and Mathematics), with only 28% of jobs occupied by women in the sector (IRENA, 2019). A recent study conducted in Brazil by Greener Consultancy (2020), in which 685 companies in the solar photovoltaic energy sector were interviewed, showed that 40% of the responding companies had no women in their workforce. In companies with up to ten employees, only 16% were female. In larger companies, with more than 26 employees, the percentage of women rose to 21%. On the other hand, according to the report "Renewable Energy: A Gender Perspective" published by the International Renewable Energy Agency (IRENA), there is an increasing interest among women in renewable energy and its applications. Despite the low representation, the renewable energy sector employs more women than the non-renewable energy sector. To cite one example, only 22% of professionals working in the oil and gas industry are women (IRENA, 2019).

According to the International Renewable Energy Agency (IRENA, 2021b), in three decades, 90% of the electricity that will be used in the world must be produced by renewable sources to prevent the increase in global warming. In the power sector, photovoltaic technology has been growing exponentially, and by 2020, an increase in installed global power of 139 GW was recorded (IRENA, 2021a), (REN21, 2021). This growth is also occurring in Brazil, a country with high solar irradiation, and in which the solar photovoltaic energy generation grew more than 60% from 2019 to 2020 (Energy Research Office—EPE, 2021). In this context, the efforts to produce electricity from renewable sources, aiming at the environmental, economic and social benefits, should continue in Brazil. According to a study by the Brazilian Solar Energy Association (ABSOLAR), more than 140,000 jobs were created in the solar photovoltaic energy sector in Brazil in 2021, most of them related to the

---

N. H. R. Chaves
Câmara de Comércio e Indústria Brasil-Alemanha do Rio de Janeiro (AHK-RJ), Rio de Janeiro, Brazil

P. Betti
Universidade Federal do Paraná (UFPR), Curitiba, Brazil

I. Zanesco
Pontifícia Universidade Católica do Rio Grande do Sul (PUCRS), Porto Alegre, Brazil
e-mail: izete@pucrs.br

installation of photovoltaic systems in residences (ABSOLAR, 2020). However, women's participation is disproportionate with the growth of the solar technology market; in fact, a downward trend in female professional participation is perceived (GIZ, 2021).

In this context, in 2019, the Brazilian Women Network in Solar Energy (*Rede MESol*) was founded with the purpose of connecting, supporting and inspiring women to actively act in the Brazilian process of energy diversification and transition (de Oliveira et al., 2020). In the MESol Network, women with scientific and technical backgrounds who work in the area of solar energy come together to share, discuss and analyse gender inequality in the workplace and to develop proposals to improve the current reality.

In order to identify the barriers and opportunities for women acting in the solar energy sector in Brazil, the C40 Cities Finance Facility, in partnership with the MESol Network, the German-Brazilian Chamber of Commerce of Rio de Janeiro and GIZ (Deutsche Gesellschaft für Internationale Zusammenarbeit) in Brazil, developed an unprecedented study on the subject (C40 Cities Finance Facility, 2021). In this study, 251 women working in the solar energy applications sector answered an online survey to gather the barriers and challenges to entering the labour market and remaining in the sector.

There is a lack of practical solutions to solve the issue and create a culture of gender equity in the solar energy field in Brazil and around the world. Therefore, the objective of this chapter is to identify and discuss actions that aim at the inclusion and retention of women in solar energy in Brazil, based on practical recommendations, to make the sector more attractive to women and thus promote gender equity as well as energy diversification and transition.

Challenges and Opportunities for Women in Solar Energy in Brazil: Recommendations for the Sector

The reduced and disproportionate participation of women, as well as the discrimination and different inequalities faced by them, demands support efforts for their insertion, permanence and continuous development in the sector. However, in addition to insufficient quantitative and qualitative data and information regarding women's participation in the national and global energy sector, there is also little understanding of actions that can be taken to minimize inequalities and better harness female potential.

According to CARE International (2015), an NGO that acts globally to address the rights of vulnerable groups (particularly women and girls), women have historically had less access to rights and opportunities than men in different countries around the world. Therefore, to adjust this imbalance and work towards gender equality, women should be empowered by promoting changes in different aspects (CARE International, 2015):

- Regarding women's own knowledge and abilities, for example, by raising awareness or building self-confidence;
- Regarding social norms, customs, institutions and policies, as in the introduction of new laws that promote gender equality;
- Concerning power relations between individuals, such as through changes in the control of household resources among family members.

These factors are interdependent to increase women's ability to access equal rights and opportunities. Therefore, interventions are needed for changes in each of these three dimensions. Empowerment starts at the individual level and must come from within, with women becoming more aware and empowered. However, it is equally important that men and boys play an active role in empowering women and work with them to change power structures and relationships towards gender equality. Otherwise, efforts will not be sustainable. It is also critical to identify practical factors involving immediate gender needs, such as providing job training so women can improve their income, and strategic factors, such as laws or social norms, to transform unequal gender social relations in the long term. If strategic factors are ignored, practical solutions are likely to have minimal sustainable impact (CARE International, 2015).

In order to support the development of strategies along these lines, the C40 Cities Finance Facility (2021) study presented recommendations aimed at the inclusion and professional development of women, black women, transgender women, mothers and other minorities in the solar sector in Brazil. Practical and strategic measures were proposed for changes at the individual, relational and structural levels in four areas considered key: (1) education, training and access to information; (2) economic opportunities; (3) health and well-being, with a focus on combating gender-based violence and supporting parenting; and (4) participation and leadership.

Based on the C40 Cities Finance Facility (2021) study and other reference sources, the following measures are presented for the promotion of gender equity and the strengthening of women's participation, with the invitation and incentive for men to also participate in the debate, understanding and promotion of these actions. We emphasize that several of these actions could, and should, be applied in combination, as there is the intersection of race, gender and social class as a conditioning factor (do Nascimento et al., 2019).

Education, Training and Access to Information

Currently, women are the majority of those completing higher education courses in Brazil (IBGE, 2021a) and are over 50% represented in graduate courses (CAPES, 2019). Moreover, they are in the (co)authorship of half of the scientific articles produced in Brazil (Elsevier, 2017). Therefore, there has been an expansion of the qualified participation of women at the bottom of academic careers. However, when evaluating the percentage distribution of graduates by gender, according to the major areas of the courses, the low representation of women in the areas of engineering (37.3%) and information technology (13.6%) becomes evident when compared to the areas of health (73.6%) and education (73.8%) (IBGE, 2019).

Up to the mid-twentieth century, it was stated which professions were for men and which were for women, and Science was culturally defined as a career unsuitable for women (Chassot, 2003). Therefore, the gap between men and women is associated with historically constructed social stereotypes. The consequence of this social construction is the aptitudes that each individual believes they have and must develop (Cunha et al., 2014). One of the ways to strengthen the female gender today is through education. Opening spaces for women to participate in scientific processes is fundamental to consolidate their inclusion and retention (Grossi et al., 2016). In addition to increasing the number of women with university degrees, women's participation in the sciences in Brazil has also grown. The number of female researchers has equalled the number of male researchers (Grossi et al., 2016). Despite this isonomy from a numerical point of view, women do not ascend in their careers in the same way as men.

The scientific career is culturally built on a "male model of career" (Velho et al., 2006). Therefore, the scientific career involves full-time commitments to work and research productivity, and women face the barrier of reconciling these demands with family responsibilities, having an excessive journey (double journey) (Silva & Ribeiro, 2014). In addition, studies show that school-age girls are persuaded into believing that the aptitude with STEM subjects is a masculine quality, which ends up discouraging them from pursuing careers in such areas (Bian et al., 2017; Falcão et al., 2019).

The study conducted by the C40 Cities Finance Facility (2021) found that almost half of women (42.6%) have at least a higher education degree, and 50.2% have some type of post-graduate degree. Many also reported that the lack of training and recognition is a barrier to professional advancement. The need for professional training seems to interfere more in getting into the job market than in remaining there. The unequal pay, as well as the need to balance personal and professional life, is highlighted in their answers as barriers encountered.

Offering specific training courses for women, especially those taught by other women, from short- to long-term courses, online and in person, presents an alternative to show the possibility of representation and inspiration to other women. MESol Network collaborated on a video[1] in 2020 towards this goal, and it had great repercussions in Brazil in social and academic media, as many girls saw the possibility of this profession, which before was portrayed as male-dominated (Pan et al., 2021).

In 2019, black or brown women between the ages of 18 and 24 had a higher education attendance rate of 22.3%, almost 50% lower than that recorded among white women (40.9%) and almost 30% lower than the rate seen among white men (30.5%) (IBGE, 2021b). This becomes evident when analysing the solar sector in Brazil, where 69% of the female workforce are white (C40 Cities Finance Facility, 2021). This data highlights the relevance of developing specific training programmes for the insertion of black women, thus contributing to social inclusion and diversification in the sector.

An effective social transformation requires the participation of all members of society within this process. Therefore, it is necessary to include, raise awareness and to encourage the participation of the male public in training courses and other actions aimed at the debate and promotion of gender equality in the sector, in order to identify their perceptions and make them allies in strengthening the role of women, and reduce chauvinism and prejudice.

[1] *Mulheres de Energia* video campaign: https://www.youtube.com/watch?v=hmMcdpiuOSk

Economic Opportunities

Even though women stand out in the educational field when compared to men, this advantage does not translate into better opportunities, higher (or equal) salaries or good working conditions when analysing the Brazilian labour market (Agência Brasil, 2017).

For instance, when looking at the average income of male and female professionals, Brazilian women receive wages 21.3% lower than those received by Brazilian men, and 24.4% lower in the case of female workers without formal employment status (IBGE, 2019). This difference is even greater when taking an intersectional reading. Inequality by colour and race occupies a central space in the debate on social inequalities in Brazil, as the black, brown and indigenous populations are the most vulnerable in the socioeconomic sphere. In this context, black and brown women are the main groups affected, receiving salaries, on average, 55.6% lower than those of white men (IBGE, 2019). They also experience multiple layers of oppression, which places them at the bottom of the social pyramid (ECLAC, 2021).

Inclusion in the labour market and income generation are central aspects of women's autonomy. It not only involves financial independence, but also presupposes autonomy to make choices. Women must also have freedom, as well as favourable and equal conditions to choose their profession, plan their futures and qualify themselves (UN Women, 2016).

Like in other parts of the world, in the Brazilian labour market, discrimination concentrates a large number of women in jobs considered as "female" occupations. These positions are seen as an extension of their "natural attributes", which is used as a justification for receiving lower wages (Cotrim et al., 2020). This distinguished occupational segregation, institutionalized by social norms and practices, makes it difficult for Brazilian women to achieve fair conditions for professional development and gender equality in the labour market.

Women also face greater challenges in taking advantage of equal employment opportunities because most of the domestic chores and care of children, the elderly and sick family members fall on them. In 2019, women spent an average of 76% more hours than men on domestic chores. This statistic is even more unequal considering the racial intersection, which means black women spend more time than white women performing these activities (IBGE, 2020). Women's responsibility for domestic and caregiving work profoundly affects their access and permanence in the labour market.

Professionals working in the energy sector in Brazil must deal with all the already mentioned challenges, enhanced by the fact that it is one of the worst areas for gender equality in Brazil (Falcão et al., 2019). In the solar sector specifically, 92.8% of the women who participated in the survey (C40 Cities Finance Facility, 2021) claim there are barriers to entering, and 94%, to remaining in the sector, which shows that despite its potential for innovation and renewal of practices, actions for greater gender equality are urgent. The biggest obstacles cited are linked to structural chauvinism and its consequences, such as lack of credibility in the quality of work developed by them, especially when it comes to the technical area, which is exemplified in the statement of one of the respondents: '*We still face prejudice! More than once I came to a client to present a solution and heard that women do not add anything technically and that they only discuss technically with men*'.

Greater equality of opportunities is a longing of women working in the solar energy sector (C40 Cities Finance Facility, 2021). This calls for a series of actions aimed at greater openness of the market to hire female professionals and equal conditions for career progression. The wide dissemination of studies such as the one conducted by the C40 Cities Finance Facility (2021) and others that expose gender inequality in the sector, as well as the holding of events to debate the issue, can help raise awareness in the public and private sectors. In addition, these opportunities can lead to the formation of partnerships and aggregate efforts to make the sector more attractive to women.

Encouraging companies and organizations that work with solar energy applications to adhere to gender equity programmes, such as the UN Women's Platform of Women's Empowerment Principles (WEPs) and similar ones, is another key action. In order to change this scenario of gender inequality, goals and strategies consistent with the reality of each organization are defined regarding these adhesions.

Establishing a fair and equivalent pay to that of men is another important agenda to be promoted in this field of action, since male professionals in the sector earned on average 31% more than their female peers in the period of 2012 to 2019, even in situations of equal education, age and time of employment (C40 Cities Finance Facility, 2021). To this end, it is recommended that studies should be conducted to identify the reasons for this wage gap and to develop mechanisms to eliminate it, such as career plans with equal pay and opportunities. Equalizing salaries will make the sector fairer and more attractive to women.

Advertising the solar energy sector as an area that is already and can be increasingly occupied by women is another necessary demand. This action will help to eliminate the male-predominant stereotype of the sector, decreasing the chauvinism and prejudice that bar the development of female professionals. In addition, it can help attract more girls and women not only to the industry, but to STEM careers. This goal can be achieved not only by increasing the visibility of women working with solar energy applications, but also by developing awareness: raising materials such as handbooks, videos, podcasts, games and social media pages on the topic, tailored for each target audience, such as elementary, middle and high school students, for example. Stimulating girls' interest in STEM fields by presenting them with a welcoming environment, in which women feel they can be part of and be successful in careers in these areas in the future, can substantially change this reality.

Health and Well-Being

Health conditions and well-being can also be influenced by gender norms, roles and relations and are directly related to women's personal and professional development. Thus, two issues strongly related to gender and fundamental to providing women's quality of life, that will be addressed in this topic, are violence and parenting.

Combating Gender-Based Violence

Violence is one of the main obstacles to ensuring human rights and fundamental freedoms for women and girls, whether it occurs in public or private life, perpetrated by the state or tolerated by it (Engel, n.d.).

According to Minayo (2006, p. 8), '*Violence, in itself, is not a health issue, but it affects health because it causes injuries, traumas, physical and emotional deaths*'. Thus, it is understood that violence has a great impact on health. For the author, to understand its effects, it is necessary to analyse the issue as a set of problems that relate to health, conditions, situations and lifestyle (Minayo, 2006).

There are multiple forms of violence, which are related to the notions of embarrassment and the use of physical superiority over another. The concept may seem neutral, but it refers to conflicts of authority, to disputes overpower, domination, possession and destruction of another being or his property. According to social norms or legal apparatuses of society in different realities such as places, circumstances and historical periods, manifestations of violence can be tolerated or condemned, licit or illicit, in visible or secret aspects, in both public and private settings (Minayo, 2006).

The specific gender type of violence can be understood as any type of aggression against a person in a vulnerable situation due to his/her gender identity or sexual orientation. However, since women are the most affected, gender violence is understood as:

> (…) expression used to refer to the various acts practiced against women as a way to subject them to physical, sexual and psychological suffering, there included the various forms of threats, not only in the intra-family sphere but also covering their social participation in general, with emphasis on their work relations, characterized mainly by the imposition or intention of imposition of subordination and control of the male gender over the female. (de Souza, 2007)

According to the World Health Organization, worldwide, one in three women (35%) has already been victims of physical violence. In Brazil, despite the legislation against gender-specific violence, the statistics are alarming, positioning the country with the fifth-highest femicide rate in the world (Agência Brasil, 2017). According to the Atlas of Violence, in one decade, from 2007 to 2017, the homicide of women increased by 30.7% in the country, with greater growth among black women, who accounted for 66% of all women murdered in Brazil in 2017 (Ipea, 2019).

Brazilian cases of violence directed at transgender people are also expressive and impactful, making the average life expectancy of these people only 35 years in the country. The Dossier on Murders and Violence Against Brazilian Transgender People (Benevides & Nogueira, 2020) reports the fact that Brazil, among 74 countries analysed, is the one that kills the most transgender people since 2008. Almost all of them (97.7%) are transgender women. The prejudice that leads to such violence also makes it difficult for trans women to access the job market. According to the UN's Trans Pages document, '(…) *they are discriminated against in selection processes, falling behind in the recruitment stages because of their gender identity*' (United Nations, 2019).

In the professional environment, a survey conducted by the LinkedIn platform in partnership with Think Eva, in 2020, concluded that 47% of Brazilian women have experienced harassment at work and that one in six has resigned after the event. Only 5% of harassed women resorted to the

Human Resources sector of the company they worked for, since 78% of them believe that the aggressor usually goes unpunished (Think Eva & Linkedin, 2020).

When analysing the statistics of women victims of domestic violence, it is observed that these women remain in the same job for less time than others, miss more work and are less focused, increasing their chances of dismissal (Estado de Direito, 2020).

In the study conducted by the C40 Cities Finance Facility (2021), gender violence was investigated among female professionals in the solar energy sector considering the following typology:

- Psychological violence: any conduct that causes emotional damage and diminished self-esteem or that harms and disturbs the full development of the victim: threat, embarrassment, humiliation, insult, etc.
- Moral violence: any conduct that constitutes slander, defamation or insult.
- Sexual violence: any conduct of a sexual nature without consent, through intimidation, threat, coercion, bribery, manipulation or the use of force: from any kind of touch that causes discomfort, such as kissing, hugging, caressing, to rape.
- Physical violence: any conduct that offends the bodily integrity or health of the victim. (Brasil, 2006, 2011).

Such definitions were presented in the survey, first, in order to raise awareness among the participants about the existence of gender-based violence and its different forms of manifestation, since many people are unaware of the term or do not realize that they experience gender-based violence.

According to the answers obtained, most women (57%) have already suffered some kind of violence as professionals of the sector in their different workspaces (companies, institutions, universities, events, meetings) with emphasis on psychological violence, which represented 47.4% of the total of situations experienced by them. Another 15.5% experienced moral violence, 10% sexual violence, 1.6% physical violence and 0.8% other violence. In addition, 8% of the respondents could not identify if they had already experienced some situation of violence in their work environment.

Among the professionals who have experienced discrimination, harassment or other forms of violence, most (65.8%) have noticed when it happened but did not feel safe enough to report them or to make a formal complaint. Only a minority, 9.6%, reported feeling safe to do so. The statement of one of the respondents indicates that impunity and the normalization of violence may be one of the reasons for the lack of security in reporting: '*I complained, and nothing happened. I don't report it anymore, they told us that men are just like that*'.

There are also cases in which women were not able to identify the violent situations they went through at the time it occurred, as well as those who declared they had dealt with the situation in other ways, such as those who took a stand and argued directly with the aggressor.

Despite gender violence being a reality in the Brazilian professional environment, including the solar energy sector, a significant amount (41%) of the respondents' workplaces do not have policies or actions aimed at combating it. Furthermore, 22.7% of the women who participated in the survey do not know if there are any actions in this direction in the organizations to which they are linked. Among the existing actions, the most frequently mentioned were the existence of a good behaviour guide (19.9%), a reporting channel (16.7%) and projects and/or other actions to combat harassment in the workplace (11.6%).

Gender violence occurs in different ways and levels of aggression. It is rooted and naturalized, and often does not mobilize energy for change. However, the less or more serious cases can be avoided if public security policies and actions in all spaces are directed to fight psychological, moral, physical and sexual violence. In other words, it is necessary to identify and combat violence in all situations, avoiding the normalization of violence and its social acceptance.

To do so, a first step is to make this commitment both individually and socially. Secondly, it is necessary to know how to identify cases of gender violence and how to address them. In this sense, it is essential to offer information to people of all genders and to public and private institutions for the recognition and confrontation of this type of violence.

In the opinion of Minayo (2005, p. 26), 'overcoming violence is a challenge that requires a great deal of social, political and subjective investment to think and rethink Brazilian culture'. For this purpose, it is possible to make available online or in-person courses, posters, brochures and videos, and promote lectures and debates about types and consequences of gender violence and healthy masculinity, for example. Dates such as "March 8th—International Women's Day" and "June 28th—International LGBTQIA+ Pride Day" can stimulate the development of a calendar in the institution for courses, film screenings with debates on the various themes involved. Such actions are necessary because many people do not even realize that they are being victimized or acting violently, since attitudes such as discriminatory jokes are quite normalized in Brazil. According to Odalia (1985, p. 85), 'violence (...) is not self-evident in all its manifestations, some of which are so subtle and so well managed that they can pass for normal and natural conditions of human living'. By generating opportunities for education on the subject for victims and aggressors, the solar energy industry will collaborate to break the cycle of violence, such as: *'We still face prejudice! More than once, I have arrived at the client to present the solution and I have heard that women don't add anything technically, that they only discuss technical issues with men'*.

In addition to understanding what gender violence is, it is also fundamental to know how to proceed in case a person suffers or witnesses its materialization. To this end, it is important to publicize public help lines, such as the phone number of the Call Centre for Women (*Ligue 180*), to create safe internal help lines for denunciation and to implement policies to combat gender violence. By developing actions for accountability in cases of aggression, the sector will contribute to eradicating it from the professional environment, making it fairer and more attractive to women and other people. The awareness that there are people and institutions available for support can encourage victims to report the aggressions they have suffered and colleagues to defend them in these situations.

On the other hand, the simple existence of a poster with a 'report hotline' and the message that the environment is against gender violence can inhibit a lot of aggressive attitudes. However, it is also important to stipulate serious consequences, such as contractual clauses that guarantee the dismissal of those people who commit aggression in the professional environment. In a transversal way, it is relevant to conduct periodical training to men and women so that they know how to deal and adequately address these situations. In this sense, awareness campaigns, courses, conversation rounds, as well as the dissemination of news about good practices, presentations about projects, institutions, channels, etc., that work with the theme, both to prevent aggression and to give support to the victims of violence, can be useful.

Since gender violence is practically taboo, the exemplary punishment of aggressors, but also the support to victims, the debate, the information and the training will contribute to eliminating it. Once eliminated in the professional environment of solar energy, the industry will contribute to a profound and urgently needed transformation that relates to respect and care for women.

Parenting Support

Being a mother or becoming a mother can be a barrier against women in society. The idea that motherhood and work are mutually exclusive is perceived in different cultures. In Brazil, among women aged 25–49 who lived in homes with children up to 3 years old, just over half (54.6%) were working outside the home. However, for men in these same conditions, the percentage is 60% higher. In households without children in this age group, the number of women who work outside the home was 20% higher and 5% lower for men compared to those who have children at home. When comparing this indicator for women from ethnic minorities with or without children, the possibility of getting a job abroad is even smaller (IBGE, 2021a). It can be seen from the data that there is a lack of structure to support women in reconciling work and family, whereas for men the data does not reflect this, and for black and brown women, the obstacles are even greater due to intrinsic racial and social problems (structural racism).

Another very recurrent problem in Brazil is teenage pregnancy. If a pregnant girl belongs to a low social class, there is a significant loss in her education and professionalization process. Without the availability of resources and support to deal with the situation and its demands (Dias & Teixeira, 2010), these young women will not be able to reach the labour market.

Motherhood is a reality for 36.7% of Brazilian women in the solar energy sector who responded to the questionnaire. Furthermore, 1.2% of respondents were pregnant. When asked if they faced or face any professional barrier for being a mother or for becoming a mother, most women in the sector (56.5%) said yes (C40 Cities Finance Facility, 2021). When analysing the employment relationship of women who are in leadership positions, it is observed that most women that are company directors are also company owners (84.3%) and most of them (56.9%) are mothers. The study concluded that these numbers may indicate that many women become "entrepreneurs out of necessity" in the sector since the number of women who leave or lose their jobs when they become mothers in Brazil is quite expressive (Machado & Pinho Neto, 2016). The decrease in the presence of women in stable and prestigious positions in strategic and high-ranking jobs is related to motherhood (Ginther & Kahn, 2009), as women with children are 38% less likely to reach these positions in all areas of knowledge (Mason et al., 2013). Also, in the scientific sector, in the period of up to 4 years after motherhood, there is a decrease in the researchers' productivity (Andrade, 2018).

Currently, actions and projects that favour parenthood in institutions that work with the applications of solar energy are insufficient. It is more difficult for women than for men to reconcile personal and family life, especially regarding the care of children, since they are responsible for a large part of the responsibilities with the home and family members. Therefore, promoting support programmes and compensation mechanisms for the maintenance of working mothers during the maternity and breastfeeding phases is essential.

Flexible working hours, the possibility of home office, breastfeeding room and day care centres in and/or near the workplace, caregivers and children's recreation at events and fairs are examples of necessary policies that should be implemented and expanded to encourage permanence and participation of women mothers in the sector.

In the Brazilian Federal Constitution, maternity leave is 120–180 days and paternity is 5 days, which drastically highlights the inequality between men and women at work. This format reinforces the sexual division of labour, as it attributes exclusive care to women. Therefore, seeking equalization of maternity and paternity leave and/or sharing is a policy that must be adopted.

Some actions that will favour the permanence of these women mothers in the areas of solar energy applications are not considering years spent on maternity leave in the accounting of productivity indicators and adding an extra year to each maternity or paternity leave. In addition, expanding the policies that allow the participation of women in events, congresses and fairs with spaces for childcare and recreation, will allow the participation of women with small children.

Enabling the reconciliation between personal and professional life will favour the permanence and development of professionals, with a consequent increase in their children's well-being.

Participation in the Workforce and Leadership

Having women in leadership positions and participating in decision-making spaces is a crucial aspect to provide working spaces and public policies that consider gender awareness. The inclusion in these spaces makes it more likely that women's demands will be considered and, thus, their needs and interests met. However, although the number of women leaders has increased in recent years, it is still far below a relevant female representation, especially when analysed from a racial perspective.

According to data from the Women in Business 2020 Survey, conducted by Grant Thornton International, women currently occupy 34% of senior leadership positions (executive directors) in companies operating in the Brazilian market. The result represents an increase of 9% compared to 2019 (Grant, 2020). This study did not bring an ethnic-racial analysis to the data. However, the research conducted by the Inter-American Development Bank (IDB) and the Ethos Institute, highlighted that in 2016, black women occupied only 1.6% of management positions and 0.4% of the executive staff of companies in the country (BID; Instituto Ethos, 2016).

In the energy sector, the presence of women in leadership positions is quite negligible. A survey carried out with the main energy companies operating in Brazil showed that only 6% of leadership positions are held by women. The survey considered female participation in CEO or other leadership positions in areas such as operations, maintenance, new business, engineering and construction (EPBR, 2021).

In the solar energy sector, female participation in leadership positions is also limited, especially in higher positions such as directors. There is a positive trend for the entry of women in management positions—6.9% for women, against 5.3% for men—however, negative for executive positions—only 0.7% of women assume such positions, against 0 0.9% of men (C40 Cities Finance Facility, 2021). Although small, this bias is statistically significant, indicating that the percentage of women in the sector decreases disproportionately as the career progresses.

This effect is known as vertical or hierarchical segregation and is popularly known as the Scissor Effect (Rossiter, 1982). The name refers to the shape of the graphic in which two complementary curves move apart or even cross, resembling an open scissors. The eventual crossover, necessarily at the 50% level, reflects the inversion of the predominance of one group over the other. Even when there is no crossing of the curves, their distance indicates that the disparity situation increases as one progresses in the career, as is the case of the proportion of women in relation to men in the energy sector as a whole, as well as in energy solar, and in management positions.

This challenging path for women to reach positions of leadership is well documented and has been called the Glass Ceiling. This is because the obstacles to greater female participation in positions of power are not formal or legal, but intangible, although they interfere profoundly in their professional ascension (Lima, 2013).

Despite the barriers to their growth, women working in the field of solar energy show great interest in assuming leadership positions. Among those who responded to the survey, 76.6% of those who are not in these positions said they wanted to occupy them. What paths can be taken to change this reality and ensure greater female participation in decision-making spaces and positions in the solar energy sector?

The first step is to become aware of the phenomenon and recognize that it is effectively a problem for the sector's development. Therefore, it is necessary to debate the reasons for such disparity and its consequences, both within private companies, as in the public sector, professional associations, events and all the spaces and opportunities to bring up this theme.

> Standing up for ourselves every day and showing that we are as capable and competent as any man in the area makes us want to give up because it is unfair that we must do every day. To reach leadership positions is even more complicated, respect and salary inequality are very present. (C40 Cities Finance Facility, 2021)

Carrying out campaigns to deconstruct gender stereotypes is another important action. Therefore, in addition to the actions that positively influence the change in the perception of gender patterns already suggested throughout this chapter, it is necessary to increase the visibility and recognition of female leaders who already work in the field of solar energy. This action also helps to create references in which professionals can project themselves and be inspired.

Establishing targets or quotas for female participation in leadership positions is another action that helps to achieve gender equity in companies and organizations. In addition, the development or support of coaching and mentoring programmes aimed at female leaders tends to help the ascension to management positions.

Achieving equal representation of women in management and leadership positions will make the solar energy sector more welcoming and promising for all professionals, not just women. Building a more egalitarian, diverse and inclusive professional environment can generate results that go far beyond the financial aspect.

Gender Equity Potential for Energy Transition

In Brazil, the energy transition is very particular when compared to other countries' reality. In 2020, renewable energy sources were responsible for 84.8% of Brazil's electricity matrix, which represents a high value compared to the 23% of the world average (EPE, 2021). Hydro energy source is the one that stands out the most, accounting for 65.2% of the internal supply (EPE, 2021). However, a large part of the national hydroelectric potential is available in very sensitive regions, in both environmental and social terms. Also, hydroelectric plants are based on centralized generation, have limited growth and are usually located far from the consumption centres, which generates high costs and energy losses.

In this scenario, Brazilian energy security is very sensible since there is a large-scale reliance on energy generated by only one type of source. Therefore, the discussion on the Brazilian energy transition requires rethinking this model even though the country's electricity matrix is much "cleaner" than other countries. Efforts must be focused on diversifying and decentralizing the Brazilian electricity matrix.

Diversifying the electricity matrix is not the only type of diversification required for the Brazilian energy transition. In fact, worldwide energy transition requires also gender diversity in the sector since gender inequalities in industry are particularly acute within the energy sector (Ernst & Young, 2015). Thus, it's necessary to support opportunities for women's advancement and highlight the multiple benefits of hiring women for new work opportunities in the energy industry, especially to scale up the energy transition and accelerate the requirements to meet agreed climate goals.

A significant and emerging growth of the solar energy sector is expected. According to IRENA (2021a), even in the pandemic situation, photovoltaic systems continued dominating new capacity installations, showing the resilience and momentum of the renewables industry. Following this development, systematic collection and distribution of data on gender in the energy workforce will encourage greater awareness about gender equity (Martinez & Stephens, 2016).

The current lack of gender parity in institutional positions can lead to harmful social impacts. It is necessary to harness all talents and foster innovation across a vast array of skills, applications and specializations. When there are women integrated into decision making, governmental transparency increases (Cornwall & Goetz, 2005).

In addition, the global energy transition, associated with gender equality, offers a chance for deep societal transformation within communication and decision making. When categorized in terms of environmental, economic and cultural aspects, women's participation can lead to more effective and inclusive outcomes, because women are more likely than men to express environmental concern, support environmental protection and enact pro-environmental behaviours (Kennedy & Dzialo, 2015). Furthermore, women on their board of directors are more likely to proactively invest in renewable energy and to reduce carbon emissions throughout their value chain (CRB, 2012).

Gender equity is an opportunity for a more inclusive workforce for the energy transition process and to societies that leave no one behind (GWNET, 2019). During this transition, there is a need to integrate social logics from diverse stakeholders whose priorities are influenced by multiple challenges (Martinez & Stephens, 2016).

The relevance of the issues of gender equality and energy transition for the entire society was recognized by the United Nations (UN), which included them in their 17 Sustainable Development Goals (SDGs). The SDGs establish priorities and goals necessary to achieve the goals of the 2030 Agenda for Sustainable Development of the UN. SDG 5 "Gender Equality" is focused on equality, empowerment and issues related to women and the benefits they can bring to society. SDG 5 is strongly interrelated with SDG 10 "Reducing Inequalities" and with the other SDGs. These SDGs impact all other sustainable goals, especially the ones related to climate change and energy, as women are more susceptible to the effects of climate change (Alston, 2014). Gender equality is, therefore, substantially connected to SDG 13 "Action against Global Climate Change" and SDG 7 "Clean and Affordable Energy", closely related to energy transition and the diversification of the energy matrix. Women can be protagonists of these goals, as the combination of both gender and energy diversification should provide the innovative solutions required to face the challenges of these areas.

Final Considerations

The Brazilian historical context marked by colonization, slavery and dictatorship contributed to building the main social problems that are currently faced in the country: racism, sexism, violence and poverty. Recognizing that these barriers are rooted in the structure of Brazilian society as a tool for political manipulation, and not as a place for the integral formation of citizens (Freire, 1987), is the first step towards the understanding of so many inequalities existing in the country. Therefore, it is necessary to leave the stereotypes rooted in so many years and to promote actions that can change these perceptions.

This discussion is raised because gender discrimination is increasingly perceived as disconnected from reality, at a historical moment when it should be as close as possible to the critical analysis of humanity's next steps, as called for by SDG 5 and SDG 10.

In recent years, Brazil has denied the environmental impacts and the series of setbacks in rights policies. In the pandemic of 2020, it is historically impoverished and marginalized populations that suffer the most from the consequences of these phenomena (Pires et al., 2020). Therefore, the discussion about the impacts of the pandemic and climate change cannot be dissociated from the social relations established in Brazilian society, such as chauvinism and gender/gender discrimination. In order to form conscientious citizens, it is essential to assume the commitment to take this knowledge in order to develop a correct perception of the problems and to foster favourable attitudes and behaviours to build sustainable development (Cachapuz et al., 2005).

In order to contemplate the strengthening of female participation and the consequent development of the solar energy sector in Brazil as a whole, this chapter presented the main barriers for women, strategies to face them and some of their possible positive impacts. Four key areas were considered: (1) education, training and access to information; (2) economic opportunities; (3) health and well-being, with a focus on combating gender violence and supporting parenthood; and (4) participation and leadership.

These recommendations aim to promote the gender agenda with the aim of strengthening women's participation in the solar sector. However, they are not fully comprehensive in themselves. They encompass the four discussion areas highlighted in the C40 Cities Finance Facility survey (2021) aiming to shed light on some of the aspects that should be addressed for building a more equitable and gender-balanced solar sector workforce.

However, more contributions focusing on discussing the solutions, instead of only the problems, should complement them. In addition, it is necessary to extend actions beyond the female audience, ensuring the participation of men in promoting gender equality.

The diversification of the Brazilian energy matrix needs to move towards the expansion of gender and race diversity so that the sector, not only in solar energy but also in all renewable energies, can take advantage of its full potential for growth and innovation.

References

ABSOLAR (Brazilian Photovoltaic Solar Energy Association). (2020). *Solar energy will generate more than 147 thousand jobs in Brazil in 2021, projects ABSOLAR* (in Portuguese). https://www.absolar.org.br/noticia/energia-solar-vai-gerar-mais-de-147-mil-empregos-no-brasil-em-2021-projeta-absolar/

Agência Brasil. (2017). *Feminicide rate in Brazil is the fifth highest in the world* (in Portuguese). https://agenciabrasil.ebc.com.br/direitos-humanos/noticia/2017-08/taxa-de-feminicidios-no-brasil-e-quinta-maior-do-mundo

Alston, M. (2014). Women's studies international forum gender mainstreaming and climate change. *Women's Studies International Forum, 47*, 287–294. https://doi.org/10.1016/j.wsif.2013.01.016

Andrade, R. O. (2018). Maternity on the curriculum (in Portuguese). *Revista Pesquisa Fapesp, 269*. http://www.revistapesquisa.fapesp.br/2018/07/19/maternidade-no-curriculo

Benevides, B. G., & Nogueira, S. N. B. (2020). *Dossier of murders and violence against Brazilian transvestites and transsexuals in 2019* (in Portuguese). Expressão Popular, Antra, IBTE. https://antrabrasil.files.wordpress.com/2020/01/dossic3aa-dos-assassinatos-e-da-violc3aancia-contra-pessoas-trans-em-2019.pdf.

Bian, L., Leslie, S. J., & Cimpian, A. (2017). Gender stereotypes about intellectual ability emerge early and influence children's interests. *Science, 355*(6323), 389–391.

BID & Instituto Ethos. (2016). *Social, racial, and gender profile of the 500 largest companies in Brazil and their affirmative actions* (in Portuguese). https://www3.ethos.org.br/wp-content/uploads/2016/05/Perfil_Social_Tacial_Genero_500empresas.pdf

Brasil. (2006). *Lei no 11.340, de 7 de agosto de 2006*. https://www.institutomariadapenha.org.br/lei-11340/lei-maria-da-penha-na-integra-e-comentada.html

Brasil. (2011). *Programa Pró-Equidade de Gênero e Raça: Assédio Moral e Sexual*. Senado Federal. https://www12.senado.leg.br/institucional/procuradoria/proc-publicacoes/cartilha-assedio-moral-e-sexual

C40 Cities Finance Facility. (2021). *Solar energy in Brazil: which are the barriers and opportunities for women professionals in the field?*. https://www.c40cff.org/knowledge-library/solar-energy-in-brazil-which-are-the-barriers-and-opportunities-for-women-professionals-in-the-field

Cachapuz, A., Gil-Perez, D., Carvalho, A. D., Praia, J., & Vilches, A. (2005). *The necessary renewal of science education* (in Portuguese).

CAPES (Coordination for the Improvement of Higher Education Personnel—Coordenação de Aperfeiçoamento de Pessoal de Nível Superior). (2019). *Master's and doctorate data* (In Portuguese). https://dadosabertos.capes.gov.br/

CARE International. (2015). *Making it count: Integrating gender into climate change and disaster risk reduction: a practical how-to guide*. CARE International in Vietnam.

Chassot, A. (2003). Scientific literacy: A possibility for social inclusion (in Portuguese). *Revista brasileira de educação*, 89–100.

Cornwall, A., & Goetz, A. (2005). Democratizing democracy: Feminist perspectives. *Democratization, 12*(5), 783–800.

Cotrim, L. R., Teixeira, M. O., & Proni, M. W. (2020). Gender inequality in the formal labour market in Brazil (In Portuguese). *Texto para Discussão*. Unicamp. IE, Campinas, 383, 1–28. https://www.eco.unicamp.br/images/arquivos/artigos/TD/TD383.pdf

CRB (Center for Responsible Business). (2012). *Women Create a Sustainable Future*. Haas School of Business, University of California Berkeley. https://www.eticanews.it/wp-content/uploads/2012/11/Report-Women_Create_Sustainable_Value.pdf

Cunha, M. B., Peres, O. M. R., Giordan, M., Bertoldo, R. R., de Quadros Marques, G., & Duncke, A. C. (2014). As mulheres na ciência: o interesse das estudantes brasileiras pela carreira científica. *Educación química, 25*(4), 407–417.

de Oliveira, A. K. V., Schneider, K., Zanesco, I., & Pan, A. C. (2020, June). Brazilian network of women in solar energy: Their training and mapping of the women active in the sector (In Portuguese). In: *Proceedings of VIII Brazilian congress of solar energy*. https://anaiscbens.emnuvens.com.br/cbens/article/view/987

de Souza, S. R. (2007).*Comments on the law to combat violence against women* (p. 22) (in Portuguese). Juruá.

Dias, A. C. G., & Teixeira, M. A. P. (2010). Teenage pregnancy: A look at a complex phenomenon (in Portuguese). *Paidéia (Ribeirão Preto), 20*, 123–131.

do Nascimento, E. C., Cabral, F. P., & Cerqueira, L. S. (2019). A interseccionalidade de raça e gênero no acesso ao mercado de trabalho: uma breve análise dos dados do IBGE ano base 2016. *Diversidade e Educação*, 68–83 (in Portuguese). https://periodicos.furg.br/divedu/article/view/9493/6178

ECLAC (Economic Commission for Latin America and the Caribbean). (2021). *The gender dimension in the Big Push for Sustainability in Brazil: Women in the context of the social and ecological transformation of the Brazilian economy* (in Portuguese). https://www.cepal.org/pt-br/publicaciones/46643-dimensao-genero-big-push-sustentabilidade-brasil-mulheres-contexto-transformacao

Elsevier. (2017). *Gender in the Global Research Landscape, Analysis of research performance through a gender lens across 20 years, 12 geographies, and 27 subject areas.* https://www.elsevier.com/__data/assets/pdf_file/0008/265661/ElsevierGenderReport_final_for-web.pdf

Engel, C. L. (n.d.). *Violence against women* (in Portuguese). Ipea. https://www.ipea.gov.br/retrato/pdf/190215_tema_d_a_violenca_contra_mulher.pdf

EPBR. (2021). *Women are only 6% in leadership positions in the energy sector in Brazil* (in Portuguese). https://epbr.com.br/mulheres-sao-apenas-6-em-cargos-de-lideranca-do-setor-de-energia-no-brasil-mostra-estudo

EPE (Energy Research Office—Empresa de Pesquisa Energética). (2021). *Synthesis report of the national energy balance 2021* (In Portuguese). https://www.epe.gov.br/sites-pt/publicacoes-dados-abertos/publicacoes/PublicacoesArquivos/publicacao-601/topico-588/BEN_S%C3%ADntese_2020_PT.pdf

Ernst & Young. (2015). *Women in Power and Utilities Index 2015.* https://www.ey.com/en_gl/women-power-utilities/could-gender-equality-be-the-innovation-boost-utilities-need

Estado de Direito. (2020). *Women victims of violence and unions, what does one have to do with the other?* (in Portuguese). http://estadodedireito.com.br/maria-da-penha-vai-ao-trabalho-mulheres-vitimas-de-violencia-e-sindicatos-o-que-uma-coisa-tem-a-ver-com-outra/

Falcão, J., Strapasson, A., Costa, K. M., Masulino, N. W. N., & Barbosa, M. (2019). Energy & Gender: An assessment on gender equality in the energy sector in Brazil. In *Technical report commissioned by the British Embassy in Brazil in collaboration with the UK Prosperity Fund: Energy Programme..* https://doi.org/10.13140/RG.2.2.19253.52964.

Freire, P. (1987). *Pedagogy of the oppressed* (in Portuguese). Paz e Terra, 2005. A importância do ato de ler, p. 44.

Ginther, D. K., & Kahn, S. (2009). *Does Science Promote Women? Evidence from Academia 1973–2001* (pp. 163–194). University of Chicago Press.

GIZ (Deutsche Gesellschaft für Internationale Zusammenarbeit Gmbh). (2021). *A mão de obra da cadeia produtiva do setor solar brasileiro.*

Grant, T. (2020). *Women's participation in leadership positions grows in Brazil* (in Portuguese). https://www.grantthornton.com.br/insights/artigos-e-publicacoes/cresce-participacao-de-mulheres-em-cargos-de-lideranca-no-brasil

Greener Consultancy. (2020). *Strategic market research of distributed generation: photovoltaic market in Brazil.* https://d335luupugsy2.cloudfront.net/cms/files/12882/1617857772Strategic_Market_Study_-_Distributed_Generation_-_2H_2020.pdf

Grossi, M. G. R., Borja, S. D. B., Lopes, A. M., & Andalécio, A. M. L. (2016). Women practicing science in Brazil (in Portuguese). *Revista Estudos Feministas, 24*, 11–30.

GWNET (Global Women's Network for the Energy Transition). (2019). Women for sustainable energy: Strategies to foster women's talent for transformational change. https://www.globalwomennet.org/women-for-sustainable-energy/

IBGE (Brazilian Institute of Geography and Statistics—Instituto Brasileiro de Geografia e Estatística). (2019). *2018 annual continuous national household sampling survey—PNADC* (In Portuguese). www.ibge.gov.br

IBGE (Brazilian Institute of Geography and Statistics—Instituto Brasileiro de Geografia e Estatística). (2020). *Outras formas de trabalho*. https://agenciadenoticias.ibge.gov.br/media/com_mediaibge/arquivos/579f8cb956b0677de8dfc870215f3951.pdf.

IBGE (Brazilian Institute of Geography and Statistics—Instituto Brasileiro de Geografia e Estatística). (2021a). *Gender statistics: social indicators of women in Brazil* (In Portuguese). IBGE Report: Coordination of Population and Social Indicators. https://biblioteca.ibge.gov.br/index.php/biblioteca-catalogo?view=detalhes&id=2101784

IBGE (Brazilian Institute of Geography and Statistics—Instituto Brasileiro de Geografia e Estatística). (2021b). *Gender Statistics: Women's occupation is lower in households with children up to three years old* (In Portuguese). https://agenciadenoticias.ibge.gov.br/agencia-sala-de-imprensa/2013-agencia-de-noticias/releases/30172-estatisticas-de-genero-ocupacao-das-mulheres-e-menor-em-lares-com-criancas-de-ate-tres-anos

Ipea. (Institute for Applied Economic Research—Instituto de Pesquisa Econômica Aplicada). (2019). *Atlas of Violence 2019* (in Portuguese). https://www.ipea.gov.br/atlasviolencia/arquivos/downloads/6537-atlas2019.pdf

IRENA (International Renewable Energy Agency). (2019). *Renewable energy: A gender perspective*. https://www.irena.org/publications/2019/Jan/Renewable-Energy-A-Gender-Perspective

IRENA (International Renewable Energy Agency). (2021a). *World energy transitions outlook: 1.5°C pathway*. https://www.irena.org/publications/2021/Jun/World-Energy-Transitions-Outlook

IRENA (International Renewable Energy Agency). (2021b). *Renewable capacity statistics 2021*. https://www.irena.org/publications/2021/March/Renewable-Capacity-Statistics-2021

Kennedy, E. & Dzialo, L. (2015). Locating gender in environmental sociology. *Sociology Compass, 9*(10), 920–929. https://onlinelibrary.wiley.com/doi/epdf/10.1111/soc4.12303#accessDenialLayout

Lima, B. S. (2013). The crystal maze: The trajectories of female scientists in physics (in Portuguese). *Revista Estudos Feministas, 21*(3), 883–903. https://www.scielo.br/j/ref/a/v7m9qdqJPRMhSmyhny7kQgq/abstract/?lang=pt#

Machado, C., & Pinho Neto, V. R. D. (2016). The labor market consequences of maternity leave policies: Evidence from Brazil. https://bibliotecadigital.fgv.br/dspace/handle/10438/17859

Martinez, R. P., & Stephens, J. C. (2016). Toward a gender diverse workforce in the renewable energy transition. *Sustainability: Science, Practice and Policy, 12*(1), 8–15. https://doi.org/10.1080/15487733.2016.11908149

Mason, M. A., Wolfinger, N. H., & Goulden, M. (2013). *Do babies matter? Gender and family in the ivory tower*. Rutgers University Press.

Minayo, M. C. S. (2005). Laços perigosos entre machismo e violência. *Revista Ciência & Saúde Coletiva, 10*(1), 23–26. https://www.scielo.br/j/csc/a/gvk6bsw36SPbzckFxMN6Brp/?format=pdf&lang=pt

Minayo, M. C. S. (2006). *Violence and health* (p. 132) (in Portuguese). Editora FIOCRUZ. Coleção Temas em Saúde.

Odalia, N. (1985). *O que é violência* (3rd ed.). Nova Cultural; Brasiliense.

Pan, A. C., Oliveira, A. K. V., Schneider, K., Chaves, N. H., & Zanesco, I. (2021). Brazil: Video inspires young women to work in solar energy. *Nature, 593*(7857), 37–37. https://doi.org/10.1038/d41586-021-01147-z

Pires, L. L., Carvalho, L., & Xavier, L. L. (2020). *COVID-19 and inequality in Brazil*. https://www.researchgate.net/publication/340452851_COVID19_e_Desigualdade_no_Brasil

REN21. (2021). *Renewables 2021: Global status report*. https://www.ren21.net/wp-content/uploads/2019/05/GSR2021_Full_Report.pdf

Rossiter, M. W. (1982). *Women scientists in America: Before affirmative action, 1940–1972* (2nd ed.). Johns Hopkins University Press.

Silva, F. F. D., & Ribeiro, P. R. C. (2014). Trajectories of women in science: "Being a scientist" and "being a woman" (in Portuguese). *Ciência & Educação (Bauru), 20*, 449–466.

Think Eva & Linkedin. (2020). O ciclo do assédio sexual no ambiente de trabalho. https://thinkeva.com.br/pesquisas/assedio-no-contexto-do-mundo-corporativo/>.

UN Women. (2016). *Gender and economic autonomy for women* (In Portuguese). SPM, MMIRDH. https://www.onumulheres.org.br/wp-content/uploads/2016/04/caderno_genero_autonomia.pdf.

United Nations. (2019). *Trans pages: Access guide to rights and services for trans people* (in Portuguese). https://nacoesunidas.org/wp-content/uploads/2019/01/WEB-P%C3%A1ginas-Trans-FINAL.pdf.

Velho, I., Santos, L. W. D., Ichikawa, E. Y., & Cargano, D. D. F. (2006). Science, technology and gender: Unveiling the feminine in the knowledge building (In Portuguese). In *Ciência, tecnologia e gênero: desvelando o feminino na construção do conhecimento* (pp. 284–284).

PART III

Women and Gender Inequality in Energy Communities

CHAPTER 8

Gender-Just Energy Communities: A Catalyst for Sustainable and Just Development

Katharina Habersbrunner and Marika Kuschan

Introduction

More than 60% of the global greenhouse gas emissions come from energy use, making energy the natural starting point for increasing our efforts to mitigate climate change (IRENA, 2019). An effective energy transition requires active citizen engagement. Renewable energy communities can play a key role in putting citizens at the centre. The participation of citizens and communities as partners in energy projects are transforming the energy system. Community energy initiatives are offering new opportunities for citizens to get actively involved in energy matters. Community energy refers to collective energy actions that foster citizens' participation across the energy system. It has received increased attention in recent

K. Habersbrunner (✉)
Munich, Germany
e-mail: katharina.habersbrunner@wecf.org

M. Kuschan
Munich, Germany
e-mail: marika.kuschan@wecf.org

N. Rocha Lawton, C. Forson (eds.), *Women and the Energy Sector*,
https://doi.org/10.1007/978-3-031-43091-6_8

years, developing a wide range of practices to manage community energy projects.

In many countries, local energy communities have already been established to increase renewable energy production and/or to purchase power grids in order to decentralize them and integrate renewables into the grids. The feed-in tariff (FIT) policy mechanism was introduced in many European countries, offering long-term contracts to renewable energy producers. As a consequence, the number of energy communities increased sharply in the beginning and remained stable in recent years. ECs are realized in different legal forms—cooperatives, stock or limited companies—and differ from large corporations to smaller entities with only a few members. One of the main goals of energy communities is to provide their members with renewable energy and energy services to use local resources and to push the decentralized, democratic and just energy transition. Additionally, this people-centred approach to future governance does not prioritize corporate profits but community well-being (Stephens, 2020).

The European Renewable Energy Directive (RED II)—part of the Clean Energy for all Europeans package—confirms the prominent role of so-called prosumers and their collective forms will play in the future energy system. RED II is aiming to stimulate the formation of 'renewable energy communities' in all the Member States. The implementation at policy level remains with member states where national energy policies and markets are still overwhelmingly centralized. This creates barriers to the decentralization and democratization of energy and, ultimately, to achieve a just transition.

The energy transition and energy communities are a fertile ground to implement climate, gender and social justice, as (changes in) energy systems are often confronted with power inequalities. Energy communities may become a 'popular and viable alternative to capitalist individualism and isolation, and the incorporation of a gender perspective in community actions and strategies has transformative potential and can open a space for a new debate on an inclusive and economically and socially sustainable development model' (Łapniewska, 2019). Considering the long operational life of energy infrastructure (e.g., public transport, energy efficiency and power plants), not mainstreaming gender will reinforce inequalities for decades, wasting limited financial, social and human resources and missing ecological chances. This input proposes a way forward to recognize energy communities as successful, gender-just and democratic legal entities including the needs, rights and talents of all gender to shape a transition towards a just new energy system (Habersbrunner et al., 2021).

The 2019 European Directive for the internal electricity market introduced new rules to engage citizens in energy production, either individually or through energy communities, 'by generating, consuming, sharing or selling electricity, or by providing flexibility services through demand-response and storage' (Directive EU, 2019). Furthermore, the revised EU RED II included new provisions to support Renewable Energy Resources (RES) self-consumption—i.e., consumers' production of their own energy. The RED II established the EU's target of ensuring that 32% of Europe's energy supply come from RES by 2030, in line with the European Green Deal's (EGD) goal of cutting overall greenhouse gas emissions with 55% by 2030. Energy communities form a powerful bottom-up approach involving local actors. This is increasingly regarded as a strong potential for delivering a more democratic and inclusive just energy transition. This energy decentralization approach aims to stimulate the formation of 'renewable energy communities' in all EU Member States, in which 'citizens take ownership of the energy transition' (RESCOOP, 2019). Indeed, as the governance and the regulatory task of embedding such entities into the national energy systems is left to member states, the challenge for new legal and regulatory tools to embed them in the energy system continues. Thus, the objective of this research is to identify how a gender perspective can contribute to an equitable energy policy and supply.

Energy Communities: Description and Relevance

Community energy projects are defined as initiatives in which citizens are either owners or have a meaningful say in the running of renewable energy plants or energy-related services. They have different legal forms and are increasingly opening opportunities for citizens to participate in the energy market and to benefit their local communities and push a decentralized energy transition. Yet, national legal frameworks and public support differ among European countries. The European electricity market and energy security have recently become heatedly discussed topics at the European Union level. In many countries, political and financial support for the transition towards renewable energy systems during the last two decades has encouraged the establishment of a substantial number of new electricity communities. Many of them can be described as 'social enterprises, demonstrate attachment to values such as equity and equality in their actions, thus they might be perceived as women[1]-friendly entities' (Łapniewska, 2019). Definition, organizational forms and financing models of energy communities might vary among countries. Community energy initiatives and energy communities come in different shapes and sizes and their role/motivation for existence is highly institutionally and policy context-dependent. The terms 'energy community' or 'local energy community' do not have common definitions and can refer to a wide range of collective energy actions. The Clean Energy Package contains two definitions of energy community: Citizen Energy Community (CEC) which is contained in the provisionally agreed recast Electricity Directive, and Renewable Energy Community (REC), which is contained in the recast Renewables Directive (European Commission, 2020a).

The definition of Renewable Energy Community (REC), regarding article 2(16) Renewable Directive (Directive EU, 2018): A legal entity:

- which, in accordance with the applicable national law, is based on open and voluntary participation, is autonomous, and is effectively controlled by shareholders or members that are located in the proximity of the renewable energy projects that are owned and developed by that legal entity;
- the shareholders or members of which are natural persons, SMEs or local authorities, including municipalities;
- the primary purpose of which is to provide environmental, economic or social community benefits for its shareholders or members or for the local areas where it operates, rather than financial profits.

RECs are also entitled to produce, consume, store and sell renewable energy, including through renewables power purchase agreements, to share renewable energy within the community and to access all suitable markets

[1] BINARY CONCEPT refers to the social classification of people into two categories, namely, male and female, and thereby hides the diversity of social and biological genders (Kuschan et al., 2020). Although we use the binary term 'women and men' in this publication, this does not refer exclusively to people who assign themselves to these genders, but to all of them. Nevertheless, the asterisk (*) stands for all genders. For a better flow of reading, we write women instead of women* and mean all people who feel themselves to be women, regardless of their biological sex with which they were born. Of course, this also applies to men or men*.

The definition of Citizen Energy Community, regarding article 2(11) Electricity Directive (Directive EU, 2019): A legal entity that:

- is based on voluntary and open participation and is effectively controlled by members or shareholders that are natural persons, local authorities, including municipalities, or small enterprises;
- has for its primary purpose to provide environmental, economic or social community benefits to its members or shareholders or to the local areas where it operates rather than to generate financial profits; and
- may engage in generation, including from renewable sources, distribution, supply, consumption, aggregation, energy storage, energy efficiency services or charging services for electric vehicles or provide other energy services to its members or shareholders.

The recitals of the Electricity Directive and the Renewables Directive provide an explanation of what CECs and RECs are:

- CECs constitute a new type of entity due to their membership structure, governance requirements and purpose (purpose being framed around provision of services/benefits for members or the local community—as opposed to profits) (Directive EU, 2019).
- Similarly, RECs constitute a new type of entity that can be distinguished from other market players based on, inter alia, size and ownership structures (Directive EU, 2018).

At their core, both definitions describe a way to organize collective cooperation of an energy-related activity around citizen ownership, governance and non-profit targets. The main purpose of CECs and RECs is to provide environmental, social and economic community benefits for its members or the regions where they operate rather than only financial benefits. Both definitions emphasize participation and control by citizens, local authorities and smaller businesses (European Commission, 2020a).

Energy cooperatives with some common elements are as a legal framework defined by the International Cooperative Alliance (ICA) and meet some of the criterion of RECs and CECs. In general, the ICA defines a cooperative as 'an autonomous association of persons united voluntarily to meet their common economic, social and cultural needs and aspirations through a jointly-owned and democratically-controlled enterprise' (ICA, 2018). Therefore, cooperatives operate under seven principles:

- Voluntary and open membership
- Democratic control of members
- Economic participation of members
- Autonomy and independence
- Education, training and information
- Cooperation between cooperatives
- Concern for the community

The earliest forms of cooperatives were established in the United Kingdom around 1840. A famine motivated a group of waivers to organize themselves to make food available to the people in the community through joint purchases (REScoop, 2020). Nowadays, cooperatives are involved in various topics such as food, housing, transportation, finance and energy and can provide sustainable business models. In the energy sector, cooperatives are increasingly becoming a suitable option as they enable communities to address proactively their specific energy needs under given conditions while contributing to their local economies. In Europe, the work of energy cooperatives and communities work is mostly based on renewable energy sources (ibid.). Nevertheless, their activities can exist in different forms. While some energy communities are more tightly linked to one exclusive economic function, such as consumption or production, others combine different activities (production, consumption and distribution) for their communities (ibid.). Country-specific regulatory frameworks determine the organizational structure. Yet, in general terms, they differ from commercial actors in energy markets in certain respects (Wierling et al., 2018). First, cooperatives enable direct participation and ownership and, therefore, involve a wider public. Second, they also seek non-commercial benefits such as fostering a sense of community and local value chains. Lastly, members are motivated to accelerate the transition to sustainable energy systems (e.g., phasing out nuclear power, regaining local ownership and control over energy supply) (ibid.). With energy communities, 'the traditionally passive consumer is becoming an energy prosumer, co-owner of renewable energy facilities and community energy participant' (Caramizaru & Uihlein, 2020). Recent statistics show that there are about 3500 so-called renewable energy cooperatives and communities, which are found mostly in North-Western Europe

(RESCoop MECISE, 2019). Estimates show that European energy communities could own about 17% of installed wind capacity and 21% of solar energy by 2030 and by 2050, nearly half of EU households are expected to produce renewable energy. Even though the first energy cooperatives can be traced back to the beginning of the twentieth century, cooperative development started mainly in the second half of the 1970s. The oil crisis of 1974 obliged energy-dependent countries to search for new and alternative energy sources. Communities living in remote and relatively low-populated areas started to produce and consume energy locally, using solar, wind and biomass (Troya Çevre, 2017). Further developments were seen especially in the 1990s in Europe and the model spread worldwide in the 2000s due to technological improvements and decreasing prices of renewables. Today, more than 20% of the renewable energy power plants all over the world are operated by communities and individual users (ibid.). Some Western European countries have valuable experiences in the implementation and successful management of energy communities and set an example for others. In countries such as Germany and Denmark, electricity development included many cooperatives and communities, where farmers and other rural communities organized electricity production and distribution in their area.

EU Energy Regulation Could Change the Energy Market

New energy legislation, agreed at EU level in 2019 (Directive EU, 2019), should boost community energy and help community projects across Europe. Energy communities across the EU have won new rights that should guarantee they can participate in the energy transition. Acknowledgement of their role and new rights to produce, consume, sell and store renewable energy are now enshrined in EU law for the first time. This is an important opportunity to encourage many more people-powered renewables projects, which are supported by their governments. The rights for community energy are indeed included since 2019 in the EU clean energy package. The EGD is as well a flagship framework of the Commission for a decarbonized development strengthening the opportunities of decentralized energy markets. 'Energy communities can take any form of legal entity, for instance, that of an association, a cooperative, a partnership, a non-profit organization or a small/medium-sized enterprise' (European Commission, 2020b).

Potential of Community-Owned Infrastructure

The decreasing price of renewable energy assets and new digital solutions allow for producing and consuming energy where it is most efficient, making energy communities feasible. Citizens increasingly will play a part in sustainability efforts; thus, they are motivated to volunteer in setting up such energy communities. Unfortunately, enthusiasm is not enough to build a successful energy community for the twenty-first century. Today, local energy communities are comprised of hundreds of individual energy producers, sharing energy with each other in real-time. Digitalized local energy communities can provide community members with new types of services and can build—if regulation allows—new forms of energy sharing at the local level. Thus, local energy communities could transform into significant actors in the European energy landscape, making our economies less dependent on fossil fuels by integrating more accessible and affordable renewables. Whilst focus has typically been put on renewable energy infrastructure, citizens' participation in the energy transition has still marginalized roles. Community-owned infrastructure remains an underappreciated approach in many countries, despite the many benefits compared to the current energy system, which primarily benefits large corporations at the expense of communities and citizens.

Intersectionality

Studies (Fraune, 2015; Radtke, 2016) show the exclusiveness of energy communities in many European countries. Members and actors are mainly older than 50 years, male and having access to financial and educational resources (Fraune, 2015). Aiming to get the (citizens') energy transition even more broadly accepted in society, it is also necessary to increase the participation of other social groups that are perceived as underrepresented. The social group that is most strongly perceived as underrepresented are people with a migration background (87%); people of younger age (65% agreement), people with physical and mental impairments (61% agreement) and people with low incomes are also perceived as underrepresented by the majority (WWEA, 2021). Sixty-seven per cent see educational attainment as the most important socio-economic factor in the decision to participate in an energy community, followed by income, gender, age and membership in local associations (WWEA, 2021).

Gender in Energy Sector and Energy Communities

Decisions within the fossil fuel energy model of the last two centuries were, and still are, mainly male-dominated, and power relations are clearly distributed at the expense of women (EIGE, 2019). The extractive energy model has been constantly genderblind (e.g., failing to include women and all gender in consultation processes regarding energy community decisions like use of lands and resources for fossil fuel extraction projects) and neglecting women's energy needs and skills. As a result, systematic gender inequalities in access to and control of energy have been created. When considering for gender equity in the energy sector, it is advisable to use an analytical framework known as gender dimensions (Habersbrunner et al., 2021). Gender dimensions are linked to different societal domains which are connected with peoples' everyday live realities that are closely interrelated (Spitzner et al., 2020). The framework, which is set up by these gender dimensions, is a relevant approach to address and to achieve gender-equitable energy policy and supply (Kuschan et al., 2020). The gender dimensions can be categorized as productive work, reproductive work, power and decision-making power, public resources and infrastructure, body and health as well as institutionalized androcentrism, respectively.

Productive Work

A productive occupation refers to the opportunities to access and perform paid work. To achieve this, the choice of occupation, the possibilities of career opportunities, fair working hours and fair and continuous payment in the European energy sector are necessary (Spitzner et al., 2020). Currently, about 1.3 million people are employed in the renewable energy sector across Europe (IRENA, 2020). However, the share of women is only 35%. Women are particularly underrepresented (28%) (Clancy & Feenstra, 2019) in STEM-related professions than, for example, in administrative professions (45%). Among other reasons, this is due to many barriers to women entering the energy sector. Chief among these are prevailing hiring practices, lack of relevant qualifications or lack of awareness of gender differences in early childhood education. However, the lack of an inclusive work environment such as flexible work hours, affordable childcare options, networking, training opportunities and gender equality goals also stand in the way of women.

Reproductive Work

Activities without financial remuneration include, above all, care work for other people. The gender care gap in Germany shows that women in particular (52.4%) spend more time on reproductive work than men (BMFSEJ, 2021). This also leads to gender differences in energy consumption behaviour, as most reproductive work takes place at home and thus increases energy consumption. However, gender-specific energy consumption is not only shaped by socially influenced behaviour, but also by biological factors such as people's comfort temperature. Other European studies show the difference in energy consumption between men and women. According to studies, women on average consume less energy than men, except in the home, which is mostly due to their care role in the household or family. A major problem here is the disregard for gendered consumption and reproductive work in existing energy policies and EU directives in the energy efficiency field. But there is also a lack of support for men in the economy to take on care work (EIGE, 2021).

Power and Decision Making

To ensure a socially and gender-equitable energy transition, a special focus must be placed on the gender perspective in decision making. Women are still underrepresented in many decision-making positions across Europe. In 2020, the proportion of women in EU-28 ministries was 29.5% and in parliaments 30.3%. In national ministries in the fields of environment, transport and energy, the proportion of female ministers was 26.8%. A look at business shows that women are underrepresented here as well. In the largest listed companies, the proportion of female supervisory board members and board members was 26.6% (EIGE, 2021). To change this, with the help of energy transition policies and programmes, more women need to be involved, experienced and empowered. Gender budgeting[2] is a good way to generate good results in this regard. But gender audits, which are seen as an important tool for gender mainstreaming, and which examine laws, regulations, taxes and specific projects for their impact on the status of women in society, also help to show the gender perspective in public policies and in the private sector (Clancy & Feenstra, 2019).

[2] Gender budgeting: Gender budgeting is a tool for determining and controlling how funds are distributed and what effects they have on women and men and on gender relations (Landeshauptstadt München).

Incentives and policies for gender diversity need to be put in place. But not only by companies, but also by governments. By means of numerical targets, such as target numbers for hiring new employees, gender balance, etc., this could be achieved (International Labour Organization, 2017).

Public Resources and Infrastructure

Transport still represents one of the largest greenhouse gas emission generators in the EU. Although emissions from the energy sector are declining, emissions from transport continue to rise and are now significantly higher than in 1990 (European Commission, 2021). According to studies, women make their journeys on foot or by public transport more often than men. Men, on the other hand, are more likely to use cars, planes, bicycles or new mobility services such as car and bike sharing. This is partly because men are more likely to have a driver's license and access to a vehicle than women. 'If everyone travelled the way women do today, energy consumption and emissions from passenger transport in Sweden would fall by almost 20 percent. In addition, the use of cars as a means of transport would already be at the level considered sustainable for 2050', says Lena Smidfelt Rosqvist, research director at Trivector. Through well-planned and inclusive infrastructure, all gender roles and needs can be addressed. Currently, gender blindness prevails, especially in the mobility and energy infrastructure sectors (Spitzner et al., 2020). Many mobility patterns are highly gendered, due to the distribution of roles in the labour market (CIVITAS, 2014).

But the real conditions of women are not included in this. Thus, a gender perspective cannot be achieved in energy and mobility policies. Therefore, most of the support programmes in the field of renewable energies and mobility do not include this aspect. In Germany, most support programmes are designed for men, as they can afford the more expensive electric vehicles due to their higher incomes on average. Also, support programmes at the local, national and EU level that aim to increase energy efficiency and reduce energy poverty do not consider gender at all (Habersbrunner & Martschew, 2020). The importance of gender inequalities in energy poverty, while recognized, is neglected in the financial aspects of support programmes. Furthermore, the energy transition requires land and buildings that are mostly owned by men. For example, only between 6.1% (Netherlands) and 34.5% (Austria) of agricultural land is owned by women (FAO, 2021). Since many project operators mostly approach property owners and community leaders, women are often excluded.

Body and Health (Sex)

One's own health is one of the basic rights of a human being. However, the term health includes not only physical health, but also well-being, social and mental health (WHO, 2006). These aspects are all closely related to energy consumption. Even today, not all European citizens have access to electricity or the amount they need, with 7% of European households unable to pay their electricity bills on time in 2018 (especially in Greece and Bulgaria) (Eurostat, 2021). Energy poverty has a direct negative impact on the health and well-being of the people affected. Those affected include mainly low-income earners (mostly women) and older people who spend a lot of time in the house (Reibling & Jutz, 2017). However, women also usually prefer a higher indoor temperature than men, as they are more efficient when the indoor environment is adequately heated (Chang & Kajackaite, 2019). Everyday tasks such as cooking and washing, which is mostly done by women, are also favoured by access to clean energy. Energy poverty is often associated with mobility poverty, which often negatively impacts the social participation and well-being of those affected. However, the impact of energy on health is not limited to household consumption. From fuel extraction to transportation to waste disposal, energy systems are associated with health risks and impacts. Energy poverty affects four different dimensions such as the physiological, economic, health and socio-cultural dimension. Above all, the dimension of health is an important factor that must be considered in energy poverty. Women are more affected by winter mortality due to low temperatures (Birgi et al., 2021). Poor and incorrect heating can have serious effects on many physical problems such as respiratory infections or cardiovascular disease. Another problem is the use of solid fuels for heating or cooking, still in European Countries. Due to the release of nitrogen, serious illnesses can be caused, even death. But mental health is also negatively affected by energy poverty (Birgi et al., 2021). Stress, low well-being due to cold and depression are often the accompanying symptoms, especially for women.

Androcentrism

Androcentrism placed masculinity at the centre and declared it to be the social norm at all political, social, economic and scientific levels. This norm claims for itself gender neutrality as well as universality, objectivity and rationality, but at the same time defines 'femininities' as deviation or addition (Smith et al., 2013). To stop a gender hierarchy, the institutionalization of androcentrism would have to be abandoned. Looking back at the fossil fuel energy model of the last 200 years, it is striking that gender blindness has prevailed here as well (Global Initiative for Economic, 2020). The barriers that exist here are mostly unnoticed by the men who dominate this sector. Only 40% of men perceive the existence of gender-specific obstacles, in contrast to 75% of women (IRENA, 2019). This leads to gender inequalities in access to and control of energy, gender inequalities in decision making and financial and political gains, and spatial inequalities in the allocation of energy and exposure to the externalities of energy production, with the female population being more disadvantaged. Gender neutrality does not prevail, e.g., technology is commonly understood as a gender-neutral tool, although in most cases it is a male activity. For this reason, 'gender-neutral energy policies' usually refer to and benefit men. Women are much more likely to face greater obstacles and limitations (Groot de et al., 2017). Thus, the use of technology cannot reinforce gender asymmetries and inequalities rather than reduce them. The same is true for other social differentiators such as education, immigrant background, ethnicity, and age. If these are neglected, further disadvantages arise for women. The advantages of an 'intersectional approach' arise all the more. It brings awareness that knowledge is situated and partial, which is a direct challenge to the universality claimed by (e.g.) tech-savvy, white, male perspectives in the field of energy research (Bell et al., 2020) (Fig. 8.1).

The Status Quo of Gender in Energy Communities: 'Relevance and Opportunities'

Energy Communities (ECs) are one possibility to encourage citizens in the energy transformation process and to push the highly needed energy transition. Citizen energy concepts can be one form of energy communities. Germany, for example, has become well-known for its bottom-up-movement of the so called *'Bürger*innenenergie'* citizen energy, which

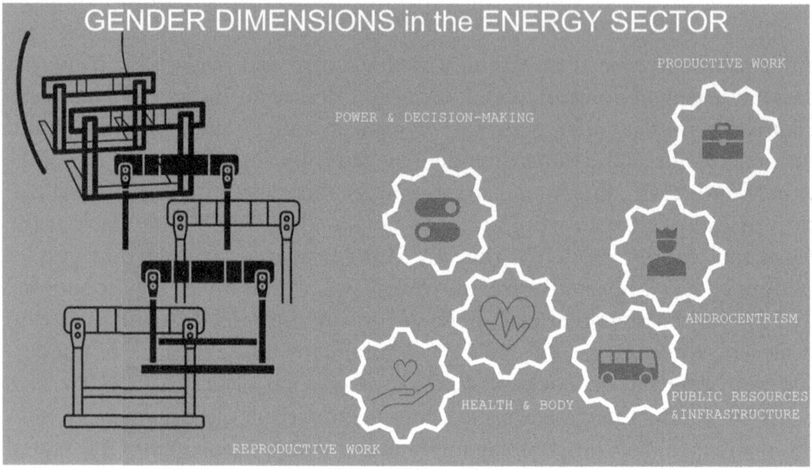

Fig. 8.1 Gender dimensions in the energy sector, WECF 2021

might be grounded in its historical connectedness to the founding of cooperatives.[3] However, ECs in all their existing diverse forms might foster or restrict the possibilities and the intention of the people to take part in the local and therefore national energy transformation. The mobilization of civil society, in its whole (the importance of the deletion of social barriers or norms not reflected for this very moment), has the potential to contribute to a democratization of this process by the development of a fundamental understanding of energy issues, de-monopolizing and pluralizing economic structures (WWEA, 2021). Even more when social groups and their specific needs are evenly reflected and addressed in the transformative process.

In principle, ECs 'cover both groups and individuals' (Renews No. 89). Furthermore, these '(EC) projects do need not to be cooperatives' (ibid.). To stick to the German example, their national energy transition was primarily started by committed citizens and later further developed by mainly community energy projects and medium-sized companies. This development is mainly based on the voluntary and very professional engagement of citizens. Up to today almost a third of the installed capacity of German

[3]Around 170 years ago, Friedrich Wilhelm Raiffeisen implemented the first cooperative model to help smaller farms out of financial needs.

renewable energies are in the hand of private property, including farmers. This exceeded 40% in 2019 (Press Release AEE, 2021), even though the share of citizen energy is dropping (ibid.).

Herein, it is recognizable that energy transformation is often understood as a process in the area of (renewable) electricity production, marketing and sale. Commonly left out in research literature, as well as media coverage, is the transformation process of mobility and heat, although the necessity of sector coupling is well known in politics, economics and research. In these sectors, the successes and development have been mainly carried out by fostering through community projects in the sense of municipal development. EC projects can therefore also involve the actions of the local public sector.

The European Commission stresses the importance and role of citizens on the continent's path to 'clean energy for all Europeans'. On a cross-border basis, two levels of participation models in the energy transition can be identified as ECs following Devine-Wright (2014): the micro and the meso level. Households, individuals and small local actors like neighbourhood initiatives work at the micro level. Regional and urban actors act at the meso level. Most community energy initiatives can be found within this level, particularly depending on their size and governance structure (Devine-Wright, 2014). The form of collective action in which ECs are viewed as such, their size, their organizational and financial models may differ between the European member states but are nonetheless 'strongly associated to the idea of energy democracy and energy justice, and thus inclusivity' (Dudka, 2021).

Since the energy transformation is not only about new technologies but about transforming societies, Fraune emphasized the need for higher attention to the dynamics and interdependencies within socio-energy systems (Fraune, 2015). In the sense of such a transformation process, the inclusive perception and consideration of marginalized citizens must also be considered. In particular, the focus is on the inclusion of all people, regardless of gender, ethnicity, religion, age or income. In many countries, energy cooperatives have been established, which build the local EC. Since cooperatives differ from private corporations, not only in the organizational form and therefore the participation scheme but also in certain values, which matter to them (Fraune, 2015), cooperatives seem to be the excellent choice and have the institutional potential to foster inclusivity. Cooperative values are per se embedded in the cooperative guiding principle such as equity and equality (all members are equal in terms of voting power, regardless of their shares in the cooperative), trust, cooperation and solidarity. These values should therefore apply to all members of a society in the same way—regardless of gender and other distinctions. Since investment and participation schemes of citizens are indispensable for the growth of acceptance and the further expansion of renewable energies, ECs, and therein cooperatives, can only benefit by attracting and engaging all social groups (Press Release AEE, 2021).

Obviously, the aim of equal participation and representation of women and other social groups in ECs is linked to the share of women in the fields of politics, economy, particularly the renewable energy industry and adjacent occupational fields. Although studies on the gender and energy nexus have been present in scientific literature for more than two decades, it has been mostly present in the aspect of traditional gender roles within household tasks (Łapniewska, 2019). However, only little research is done on ECs from a gender perspective. Although the authors represent encourage an intersectional approach on the gender-energy nexus, the eligible literature is concentrated on the binary concept of gender. Gender-Mainstreaming provides a binary glance at energy in terms of accessibility, use of energy services, investments or decision making in policies. Opportunities to actively participate in shaping the energy agendas in politics and economy are still rare for women, despite the influence of women on the energy value chain (in households, local communities and energy markets). This results in a reduced effectiveness of energy strategies, policies, programmes and other energy-related activities (Habersbrunner et al., 2021). The triple burden of women,[4] which corresponds to the traditional assigned gender roles are production, reproduction and community management, especially community tasks, include the possibility of increasing the influence in the restructuring process of the industrial energy regime by taking women's power to locally anchored ECs (Łapniewska, 2019).

ECs operate on three functional levels: the constitutional, the collective and the operational level, which have distinctive modalities to take into consideration. Furthermore, representation, active influence and power margin (participation) as well as financial shares (distribution) are

[4]The triple burden refers to the division of labour between genders. Typical roles of women are reproductive, productive and community managing tasks. In most societies, low-income women perform all three roles, while men primarily engage in productive and community activities that typically generate earnings, status or power (EIGE, 2021).

important to consider. It is thought to be beneficial and necessary to integrate gender and social justice targets both when formulating the goals and other EC policies, as well as to use inclusive or gender-responsive communication throughout (language, media) to generally attract more women and other social groups. Therefore, it is necessary to reflect on the differences, which might result from the operation of the EC, for different social groups and not as society as a whole. However, to build reliable considerations, it is recommendable to think about the six gender dimensions, which are explained above, though other distinctions exist. Several factors and conditions can be decisive for a different gender-related perception and social benefits of ECs. On the collective level, increased visibility of others of the same social attributes, as well as the own social representation in management levels and advisory boards are of importance. From an organizational point of view, for instance for member assemblies, it must be considered that women generally suffer from time poverty (triple burden) as well as women have in general a smaller financial backup. The Gender Wealth Gap consists of Gender Pay Gap, Gender lifetime-earnings gap and Gender pension Gap (Eurostat, 2022). In regard to the operational level, the authors find it necessary to strive for equal staffing at all EC levels (members, staff, management and supervisory board), to plan for financial means to achieve targets regarding gender and social justice, to integrate aspects of gender and social justice in all phases of the daily operational level on all EC levels (members, staff, management and supervisory board). To name a few measures on the operational level to integrate aspects of gender and social justice in the EC: Gender quota in members, staff, boards, using and monitoring Gender indicators; Gender disaggregated data; Gender training and awareness raising for members, staff, boards; special offers addressing the needs of women or other social target groups; financial mechanisms on accessing RE Energy with special attention, credit lines and conditions for women or other social target groups; sexual harassment policy; follow guidelines on inclusive communication throughout.

Androcentric structures are reproduced and not reflected, where and whenever predominantly men in politics decide on framework conditions for the energy transformation in general or citizen energy in particular, and when at the same time, be advised by mostly men from the media, business or civil society (WWEA, 2021). Through 'selective perception', a very common cognitive distortion, androcentric thinking and decision making is reinforced and strengthened since it makes one see, listen or focus the own attention on a stimulus based on personal expectations, without considering the rest of the information.[5] The selection of perceived sensory impressions is influenced by various filters in which experiences, expectations, attitudes and interests play a major role. Thus, it is major to raise awareness among all genders and to have gender trainings mandatory.

Resistance or Reproduction of Gender Inequalities in Energy Communities?

It is known that the energy sector in general is largely dominated by men. Yet, which kind of social groups are represented in ECs? How are cooperative values reflected in day-to-day business? In the following section, the authors reflect upon three studies particularly three studies conducted in Germany and Europe, respective, to draw a comprehensive picture of the realities of women in European ECs, namely: Fraune, 2015, Łapniewska, 2019, WWEA, 2021. These studies are used to draw conclusions whether gender inequalities are reproduced, or even extended, or whether gender inequalities are reduced. Especially by reflecting the situation in almost a decade, it can be assumed that these results are manifested and characteristic for the majority of European ECs. And that they are valid both for newer and more dignified ECs.

Fraune tested hypotheses 'on the basis of empirical evidence generated by a pilot study of gender relations within citizen participation schemes in (renewable electricity production) RES-E in Germany' (Fraune, 2015). Among other findings, Fraune's exploration of causal mechanisms showed for citizen participation schemes and the gender wealth gap (1) that women on average spend a lower total investment volume per citizen participation scheme in RES-E than men (2) the ownership of women in citizens' associations operating solar power plants is lower and in general, it can be stated that (3) the female ownership rate in citizen participation schemes in RES-E is lower than the male counterpart. Regarding voting and control rights it was found that (1) in GmbHs the investment sum by

[5] Selective perception optimizes our cognitive resources and focuses them on something that we expect to happen. This is based on the ability to recognize patterns, a fundamental function of the human brain. The brain is constantly on the lookout for patterns in order to better classify new information into existing information. Selective perception is the—mostly unconscious—search for a specific pattern.

women is significantly lower than that of men and subsequently have less voting and control rights at their disposal than men within GmbHs (2) in GbRs as well as in eGs, women's ownership rate is lower than that of men and subsequently inferring that women have less voting and control rights at their disposal, (3) women's ownership rate is significantly higher in eGs than in GbRs. For the representation of women in decision-making bodies, Fraune conducted female board members are significantly underrepresented in managerial positions in citizens' associations operating renewable electricity production plants (ibid.)

Łapniewska tested hypotheses on 'the basis of deductive reasoning using a pilot survey that was carried out in the first half of 2015 at electricity cooperatives in Europe' (Łapniewska, 2019). The survey consisting of 31 questions included 51 responses from 45 cooperatives from 11 European countries. The majority of European electricity cooperatives have not included regulations relating to gender equality in their constitutional documents (see previous chapter). Neither do the majority of them incorporate aspects of gender equality into their strategies, plans and actions. It was also found that the size of a cooperative and its adopted governance model correlates only partially with the involvement of women as cooperative members, employees or as members of boards. It was also confirmed that 'traditional roles attributed to women in society and cultural determinants of a region/country are perceived as affecting women's careers in energy cooperatives' (Łapniewska, 2019).

WWEA conducted a study consisting of two core elements as surveys expanded by in-depth interviews to object or prove hypotheses known as 'explanatory sequential mixed-method design' (WWEA, 2021). There were 29 responses received from energy communities of Nordrhein-Westfalen (NRW) (a county in Germany), as well as 347 responses from shareholders of those or other energy communities. They found that only 29% of the shareholders* of community energy surveyed in NRW are women, who hold 27% of the shares. Considering the corporate form, ECs have higher female shares in NRW (33%) compared with only 14% of women in GmbHs & Co. KGs (WWEA, 2021). If taking the form of energy into consideration, solar EC projects have a share of 37% against 21% in wind energy projects.

Though other similar studies exist (Radtke, 2016; Yildiz et al., 2015), the conclusions which can be drawn from the selected literature build nonetheless an identical scenery. It seems clear, those distinctions in the involvement of citizen's participation schemes like in EC projects or energy cooperatives do exist (Fraune, 2015; Łapniewska, 2019; WWEA, 2021) from, at least, a gender perspective. The exemplary results give an indication, that there is an additional reasoning besides the suspected technical and economic aspects. Nor can the differences in investment volume, ownership and shareholder rates, or in the female underrepresentation in membership numbers and decision-making bodies be explained alone by individual reasons, for instance male or female preferences and attitudes (Fraune, 2015). It is likely reasons go deep with social, cultural and thus political conditioning.

The technical argumentation 'indicates a stereotypical perception of gender and internalised (and conscious) gender prejudices, that is, the perception of energy as a technical, and thus "masculine", issue (...)' (Łapniewska, 2019). So, the reasoning for a male* domination in the energy sector is viewed as a consequence of the well-known gender imbalance in the enrolment and graduation in STEM fields. However, participation in ECs does not require specific technical skills, neither do all of the ECs employ technical staff (ibid.). Women* in the energy sector are mainly occupied in the administrative branch. The observed gender-specific occupational segregation, which is typical for the (renewable) energy sector, is also present in ECs: the well-known glass ceiling, which prevents women from being promoted or being present, e.g., in decision-making bodies, can be expanded to other areas of vertical and horizontal occupational segregation (known as 'glass walls' and 'sticky floor').

Within the energy sector, as well as in ECs, women are still more often employed in poorly paid professions. In addition to the understanding that women also work more frequently in part-time and mini-jobs and suffer from a gender-specific wealth gap, it seems crucial to shed light on the economic reasons for the existing gender gap in ECs. The studies of Fraune and WWEA concluded diametral arguments, in this case. Fraune is demonstrating that 'the generally low involvement of women appears to be dependent on factors other than (...) financial factors like capital-intensity of investment' because the 'observed gender investment gap (...) is even higher that the gender-specific wealth gap' (Fraune, 2015). On the other hand, in the WWEA study, the WWEA study, which research about the barriers for a higher participation of women in ECs, concluded that the barrier of different female/male participation was the lack of financial resources (43.7% of women and only 24.5% of men) (WWEA, 2021). Fraune argues that 'the structural outcomes of this gendered distribution of labour are of special importance in terms of (...) participation schemes' in ECs and are not independent of an institutional context, which connects to social and cultural conditions, as well as political framing (Fraune, 2015).

Cooke elaborates on the connections of 'the distribution of paid and unpaid work, and thus relative gender, class and other group inequality', and roots it to structures 'by mainly governmental policies: education policies, labour regulations, social insurance programs, public and private transfers, and tax and family law' (Cooke, 2011 found in Fraune, 2015). Inconsistency in framework and policymaking in the economy and politics emerge in different social outcomes in representation, participation and distribution (of resources) in ECs. It has already been shown that ECs (in most cases) do not meet the egalitarian claim to enable open participation and membership for all parts of society (Radtke, 2016; WWEA, 2021; Yildiz et al., 2015). For instance, the German citizens' energy tends to be dominated by a homogeneous group: Many of the members and actors were male (80%), often had a high level of education and above comparatively high financial resources, were already at an advanced age and exercised civil society engagement in more than one organization (Radtke, 2016). Within national and European energy transformation processes energy communities are not up to neither overcome gender barriers nor reduce gender inequalities at all, though rare lighthouses exist. Moreover, the influence of both cultural and institutional aspects not only reproduces but also strengthens gender inequalities since they reinforce each other. 'It is this interdependency of institutional and cultural factors that make gender relations a specific matter within energy transformations' (Fraune, 2015).

The International Renewable Energy Agency (IRENA) concludes that the (renewable energy) sector is in dire need of women in order to tap a wider and deeper pool of talents and skills and thus promote the future prospects of the sector (IRENA, 2019). As a consequence, energy communities should welcome and foster women's participation and representation and all gender dimensions need to be considered when planning programmes and actions. Thus, participation schemes must ensure that all citizens of different social groups—such as female and male, rich and poor, high and less educated citizens—are able to participate equally in ECs to foster decentralized, renewable and democratic ECs in Europe.

Good Practice in Europe and Beyond: Spain, Turkey, Germany and Uganda

The case studies from Spain, Turkey, Germany and Uganda show good practice for gender-just and inclusive energy communities. They represent a good variety of existing ECs considering gender and women's empowerment.

Good Practice Spain: Gender-Responsive Communication as Well as Setting Up and Implementation of Gender Equality Plan

Som Energia is a Spanish non-profit energy cooperative, which was founded in December 2010 (Som Energia, 2021). The cooperatives' activities are mainly in the production of renewable energies with the aim to promote a change of the current energy model to achieve 100% renewables. As of today, Som Energia is producing 24.60 GWh/year. Som Energia is aware of gender discrimination and oppression, at all levels, in all areas of society. As a member of the Social and Solidarity Economy, this cooperative has, as one of its principles, the equality of all people, without discrimination based on gender, and is committed to implement measures to minimize gender differences. The cooperative is aware that a gender perspective must be transversal, since it affects all areas, including communication. It is also aware of its own potential and impact with more than 75,000 members and 120,000 electricity contracts.

Thus, Som Energia has published a gender guide for its members. The target group is its own staff aiming for awareness raising and training of gender perspectives in their oral or written, communication when dealing with members, clients and citizens. Although it is an internal document, the cooperative makes it available to anyone who wants to use it. As the guide explains, 'our expressions and communication must be impregnated with a gender perspective, using inclusive and non-sexist language that makes women visible and at the same time serves to create and recognize female references, break stereotypes and neutralize sexist prejudices' (Som Energia, 2021). In order to create the guide, Som Energia made an analysis of the specific cases encountered by people who dealt with requests for information, contract management, incident resolution and other queries by telephone or mail. After grouping the most common cases, professionals, trained in gender and communication, prepared the document. The cooperative is convinced that applying the gender perspective goes beyond the non-sexist use of language, and that personal attention, in oral and written customer management, requires a deeper awareness. The guide also includes a series of other gender traps and indications to be taken into account dealing with the cooperatives' customers, such as not reproducing sexist stereotypes, not assuming that the contract holder will be a man, or that the technical staff, electricians and engineers are necessarily men; not assuming that a couple (a man and a woman) live in the house, or, in telephone conversations, not assuming the gender of a person by their voice, among others. Moreover, Som Energia anchors its engagement for social inclusion and gender equality in official documents. The cooperative declares its commitment to the establishment and development of policies with equal treatment and opportunities between women and men, without discriminating directly or indirectly on the basis of gender, as well as the promotion and encouragement of measures to achieve real gender equality in the scope of its organization, establishing equal opportunities between all genders as a strategic principle of its Corporate and Human Resources Policy. In each of the work areas in which the activity of this cooperative is developed (from recruitment to promotion, including wage policy, training, working conditions and work-life balance), the principle of equal opportunities between women and men is assumed, paying particular attention to indirect discrimination. With regard to communication, both internally and externally, all decisions taken are reported and verified in accordance with the principle of equal opportunities between all genders. The mentioned principles will be put into practice through the implementation of the 'Equal Opportunity and Equality Plan' leading to improvements with respect to the present situation, setting up corresponding follow-up systems, with the aim of advancing towards real gender equality in the cooperative and, by extension, in society as a whole (Som Energia, 2021). Som Energia addresses gender dimensions like productive work, reproductive work, infrastructure, decision making and androcentrism overall and applies gender tools to raise awareness and share skills and arguments.

Good Practice Turkey: Community-Based Energy Generation with Renewable Energy, Planned and Established Mainly by Women
The renewable energy business is becoming very important in Turkey as they have good climate conditions for solar and wind energy (Yenkoop, 2021). Therefore, energy cooperatives in Turkey form a suitable option to include interested citizens and thus pushing a decentralized energy transition. Three women (Nazan Unverir Wish Cakilcioglu Derya Özsoy, Filiz

Kircin Rock) started their idea of the first energy cooperative in Turkey. They prepared the legal, technological and economic steps which were a big challenge. With expertise, women's power, cooperation with other Turkish and international cooperatives and gender stakeholders they managed to establish the first energy cooperative in Turkey, whose board entirely has female members. They are running photovoltaic power plants and got 2020 the 'energetic woman of the Year' award, initiated and traditionally organized by the association for energy economics. The board members of the cooperative are visible role models, they are aiming to qualify more women for energy cooperatives, to overcome structural and personal barriers and to support for further energy cooperatives in Turkey. They use gender instruments like gender targets (e.g., quotas), gender-just communication and offer gender trainings for energy stakeholders.

Good Practice Germany: Heidelberg Becomes Sun City—Getting More People Involved in the Energy Transition by Heidelberger Energiegenossenschaft

In September 2018, the project 'Heidelberg Becomes Sun City—Getting More People Involved in the Energy Transition' was launched. The aim of the communication campaign is to raise awareness that more and diverse people join in and increase the impact. Citizens invest in the Heidelberg Energy Cooperative, which uses the money to build renewable energy power plants that produce green electricity and sell it to the citizens. A great idea for a democratic and bottom-up energy transition. It is really important to include the broad society. So far, the involved citizens have been a pretty homogeneous group: male, white, older than 50 years, access to funding and education. The project 'Heidelberg Becomes Sun City' was exactly addressing this point Bündnis Bürgerenergie (2020). Campaign activities, information events and flyers were used to draw attention and achieved the following results:

- Membership growth in 2018 and 2019 by 35% each, before that only by 16% per year.
- Younger female members
- More investments of citizen's capital for new projects
- Significant increase in electricity customers, also more female customers
- Increased awareness for a broad citizenship

- More visibility and reputation of the energy cooperative in politics and administration as an energy and social actor and the opportunity for successful, quick and affordable climate protection
- Establishment of an ambassador group, training for women as energy ambassadors

The campaign is still ongoing and has shown the need for inclusive communication concepts to address the broad society (Heidelberger Energiegenossenschaft, 2020).

Good Practice Uganda: Gender-Just Energy Communities

In Uganda, women are generally responsible for the household; men tend to pursue paid employment, primarily in agriculture. However, incomes are low and product quality often mediocre. These factors result in a lack of employment opportunities, particularly for young people. While men increasingly seek work away from rural areas and in towns and cities, rural women more frequently remain at home, farming the land and caring for family members. Farmland and cooperative shares are owned almost exclusively by men, while everyone—men and women—shares the work on the land (women are largely responsible for cultivating food crops, men for cash crops). Women Engage for a Common Future (WECF) is working with local partners in Uganda on the implementation, operation and productive use of renewable energy systems like solar, biogas and briquetting for local agricultural cooperatives. Advice, expertise and the development of gender-responsive funding plans are provided in conjunction with Ugandan credit cooperatives including Savings and Credit Cooperative Societies (SACCOS) and Village Savings and Loans Associations (VSLA) organized primarily by women. The project activities are able to guarantee good levels of quality management and sustainability through expertise and networking. Technical expertise on the technologies used and the management of energy cooperatives was provided principally by solar energy businesses and international energy cooperatives.

The following goals and targets have been achieved via trainings and gender-just communication

- Access to clean, self-produced energy through solar, biogas and briquettes for women (Sustainable Development Goal (SDG) 7).
- Reduction of greenhouse gas emissions and climate change mitigation (SDG 13).

- Training of women and girls specifically as entrepreneurs, energy managers and energy ambassadors (SDG 5).
- Less (energy) poverty amongst women (SDG 1).
- Business models, access to finance and secure jobs for women and young people in rural areas (SDG 4, 8, 9, 10).
- Qualification of young women and men in construction, use and operation of energy technologies, as well as the dissemination of needs-based technologies (SDG 4, 5, 6, 7, 9).

The cooperative approach uses existing structures and invests in people and communities to strengthen resilience, develop and keep the knowledge in the communities and reduce vulnerabilities, injustice and poverty. The goals could be achieved with needs assessment, targeted training, support for business development, considering the gender dimensions (e.g., high workload and unpaid work of women in households and communities, less financial means) and applying gender tools (gender analysis, gender targets, gender training, women's empowerment, gender-just communication). The impact is shown via policy recommendations to local and national policymakers.

Recommendations

Gender plays a decisive role in advancing a successful energy transition. Yet, energy decisions regarding the needs and preferences for production, transmission and consumption lack the foresight to include the key criterion of gender. This is also the case for energy communities' policies and decentralized energy concepts. Gender mainstreaming across energy policies, directives and practices are required to engender the energy transition and untap the potential for a successful and accelerated energy transition.

Following are some key recommendations for energy communities and national energy European policies:

EU and National Level

- Redefining the energy transition so its focus goes beyond technical applications and economic profit and includes social aspects, needs and practices of all citizens who are directly or indirectly affected.

- Including not only environmental aspects but also social aspects in the revision of directives (e.g., Energy Taxation Directive, Energy Efficiency Directive).
- Obligation to have gender action plans and inclusive indicators (for gender-responsive monitoring and evaluation) in each energy-related division in politics and the economy (e.g., gender impact assessment and gender budgeting).
- Applying and reinforcing gender tools and instruments:

  - Gender equality goals like binding quotas at the decision-making level in energy companies and political institutions.
  - Introducing incentives to companies with inclusive and family friendly work environment (flexible work hours, childcare, training, scholarships for women, etc.).
  - Providing and strengthening expertise among all genders.
  - EU institutions apply gender-just communication and become role models in the energy field for empowering not only female but gendered voices in the debate about energy transition, campaigns with the aim to attract more women and girls to STEM.
  - Gender-just funding and grant programmes.
  - Energy consultations tailored to female consumers and all gender.

- Gathering sex-disaggregated data with an intersectional perspective and regular monitoring (analysis and audit of energy production, distribution and consumption).
- The European Union can give more directives to local/national governments and emphasize cooperative values, decentralization, new economy, energy and climate justice with a gender-sensitive approach.
- Improving the approach of making prosumer key actors of the energy transition within the RED II through a gender-responsive prosumer model, encouraging all citizens to participate in and benefit from decentralized concepts.
- Energy communities are often limited to not-for-profit status and are relatively small compared to traditional energy players, communities face issues regarding access to funding and finance. European finance institutions should help to cover this risk and offer guarantees and support for energy community initiators in cooperation with local governments.

- National governments should offer free technical and legal advice to gender-just energy communities.
- Market access rules and grid usage rules should be in favour of energy communities. Communities managing their own grids can integrate more renewables in local areas, allowing them to act as an aggregator and offer flexible services to the regional or national grid and thus more cost-efficient grid services for society as a whole.
- Introduction of a common European definition of energy poverty covering its multifaceted nature and intersectional factors, including gender.

Energy Companies and Communities

- Adoption of Gender tools: Gender training, case studies from other cooperatives, focus of umbrella organizations on Gender equality, active recruitment, technical formation and promoting deeper participation by female members with support from policy and energy agencies, gender-just communication.
- Roles models: Women already involved in cooperative activities might attract new women members by showing the example of their engagement.
- There is a strong interest in ECs to undertake measures to address gender equality and to implement gender-sensitive activities. ECs lack sufficient knowledge about what a gender-responsive approach might accomplish (both in terms of achieving their targets and in normative terms). The establishment of a platform for exchange, case studies, training sessions on gender equality or consulting workshops might be valuable.
- Defining gender targets and involvement indicators in the official constitution documents or all documents.
- Mobilize people to advocate for and support local and democratic renewable energy communities.

We urgently call for inclusive and holistic policies that focus on society and its diverse preferences, needs and practices according to gender, age, ethnicity and cultures rather than mere technical and economic solutions. The shift towards decentralized energy concepts and markets is needed, acknowledging that the main sources are decentralized solar and wind energy. Investing in people and energy communities strengthens resilience, reduces vulnerabilities and inequalities, and increases expertise and power. After the initial investment in solar and wind, renewable energy is free, abundant and very reliable. The sooner households, companies, schools or communities are able to produce renewable energy jointly, the better off they will be in the future (Stephens, 2020).

Energy communities have a legal framework by the Renewable energy directive and need to be transposed ambitiously. Applying and using the intersectional approach for energy communities is likely to increase the success of future energy policies and projects due to the collective power of higher number of involved stakeholders, higher expertise and interests, and acceptance of the energy transition.

Conclusion

It is indisputable: Women are indispensable for implementing a community-driven energy transition. In fact, they as well as other social groups are facing hidden barriers in the energy transition and are hence structurally hindered to take up participation and representation as well as recognition and distribution. Additionally, women are key players in their community and take on a managing role. They are decisive to move the energy transition forwards efficiently and at the required pace. Without equitable participation of women and marginalized groups, the implementation of renewable energy communities will be slowed down, and successful energy transition will be achieved too late. The renewable energy system of the future is naturally decentralized. Decentralization, in energy production for primarily local energy consumption, underscores the paradigm shift in the energy system. The constantly excessive demand for fossil fuels of international origin from major energy suppliers is drying up for on-time produced energy from local energy communities. The change from the inherited, one-sided condition requires a break from the still technocratic framing and expands the energy transition process by adding social and personal factors.

Against the background of both needs to (i) accelerate the energy transition by decentralization, democratization and higher acceptance within all European societies and (ii) strengthen the share and importance of ECs in the just transition process, this paper summarizes that most of the ECs do not live up to their claim of 'democratic and inclusive institutions'.

Moreover, we argue that marginalization of social groups, gender stereotypes and social inequalities are not only reproduced and reinforced by political institutions and energy corporations but also by most ECs. The success of social transformation processes depends to a considerable extent on the existence, skills, interests and networking of key actors. This is also true of the transition to a decentralized, renewable and democratic energy system. Women are the greater advocates of the energy transition and in this role are active on the ground as initiators, process designers, multipliers and conflict resolvers (Schlüsselakteure, 2020). Seeing women as key actors is one of the basic prerequisites for the formation of wills throughout society and the development of a high and positive group dynamic among social actors (Hohmeyer et al., 2017), e.g., the founding of energy communities. Informal, women are therefore already successfully active at the local level as drivers of a system change. Formally, they and other marginalized groups still miss their respective share of participation and representation, recognition and distribution within energy communities. Socially and gender-just energy communities push energy democracy and social transformation: They are institutional instruments to connect renewable transformation with redistributing political and economic power, wealth and ownership to create a more just and equitable world (Stephens, 2020).

Thus, recommended measures for political institutions on the national and EU level for energy communities should be initiated immediately, implemented on a permanent basis and be subject to consistent monitoring and evaluation. Not (only) out of ethical-moral principles, but especially out of obvious and logical consequences as shown in this research. The only possibility for a successful energy transition, towards a renewable, decentralized and democratic energy system, is a gender-responsive and social-inclusive implementation. Only an energy transition supported by the entire society makes it possible to implement the energy transition in a forced manner and to achieve the set ambitious goals within the Paris Agreement in a timely manner. Energy communities have the legal, social and economic potential to address and reduce underlying exclusion and injustice (Robinson, 2018), catalysing a successful, broadly accepted and supported energy transition and sustainable development that will make our future liveable, socially and ecologically acceptable.

References

AEE Press Release. (2021). https://www.unendlich-viel-energie.de/mediathek/grafiken/eigentuemerstruktur-erneuerbare-energien

Bell, et al. (2020). Toward feminist energy systems: Why adding women and solar panels is enough. https://www.sciencedirect.com/science/article/abs/pii/S221462962030133X?via%3Dihub

Birgi, et al. (2021). Gender and energy poverty – Facts and argument. https://www.empowermed.eu/wp-content/uploads/2021/05/2104.Empowermed-Energy_Poverty_and_gender.pdf

BMFSEJ. (2021). Women and men in Germany. https://www.bmfsfj.de/bmfsfj/meta/en/publications-en/women-and-men-in-germany-160836

Bündnis Bürgerenergie. (2020). Raus aus der Nische! Rezept zur schnellen Energiewende. Heidelberg wird Sonnenstadt. https://nkmdn.de/kannstduauch

Caramizaru, A., & Uihlein, A. (2020). *Energy communities: An overview of energy and social innovation, EUR 30083 EN.* Publications Office of the European Union.

Chang, T. & Kajackaite, A. (2019). Battle for the thermostat: Gender and the effect of temperature on cognitive performance.

CIVITAS. (2014). Smart choices for cities. Gender equality and the energy transition in the EU. http://civitas.eu/content/civitas-policy-note-gender-equality-and-mobility-mind-gap

Clancy, J., & Feenstra, M. (2019). *Women, gender equality and the Energy Transition in the EU.* Publications Office of the European Union. https://www.europarl.europa.eu/thinktank/en/document.html?reference=IPOL_STU(2019)608867

Devine-Wright, P. (2014). Towards a comprehensive understanding of energy and communities: Advances, experiments, gaps and challenges. Presentation, 1st Symposium on Community Energy Systems, 10 March 2014. Lille.

Directive EU. (2018). Directive of the European Parliament and the Council of 11th December 2018 on the promotion of the use of energy from renewable sources (recast). https://eur-lex.europa.eu/legal-content/EN/TXT/PDF/?uri=CELEX:32018L2001&from=EN

Directive EU. (2019). Directive of the European Parliament and the Council of 5th June 2019 on common rules for the internal market for electricity and amending Directive 2012/27/EU 2019, in Official Journal of the European Union, 14 June 2019. https://eur-lex.europa.eu/legal-content/EN/TXT/PDF/?uri=CELEX:32019L0944

Dudka, A. (2021). Women in energy communities: How inclusive are energy communities towards women? PDF. https://energytransition.academy/sites/asset-h2020.eu/files/public/content-files/articles/05_Aurore%20Dudka.pdf

EIGE. (2019). Gender statistics database. https://eige.europa.eu/gender-statistics/dgs/indicator/bpfa_k_offic_k1__wmid_env_nat_ministers

EIGE. (2021). Women's triple role definition. https://eige.europa.eu/thesaurus/terms/1442

European Commission. (2020a). Energy communities in the clean energy package: Best practices and recommendations for implementation. https://op.europa.eu/en/publication-detail/-/publication/4b7d5144-91c9-11eb-b85c-01aa75ed71a1/language-en

European Commission. (2020b). Energy communities. https://ec.europa.eu/energy/topics/markets-and-consumers/energy-communities_en

European Commission. (2021). Transport emissions. https://ec.europa.eu/clima/policies/transport_en

Eurostat. (2021). Arrears in utility bills. https://ec.europa.eu/eurostat/web/products-eurostat-news/-/DDN-20200120-1

Eurostat. (2022). Gender pay gap statistics. https://ec.europa.eu/eurostat/statistics-explained/index.php?title=Gender_pay_gap_statistics

FAO. (2021). Gender and land rights database. http://www.fao.org/gender-landrights-database/data-map/statistics/en/

Fraune, C. (2015). Gender matters: Women, renewable energy, and citizen participation in Germany. *Energy Research & Social Science, 7*, 55–65.

Global Initiative for Economic. (2020). Renewable energy and gender justice. https://www.gi-escr.org/publications/renewable-energy-and-gender-justice-briefing-paper

Groot de, J., Mohlakoana, N., Knox, A., & Bressers, H. (2017). Fuelling women's empowerment? An exploration of the linkages between gender, entrepreneurship and access to energy in the informal food sector. *Energy Research & Social Science, 28*, 86–97. https://www.sciencedirect.com/user/identity/landing?code=powq9elq_FYDsZVLAj5zq1xPlVOilNGgb8YgK7-0&state=retryCounter%3D0%26csrfToken%3D15d180e9-d9d0-4c15-8181-582709a0b52c%26idpPolicy%3Durn%253Acom%253Aelsevier%253Aidp%253Apolicy%253Aproduct%253Ainst_assoc%26returnUrl%3D%252Fscience%252Farticle%252Fpii%252FS2214629617300981%253Fvia%25253Dihub%26prompt%3Dnone%26cid%3Darp-6e0e3ca9-7890-4b00-9cfb-dca416f5127d

Habersbrunner, K., Kuschan, M., Lins, C., Rühlemann, A. (2021). Gender dimensions of the energy transition as an important pillar for successful climate policy. In Heidegger et al. Why the European green deal needs ecofeminism, WECF and EEB. https://www.wecf.org/de/wp-content/uploads/2018/10/EGD_Report_2021.pdf

Habersbrunner, K., & Martschew, E.-C. (2020). Report on gender aspects of existing financial schemes for energy poverty measures. https://www.wecf.org/de/wp-content/uploads/2018/10/EmpowerMed_Report-on-Gender-Aspects-of-Financial-Schemes-FINAL_kompr..pdf

Heidelberger Energiegenossenschaft. (2020). Heidelberg wird Sonnenstadt. https://www.heidelberger-energiegenossenschaft.de/

Hohmeyer, O., Beer, M., Schirrmacher, J., & Bichler, M. (2017). Schlüsselakteure bewegen kommunalen Klimaschutz: Grundlagenpapier zum theoretischen Hintergrund. Online: http://schluesselakteure.de/wp-content/uploads/2017/10/20170927_Grundlagenpapier_final.pdf

ICA. (2018). International cooperative alliance. Retrieved February 20, 2021, from https://www.ica.coop/en/cooperatives/cooperative-identity

International Labour Organization. (2017). Gender diversity journey: Company good practices. https://www.ilo.org/actemp/publications/WCMS_578768/lang%2D%2Den/index.htm

IRENA. (2019). Global energy transformation. A road map to 2050. https://www.irena.org/publications/2019/Apr/Global-energy-transformation-A-roadmap-to-2050-2019Edition

IRENA. (2020). Renewable energy and jobs: Annual review 2020. https://www.irena.org/publications/2020/Sep/Renewable-Energy-and-Jobs-Annual-Review-2020

Kuschan, M., Habersbrunner, K., & Rühlemann, A. (2020). Frauen. Energie. Wende! Warum wir eine geschlechtergerechte Energiewende brauchen. WECF. https://www.wecf.org/de/wp-content/uploads/2018/10/Frauen.Energie.Wende2020.pdf

Landeshauptstadt München. (2014). Gender budgeting. PDF. https://www.muenchen.de/rathaus/Stadtverwaltung/Direktorium/Frauengleichstellung/themen/Geschlechtergerechtigkeit/genderbudgeting.html

Łapniewska, Z. (2019). Energy, equality and sustainability? European electricity cooperatives from a gender perspective in Energy Research & Social. *Science, 57.*

Radtke, J. (2016). *Bürgerenergie in Deutschland.* Springer Fachmedien.

Reibling, N., & Jutz, R. (2017). Energiearmut und Gesundheit. Die Bedeutung von Wohnbedingungen für die soziale Ungleichheit im Gesundheitszustand.

RESCoop. (2019). Europe's Green New Deal – What's in it for energy citizens? https://www.rescoop.eu/toolbox/europes-green-new-deal-whats-in-it-for-energy-citizens

REScoop. (2020). Energy communities: A practical guide to reclaim power. https://www.rescoop.eu/toolbox/community-energy-a-practical-guide-to-reclaiming-power

RESCoop MECISE. (2019). Mobilising European Citizens to Invest in Sustainable Energy, Clean Energy for All Europeans, Final Results Oriented Report of the RESCOOP MECISE Horizon 2020 Project.

Robinson, M. (2018). *Climate justice: Hope, resilience and the fight for a sustainable future.* Bloomsbury.

Schlüsselakteure. (2020). Das Modell der Schlüsselakteure. http://schluesselakteure.de/wp-content/uploads/2018/05/Steckbrief-Modell-Schl%C3%BCsselakteure.pdf

Smith, et al. (2013). Energy and human health. https://www.annualreviews.org/doi/10.1146/annurev-publhealth-031912-114404

Som Energía. (2021). Cooperativa de Consumo Verde. https://www.somenergia.coop/es/

Spitzner, et al. (2020). Interdependente Genderaspekte der Klimapolitik Gendergerechtigkeit als Beitrag zu einer erfolgreichen Klimapolitik. https://www.umweltbundesamt.de/sites/default/files/medien/1410/publikationen/2020-02-06_texte_30-2020_genderaspekte-klimapolitik.pdf

Stephens, J. (2020). *Diversifying power. Why we need antiracist, feminist leadership on climate and energy*. Island Press.

Troya Çevre. (2017). Enerji Kooperatifleri El Kitabı: Topluluk Bazlı Enerji Üretimi.

WHO. (2006). Constitution of the World Health Organization. https://www.who.int/governance/eb/who_constitution_en.pdf

Wierling, A., et al. (2018). Statistical evidence on the role of energy cooperatives for the energy transition in European countries. *Sustainability, 10*(3339).

WWEA. (2021). Frauen in der Bürgerenergie. Durch Offenheit zur Vielfalt. https://wwindea.org/wp-content/uploads/2021/07/WWEA_Frauen_in_der_Burgerenergie.pdf

Yenkoop. (2021). Community-based energy generation with renewable energy, can contribute to women's empowerment? https://www.yenkoop.com/kadinlar-ve-enerji/

Yildiz, Ö., et al. (2015). Renewable energy cooperatives as gatekeepers or facilitators? Recent developments in Germany and a multidisciplinary research agenda. *Energy Research & Social Science, 6*(2015), 59–73. https://doi.org/10.1016/j.erss.2014.12.001

CHAPTER 9

Engendering the Energy Transition: Inspiring Examples of Gender-Just Citizen Energy Communities in Europe

Antonia Proka

Introduction

Access to energy is a basic condition for human development and prosperity. Energy has been described as 'the life blood of our society'. The availability of cheap electricity, through fossil fuel combustion, has transformed the contemporary world and the way we live in it. The current fossil fuel-based energy system comprises undeniably a very critical infrastructure in our societies, because all societal sectors, such as housing, mobility, food production or health care, depend on a reliable provision of energy (Verbong & Loorbach, 2012; Proka, 2021). By the end of the twentieth century, it became the task of the energy industry to provide for it. And thus far, energy has mostly been a 'taken for granted necessity confined to the world of engineers' (Becker & Naumann, 2017 as cited in Szulecki &

A. Proka (✉)
REScoop.eu vzw, European Federation of Citizen Energy Cooperatives, Brussels, Belgium
e-mail: antonia.proka@rescoop.eu

N. Rocha Lawton, C. Forson (eds.), *Women and the Energy Sector*,
https://doi.org/10.1007/978-3-031-43091-6_9

Overland, 2020). More recently, this fossil fuel–dominated energy industry has ended up threatening the prosperity it once helped to create. The energy system is associated with multiple interconnected challenges: climate change and air pollution; growing demand and exhaustibility of natural resources; dependency on unstable political regimes for supply; uncertainties about reserves and high volatility of resource prices; disparate social vulnerabilities and widespread injustices among and within communities around the world (Verbong & Loorbach, 2012; IEA, 2014; Stephens, 2019; Proka, 2021).

In response to such a predominantly unsustainable energy system and neoliberal policies of the 1980s and 1990s, the advances in information and energy technologies, along with the dropping prices of renewable energy technologies around the turn of the millennium, brought about an abundance of alternative models of organising the societal function of energy provision (Vansintjan, 2015; Proka, 2021). Citizen-driven renewable energy initiatives begun to challenge the unsustainable status-quo, which facilitates an unjust division of cost and benefits prioritising business interests and privatising profits at the cost of the environment and community interests, and in fact, with their growth and expansion manifest in practice that an alternative way is possible (Proka, 2021). The vision for a just and equitable renewable energy future that connects environmental sustainability, community resilience, social justice and economic equity has been described under the term 'energy democracy' (Burke & Stephens, 2017). As Angel (2016) notes, it 'is not a future utopia to be won but, rather, is an ongoing series of multiple struggles over who owns and controls energy and how, where and for whom energy is produced and consumed' (p. 4). While the energy sector has previously been seen as purely technical and not requiring public involvement, the scale of the challenges, as well as the competing emerging alternatives, can no longer allow the framing of the choices involved in designing energy transition pathways as non-political. Technocratic expertise is not sufficient for making these decisions. To address the environmental and social injustices associated with our energy system, a broader systems perspective should be taken; a perspective that considers the particular way in which technological solutions get linked with social structures, and articulates distinct cost-benefit distribution (Proka, 2021).

In the early 2010s, the concept of energy justice emerged as a conceptual, analytical and decision-making framework to address justice issues in energy access, use and policy making (Feenstra & Ozerol, 2021). Sovacool and Dworkin (2015) have defined energy justice as 'a global energy system that fairly disseminates both the benefits and costs of energy services, and one that has representative and impartial energy decision-making'. The work of energy justice scholars has been focusing on disparities among socioeconomic groups in terms of impacts from energy infrastructure, and access to energy and energy decision-making power (Jenkins et al., 2016; Allen et al., 2019). From an energy justice perspective, energy transition processes must ensure fairness via equal distribution, full recognition of rights and labour contributions, equal participation in decision-making procedures, and equal capabilities in renewable energy outcomes (Johnson et al., 2020). The adopted in 2015 Paris Agreement notes: 'Parties should, when taking action to address climate change, respect, promote and consider their respective obligations on human rights… as well as gender equality, empowerment of women and intergenerational equity'.

The issue of gender equality and empowerment is also recognised by the UN as one of its Sustainable Development Goals (SDG 5), as 'an integral part of human rights and a fundamental criterion for democracy (…) Providing women with equal access to education, health care, decent work, and representation in political and economic decision-making processes is fundamental for social cohesion and prosperity' (UN, 2015).

The energy sector has always been male-dominated. Despite the progress made, a gender gap still exists concerning female participation, visibility and representation in management and top-level positions in the energy field. According to an IRENA (2019) study, women represent only 32% of the renewable energy workforce and 22% of the oil and gas industry workforce. Considerable deficits in the engagement of women in community energy also exist, where male participation often rises to even up to more than 70% (e.g. WWEA, 2021; Radtke, 2016). Literature suggests that beyond individual preferences and investment attitudes, cultural, social and political factors also influence an individual's agency to participate in citizens' initiatives (Fraune, 2015).

Integrating a gendered lens into the analysis and policy making for energy transformation offers multiple possibilities for accelerating change (Allen et al., 2019). For decades, already, energy scholars have been examined how women are disproportionately affected by inefficient and unsafe indoor energy sources, how natural resource jurisprudence reinscribes masculinist hierarchies, and how women are under-represented in science, technology, engineering, and mathematics (STEM) fields (Ryan, 2014). Energy researchers in the EU have analysed the gendering of energy usage, pollution, and policymaking across geographies and economies (Clancy et al., 2017).

Ecofeminist voices have for years been calling attention to women's potential to bring about an ecological revolution. For ecofeminists, there is a parallel between the exploitation of nature by (hu)man and the inherent exploitation of the majority of women in society, through a gendered division of labour that allows for unpaid reproductive work. For them, it's clear that as long as this systemic injustice goes unchecked, a real energy transition will be highly unlikely (Steinfort, 2018). At the same time, after the 2019 European elections, the issue of gender equality rose up the political agenda with the backing of the European Commission President von der Leyen, culminating in the Gender Equality Strategy 2020–2025 (European Commission, 2020). As President von der Leyen notes, 'In business, politics and society as a whole, we can only reach our full potential if we use all of our talent and diversity. Using only half of the population, half of the ideas or half of the energy is not good enough'.

The energy transition offers a unique opportunity to transform the energy sector in all aspects. Without careful consideration of social justice and who in society is benefiting and who is being excluded, humanity could move to a 100% renewable energy world that continues to perpetuate, rather than reduce, vulnerabilities, injustices, and inequalities around the world (Stephens, 2019; Proka, 2021). Citizen-driven renewable energy communities, by acknowledging their own challenges and embracing diversity, can democratize, decentralize and diversify the energy sector contributing to a socially and environmentally just energy transition.

This contribution focuses on inspiring examples from the European community energy context that illustrate alternative cultures, structures and practices that have the potential to radically transform the energy system. After the description of the methodology applied, the chapter begins with a short discussion of the background of energy democracy and how it links to gender justice. Then, following a short discussion of the benefits of gender-just energy communities and the current status, some frontrunning initiatives are being presented. In the discussion and conclusion, some first recommendations for mainstreaming gender justice in the energy transition are made.

Methodology

This chapter reports on the author's research focusing on the gender (in) justice in the cooperative energy movement and the ambition and efforts taken by energy cooperatives to promote gender justice in their sector and engender the energy transition. The investigation began with literature review on the topic of gender justice in energy cooperatives. The focus has first been on both academic articles as well as professional publications; then, official publications and material published online by front-running energy cooperatives were also analysed.

The energy cooperatives discussed in this contribution have been selected on the merit of the good practices they exhibit on the topic of gender justice. The selection was, furthermore, driven by the wish to cover a variety of geographical regions.

After the investigation of the online material, the directors or employees of the energy cooperatives presented in this chapter were contacted, and the author exchanged on the specific issue. After one-to-one conversations/interviews on the challenges and the good practices employed, access to any relevant official data was requested and granted. Following, a draft paragraph on the case was prepared by the author and shared with the interviewee for review, verification of accuracy, and any possible corrections. Overall, only minor adaptations have been made.

A Just Transition Towards Energy Democracy

The concept of energy democracy has emerged from social movements that orient their action towards system transformation and is less grounded in academic theory (Burke & Stephens, 2017). Recently, more scholars have started looking at the issue. Becker and Naumann (2017), for instance, list *decentralized energy generation*, *public and cooperative ownership* and *energy sovereignty* as the dimensions of energy democracy.

In their review, Szulecki and Overland (2020) propose three understandings of energy democracy as (1) a *process* (and existing grassroots initiatives' and a transnational social movement that challenges unsustainable energy incumbents), (2) an *outcome* (the transition towards decentralised renewables sees in parallel a reconfiguration of social relations), or (3) a normative *goal* which communities and other stakeholders aspire to be achieved in the future.

When understanding energy democracy as a process, the social movement appears as the main driver, epitomizing the concept while focusing on making it happen. Maybe one of the most popular slogans in this part of the literature is: *resist*, *reclaim* and *restructure* (Burke & Stephens, 2017; Stephens, 2019). Energy democracy entails '(1) resisting the agenda of large energy corporations, (2) reclaiming to the public sphere parts of the energy economy that have been privatized or marketized, and (3) restructuring the global energy system in order to massively scale up renewable and low-carbon energy, aggressively implement energy conservation, ensure job creation and local wealth creation, and assert greater community and democratic control over the energy sector' (Sweeny, 2014, p. 218).

Understood as an outcome energy democracy involves the changes in how energy systems are organised. From this perspective, a shift to renewable sources, which are more intermittent, distributed and scalable, should lead to a reconfiguration of cultures, structures, practices and social relations. In other words, these scholars claim the technological transition comes first, thereby enabling political and social change, as in the case of Europe, most visibly in the context of Germany's *Energiewende* (e.g. Kooij et al., 2018 as cited in Szulecki & Overland, 2020).

Last, energy democracy can also be understood as a goal—an ideal to which communities can aspire, and a principle guiding policies and action towards a just and democratic energy system (Szulecki & Overland, 2020). Like the 'outcome' understanding, the 'goal' is also something that comes in the future. However, from this perspective, energy democracy is not brought about by technological shifts alone, but by political change informed by democratic ideals. The difference from the 'process' understanding is in its prescriptive character.

In their review of existing activist and academic literature Van Veelen and van der Horst (2018), the goals of energy democracy suggest that: 'the electricity system (…), as well as our economy and society should become more inclusive, equitable and low carbon. Secondly, political power and decision-making should be more devolved to the local level. (…) Access to the electricity grid should be widened, especially for new and small renewable electricity producers (and) the ownership base for various aspects of our electricity system should be broadened, (while) greater citizen involvement and ownership should be achieved through voluntary means'. Nevertheless, as Burke (2018) highlights energy democracy as both a movement and an organizing principle is not a single vision, but a diversity of energy democracies that diverge in 'problem framings, the form and specificity of solutions, the critical stance, the historical positioning, and importantly, the scale, agency and model of social organization' (p. 12).

Undeniably, the calls for energy democracy count on the hope that activism, grassroots democratic organizing, local governing structures, and public participation have the power to change the status quo with its existing hierarchies and relationships (Feldpausch-Parker et al., 2019). However, while democratic principles suggest participation open to all, decision-making at both collective and individual levels can be unjust and reify problematic power dynamics. This highlights why simultaneously attending to justice, participation, and power is crucial for energy democracy. As Feldpausch-Parker et al. (2019) note, 'the concept of justice should highlight the importance of equity; the concept of participation should highlight the importance of group decision making; and the concept of power should highlight the importance of recognizing extant structures of power and possibilities for resistance'.

In that, justice in the frame of energy democracy calls attention to the distribution of costs and benefits in relation to energy decisions, the decision-making processes and who participates in it, the nature of the relationships involved and the role of structural inequities (such as racism, colonialism, sexism, classism, etc.) on whom is served by energy decisions (Feldpausch-Parker et al., 2019). It is important that there is an open and meaningful participation in the decision-making processes. Attending to justice, participation and power at the same time creates awareness of the existing inequalities which need to be recognised and tackled for a just and democratic energy transition.

The Concept of Gender and Gender Justice

Gender can be described as a form of socially constructed difference associated with different inequalities and hierarchies traditionally between women and men (Johnson et al., 2020). Gender roles are shaped through the process of socialisation and change over time. They may vary across cultures but are always related to the issues of power and resources. Recent literature captures the intersectional nature of gender (Kaijser & Kronsell, 2014 as cited in Johnson et al., 2020). As Johnson et al. (2020) explain, 'an intersectional approach to gender is to analyse social inequalities and systems of power with the awareness of their interconnected character, such as how gender inequality interacts with other inequalities of ethnicity, race, class and age' (Azocar & Ferree, 2016). An intersectional approach when tackling gender inequalities is crucial. While women are often treated as a homogenous group, single mothers, women over 65, migrant women, and domestic and service sector workers, are more heavily impacted by the structural injustices of the current energy model. In fact, this is aligned with what the European Parliament and the Council of the European Union describe as vulnerable people in the Directive 2013/33/EU (Recast Reception Conditions Directive). Here, in line with this European Directive, with vulnerable, we refer to women that belong or are perceived to belong, to groups that are in a disadvantaged position or marginalised, such as single mothers, elderly women, migrant women and women with serious illnesses.

The discussion about gender justice in the energy transition is about much more than merely increasing the representation of women in decision-making positions. We need to ask some fundamental questions. Who are the energy users? What are their social positions and aspirations? Who is consulted during the process of implementation, and how is the energy supply organised? Gender justice is, therefore, not a female issue; it's an issue that should engage everyone. It is essential to shaping positive broader socioeconomic development results.

A just transition towards energy democracy must reflect the needs and labour of people of diverse genders. We need everyone's perspective on key decisions, from investment priorities to project design. Beyond the justice and fairness argument, engaging women as well as non-binary people in the energy transition can bring several additional benefits. The contributions of everyone will be central to ensuring that the energy systems of the future address the needs of modern societies and truly leave no one behind.

The Value of Gender-Just Energy Communities

The advantages of taking gender considerations on board when developing renewable energy projects, and in general, when one is engaged in the energy transition are multiple.

First of all, studies suggest that projects taking gender issues into account are more likely to achieve their objectives, compared to projects that do not. When the gender situation is assessed and gender-needs are taken into account, projects tend to have a better impact on the ground (Cecelski & Dutta, 2011).

Through the engagement of women, new perspectives are added to the workplace, and overall collaboration tends to improve (Moodley et al., 2016; IRENA, 2019). A better gender balance in male-dominated professions appears to be contributing to the improvement of working conditions for both men and women, thereby offering positive effects on well-being, work culture and productivity (WISE, 2017; IRENA, 2019). Increasing the number of qualified women in an organisation's leadership yields better performance overall, including better financial results (Noland et al., 2016).

Nevertheless, critics stress that 'increasing diversity does not, by itself, increase effectiveness; what matters is how an organisation harnesses diversity, and whether it is willing to reshape its power structure' (Ely & Thomas, 2020). Societal norms can be a challenge to gender justice. Specific constraints for women's participation need to be understood and planned for in order to realise the benefits. According to a 2015 survey of Harvard Business School alumni, 76% of people in senior executive positions believe that 'a more diverse workforce improves the organization's financial performance'. Yet, scholarly researchers have rarely found a direct link between increased diversity and improved financial outcomes. As described above, they have found that it leads to higher-quality work, better decision-making, greater team satisfaction, and more equality *under certain circumstances.* For instance, the scholars note that diverse teams realize performance benefits when: 'team members are able to reflect on and discuss team functioning; when status differences among ethnic groups are minimized; when people from both high- and low-status identity groups believe the team supports learning; and (...) when teams orient members to learn from their differences rather than marginalize or deny them'. They also stress that without conditions that foster inquiry, egalitarianism and learning, diversity may either be unrelated to or undermine team effectiveness.

It is, thus, important that when one tries to integrate gender justice principles in the functioning of an organisation, attention also goes to the creation of trust and egalitarianism, together with a desire to learn and support each other. In parallel, structures of discrimination and subordination need to be dismantled. Alternatively, any conversation about diversity will happen (if at all) only in reaction to diversity-related crises.

The Present Challenges and Current Status

The energy sector has been excluding the perspectives of women, minorities and vulnerable and disadvantaged individuals and communities from its decision-making processes (Pearl-Martinez & Stephens, 2016). As the IRENA (2019) study showed the representation of women in the renewable energy field rises to only 32%, still better than the 22% of the oil and gas industry workforce. When studying the international wind energy sector, the situation is even worse. Female participation rises to only 21% of employees, and less than 8% of senior management positions (IRENA, 2019).

The case differs across different countries. A study from AEMENER, the Spanish Association of Women in Energy, for instance, found that in 2018, about 28.5% of the employees of the energy companies participating in their survey were women (AEMENER, 2018). The percentage of women in public services was higher than those in the private sector, yet, the latter appeared to be rising faster than that in the public.

Focusing on the community energy sector, the engagement of women is still significantly low. A study from 2012 showed that the participation of women in citizen energy initiatives in Germany rose to about 20%; this number was later confirmed by another study in 2016 (Yildiz et al., 2015; Radtke, 2016). Similarly, a study from Austria dating back to 2016 registered a 29.6% female participation in the renewable energy sector, compared to 19.3% of the energy sector as a whole. The study also found there were twice as many female executives in the renewable energies sector as in the rest of the energy sector (ÖGUT, 2016).

More recently a survey carried out by the World Wind Energy Association between October 2020 and January 2021 showed that the share of women in community energy in the German region of North Rhine-Westphalia is currently a bit less than a third (29%), which marks an increase compared to the previous studies (WWEA, 2021). These women hold about 27% of the shares in the energy communities in the region. A first study on the leadership positions of citizen energy communities in Greece showed that, in 2020, 93% of all the recently created communities had less than two women participating in their Boards; specifically, 42% had none, 34% had one, 17% two, 5% three, and only 1% had four or five (Vasilakis et al., 2020).

And certainly, beyond participation, previous research has also shown significant gender differences in the average ownership rate, the average investment sum and decision-making bodies (Fraune, 2015). Also, while energy communities employ more women, they often struggle to attract female members, especially vulnerable women.

There are different factors that can explain this situation. Uneven sharing of the household responsibilities and childcare, to start with, keep women busy at home. Household obligations have been recognised as one of the biggest barriers to women participating in the energy value chain. Furthermore, women often have confidence issues, which refrains them from taking the lead, even if they may be more technically qualified and competent to do a task than a male colleague. Policy plays also a role. Gender-neutral does not automatically translate into gender-equal. Winther et al. (2018) underline that gender-neutral and generic interventions to involving communities can result in benefiting the groups that are already in a more privileged position, perpetuating in this way the structures of inequality.

International non-governmental organisations (NGOs) and women and gender justice associations, along with grassroots organisations, stand up for women's rights and call for gender equality and women's empowerment in the renewable energy sector. In this context, gender quotas are discussed as a valuable tool to correct hidden male biases. As research by the London School of Economics showed 'quotas weed out incompetent men rather than promote unqualified women' (Besley et al., 2017). Also, echoing Pateman's words, it is underlined that 'we need a conception of universal, participatory citizenship that is grounded in the recognition of sexual difference, so that women, to become full citizens, do not have to attempt to become paler reflections of men, but can actively participate as women' (1985, p. 14). In line with this, appropriate tools should be used to empower women to reach their full potential, and in doing so, the current challenges women face should be considered. For instance, as, for reasons described earlier, women have less time compared to men, this should be considered when organising trainings for them.

Inspiring Examples of Citizen Energy Communities in Europe

Community energy encompasses a variety of collaborative actions that involve citizens' participation in the energy system. Energy cooperatives are linked to the call for energy democracy. A citizen energy cooperative refers to a business model where citizens jointly own and participate in renewable energy or energy efficiency projects. While joining an energy cooperative is not the only way in which citizens can participate in the energy transition, cooperatives are the vehicle for making energy accessible to all. This is the case because energy cooperatives are not only *open to all*, but they facilitate the meaningful participation of everyone by reducing the financial threshold (via low membership fees), and by providing education and trainings for everyone to become an active agent in the energy transition.

Energy communities do not necessarily need to have the legal statute of a cooperative, but rather distinguish themselves by the way they do business. Specifically, energy cooperatives implement the seven principles that have been outlined by the International Cooperative Alliance.

- Voluntary and open membership
- Democratic member control
- Member economic participation
- Autonomy and independence
- Education, training and information
- Cooperation among cooperatives
- Concern for Community

Guided by these principles when facilitating the participation of citizens in the decision making and the distribution of costs and benefits of the energy transition, energy cooperatives reflect the energy justice definition of Sovacool and Dworkin (2015). And while all principles can be supportive, especially three of the cooperative principles are clearly aligned with the effort to transition to a gender-just energy system. Firstly, the call for a *Voluntary and open membership* means that participation is open to all persons who are able to use the cooperative services and willing to accept the responsibilities of membership, without gender, social, racial, political or religious discrimination. Second, the *Democratic member control* stresses the fairness in the decision making that is achieved through the one person, one vote principle. And lastly, *Education, training, and information* focuses on the self-empowerment that can be achieved through the socio-economic power of membership.

The values and activities of energy cooperatives are, thus, framed by the previously mentioned cooperative principles. In what follows examples from the European community energy context are presented. These examples illustrate how this alternative culture translates into concrete practices and structures that are aligned with the principle of gender justice.

The Socio-political Context

Before focusing on the specific cases, it's worth mentioning that while progress is registered in several European countries, the status of development differs. The overall socio-political landscape may also play a role in the progress that is registered. In Spain, for instance, the national government has introduced a Ministry for Ecological Transition, which merges the former Ministries for Environment and Energy (the Ministry is also led by a female minister). The Spanish National plan that describes the Ministry's roadmap toward the country's climate and energy 2030 goals has been ranked at the top of the EU Member States National Energy and Climate Plans, based on national targets, the comprehensiveness of policies, inclusiveness and participation (Lieu et al., 2020; Duwe et al., 2019; WECF, 2021). Bulgaria too showcases the good practice of integrated cooperation on the matter. Specifically, the Ministry of Labour and Social Policy, the Ministry of Energy, and the Energy and Water Regulation Commission cooperate to design and establish a socially just policy for vulnerable consumers to secure their affordable access to energy and water services (WECF, 2021). At the same time, other EU countries mark rather slow progress. An area where most EU countries perform quite bad from a gender justice perspective is the interface between renewable energy and mobility. In Germany, for instance, the introduced electromobility grant programs are mostly benefiting men as they tend to receive higher salaries and thus can afford more expensive e-vehicles, and that, on top of the fact that men most often use company cars (ibid.). Alternative, gender-just clean and sustainable mobility models do not receive equal support and attention.

The Citizen Energy Communities

The Netherlands—Zuiderlicht, i.e. Southlight was established in Amsterdam, the Netherlands, in 2013. The cooperative, led by a female director and a gender-balanced team develops community energy projects, installing solar panels, for instance, on roofs of schools, sports clubs, municipal offices, apartment blocks, among others. The cooperative is also involved in wind energy projects.

Beyond the development of projects, *Zuiderlicht* also participates in and co-organises community activities such as Energy breakfasts, Energy info evenings, and celebrations (Fig. 9.1). The goal of the cooperative is to help increase awareness about the necessary energy transition in a positive manner, build and empower the local communities to take energy into their own hands. In its events, the cooperative is represented by a gender-balanced team. In fact, the *Zuiderlicht* team is quite gender-balanced overall and the same holds for its membership. Note that one can join the cooperative by investing as little as €1.

Beyond the development of projects, *Zuiderlicht* celebrates victories and at least twice a year it organises gatherings with energy lectures and performances: inviting members and partners. The goal of the cooperative is to accelerate the energy transition and help increase awareness about the necessary energy transition in a positive manner, build and empower the local communities to take energy into their own hands. In its events, the cooperative is represented by a gender-balanced team, reflecting its overall quite gender-balanced team and membership.

Fig. 9.1 Zuiderlicht: energy (and coffee) with a positive story ©Zuiderlicht

Germany—Bündnis Bürgerenergie (BBEn), i.e. the Alliance for Citizens' Energy, wishes to safeguard that citizen energy projects will remain an essential pillar of Germany's power supply. The organisation supports networks of actors and regions involved in citizens' energy, in order to shape innovative ideas on a decentralised energy transition. BBEn is an association with democratic decision making.

For BBEn, the issue of language and female representation in its activities, including the communication around them, is crucial for the empowerment and engagement of more women in community energy. For this reason, it puts more women to the front both as organisers and as speakers in its events, but also it uses more visuals that represent women in its communication. Regarding the former, BBEn makes sure that the timing of its activities allows for the participation of women, also considering their responsibilities with children. Regarding the latter, BBEn has also developed a scrolly telling to inspire more women to get engaged in community energy (Fig. 9.2).

Fig. 9.2 Aylin from the scrolly telling of BBEn ©BBEn

Greece—Electra Energy, founded in 2016, is a social cooperative based in Athens, Greece. Electra Energy focuses its efforts on the transition to a decentralized, efficient and sustainable energy system with citizens and local communities at its core. Its main activities are research on topics relevant to the community energy movement, renewable energy project development, capacity building and advocacy focusing on strengthening the policies that can enable the flourishing of community energy initiatives. Electra Energy was among the initiatives that were invited in 2018 to support the Greek Ministry of Energy and Environment during the preparation of law 4513/2018 about energy communities. In the same year, it also coordinated and contributed to the publication of the first guidebook about energy communities in Greece, with the support of other relevant organizations and academic institutions.

The director of Electra, Miriam Rodriguez-Ruiz, is involved in the Spanish Association of Women in Energy (AEMENER). With her involvement in AEMENER, Miriam aims to empower women and girls and support the mainstreaming of gender justice through education and mentoring programs. The coaching engages both female and male tutors and targets both female and male participants. It is crucial to underline that the role of both men and women is crucial for the implementation of gender justice in the field. Electra is one of the initiatives that have a clear reference to gender equality in its statutes. Specifically, Article 3 mentions:

> In particular, the Cooperative in the context of promoting sustainable development will carry out activities which:
>
> (a) promote the sustainability of the environment;
> (b) contribute to social and economic equality;
> (c) promote gender equality;
> (d) protect and develop common goods

The cooperative, in the context of the European Citizen Energy Academy (EUCENA) project funded by the European Climate Foundation, has also been providing support for the development of the first female energy community in Greece: *WEnCOOP* energy community. *WEnCoop* members are female entrepreneurs in the North of Greece. It is worth mentioning that EUCENA has a specific KPI focusing on the engagement of women in citizen energy communities.

Belgium—Ecopower is a cooperative from Flanders which was founded in 1991. It currently numbers more than 60,000 shareholders and it so far helps avoid more than 65,000 tons of CO_2 per year as it is proudly being noted. The cooperative produces renewable electricity through wind and solar energy and has also a factory producing wood pellets for domestic heating. A cooperative share of *Ecopower* costs €250.

In the case of *Ecopower*, while no specific text exists in their statute about gender balance, a concrete reference is present in the internal regulation. Specifically, it is noted that:

> The composition of the governing body shall include the following targets
>
> - balanced composition between internal members and external members
> - complementarity, diversity of expertise and network of Board members
> - sufficient involvement so that board members can make their expertise available to Ecopower
> - a balanced gender distribution

Also 'gender neutrality' is used in the labour regulations document. It is stated that 'equal pay for male and female workers must be ensured for all elements and conditions of the wage, including job evaluation systems'. This derives from the official national rules in Belgium.

The organisation is almost gender equal. This is because the seven workers of the pellet factory are all male. The rest of the employees are balanced: out of 42 employees, 21 are male and 21 females. Also, the management positions are distributed fairly. At the level of coordination, five are men (general coordination, legal, customer service, projects, pellet factory) and three women (finance, communication, ICT). At the highest seniority level, nevertheless (i.e. coordination and decisions), the presence of males is more dominant 14 men versus 8 women.

When it comes to the board of directors, while originally three out of nine members were female, currently the balance has shifted towards more male presence due to resignations (and the passing away of one board member). New members will soon be appointed. In the past, to ensure a more gender-balanced board, they actively searched for more women.

Regarding membership, a survey has shown that 80% of the respondents are male and 20% are female. More recently, this shifted to 78% male and 22% female. While of course the electricity generated is used by men and women, the cooperative has many more male members than females. In the past, in the case of a couple the names of both partners were recorded in their accounts, but nowadays no double fields can be recorded. This may obscure the real numbers. Nevertheless, it is noted that, in general, in members' meetings the participation of men is higher than that of women.

This situation may reflect the Belgian society, as it is being noted. When it comes to the organisation, to improve the situation, Ecopower applies specific policies that aim at improving the work conditions of women in the cooperative, including their work-life balance, enabling for instance, breastfeeding, and supporting overall flexibility. The cooperative has been analysing its situation and it is eager to learn what more could be done to further increase taking gender justice on board of its operation and wider activities.

Spain—GoiEner was created in 2012 by a team of 30 people. The initiative started as retailers (consumption side) because the generation was very difficult at that point of time. The first invoices were issued in autumn 2013. The membership has been steadily growing by around 1500 members per year. Nowadays, GoiEner numbers 14,618 members, 19,011 electricity supply contracts, and employs 53 people.

In terms of gender, their employees are quite balanced (M 27 / F 26) and so is their clientele (55% M/ 45% F). Nevertheless, the membership on the generation's side is still male dominated (about 80% /20%), and more male than female volunteers are engaged. Also, while the employees of the coop are gender balanced, the subgroups are not. The IT group has more males; the clients' support has more females, etc. The 50/50 rule does not apply at this stage in these sub-groups.

When it comes to their hiring process, they never seek a female or a male exclusively; the decision is made after evaluation. And if they choose a female instead of a male (or vice versa), the qualifications of the two need to be the same.

Their vision is to become *a* 'Community Cooperative that is a benchmark in the Basque Country in the energy sector with diversified activity, professional and voluntary work, which places life at the centre of its activity and is linked to other agents of transformation'.

GoiEner's mission is to build an energy model that respects the environment and is fair,

- marketing 100% renewable electricity, promoting generation projects, advising, participating in campaigns, and collaborating with organisations that share our objectives,
- being close, generating trust, accepting diversity, communicating with transparency, facilitating participation, and taking into account the needs and care of people.

Back in 2018, they had to take a decision of whether they would support or not the strike of 8 March. This was the first step for them, because as the director of the cooperative Erika Martinez puts it, 'what is the point of the support of an International Day if you do not do anything the rest of the year?' In the meantime, nowadays, new laws are being voted in Spain and the topic of gender justice is gaining more space in the public discussion.

They organised a gender group which was later renamed as *Care group* (for ones' self, the organisation and the others). Everyone participated in that group (male and female, employees and volunteers); the core group comprised seven employees, three volunteers and one partly employed specialised in the topic.

At that time, they looked for help in *another coop* that had already worked on their gender plan. To ensure that the plan does not stay on the shelf, but it is useful and transformative, for its implementation they asked for and received support from the *Basque country*.

The process started in March 2019 and went on until December of the same year with info *training on Feminism and Ecofeminism* followed by interviews of their employees and volunteers. The model will be applied, and the cooperative has four years (until 2024) to make any corrections necessary. Then, the authorities of the Basque country will evaluate and potentially promote them as a novel concept. Thus, GoiEner has invested nine months, public funds as well as their own cooperative funds in the development of the gender plan.

The gender plan covers issues regarding communications (the language used—a public speaking course for women); decision making and participation (in meetings and activities); the (envisioned) future the in coop (i.e. are there different opportunities for males and females in the coop? are people reaching their aim?), and the overall management of volunteers (e.g. would a change in the timetable support them better?), among others.

Some Elements of GoiEner's Gender Plan

First of all, the targets established by the coop are *not mandatory*. *Qualitative indicators* have been designed to map the progress. There are different areas of action.

Changing the discourse: In the context of the coop activities, *socias* (*soci@s*) is being used (in Feminine form) which, as noted, is aligned with the Spanish word for people ('personas'). 'Sometimes to integrate change you need to break the rules,' as the director of the cooperative Erika notes.

Assessing overall equity and treatment: While employees mentioned no problem, the research showed that female volunteers were not always feeling well, e.g. unofficial communication in the weekends was not always good for them.

Comms training programme—mentoring: A public speaking training is planned in order to support more women in taking the floor to talk about the coop in public.

Mapping and controlling participation during meetings: A fixed member of the Care working group controls how many times do men and women take the floor during meetings and tries to ensure balanced participation. The next step will be to count the exact time men and women speak during meetings. (It is mentioned that women when in agreement say nothing, while men usually add a few words (to say what has already been said).)

Timing and setup of meetings: The timing of the meetings (evenings) was not challenged by the members. Also, during the assembly, they investigated whether there was a need for special care for kids, and no interest was found; only one coop member joins and the other partner stays home with the kids. During COVID-19, hybrid meetings took place, in which the participation of men and women was more or less the same as before. The cooperative will go back to physical meetings as soon as possible due to the tiredness of the online meetings. Of course, the coop has not escaped criticism. But as Erika mentions in a coop of more than 14,000 people, not everyone will agree with everything, and not everyone is always interested in everything.

Mujeres con Energía, i.e. women with energy, is another female-driven energy initiative from Spain. This cooperative was born in 2018 and since then they have participated in several events, mentoring programmes, competitions, also receiving numerous national and international awards. *Mujeres con Energía, Enerxía* has received support from existing cooperatives, such as *Som energia, La Corriente, Coopernico*, research projects, such as PROSEU, and also NGOs, such as the ONGAWA, *Engineering for Human Development*. Back in 2018, a meeting took place in Madrid between women involved in energy cooperatives. The meeting revealed a lack of women in the management of energy cooperatives as well as in their responsible bodies. Indicative is that out of the 17 cooperatives that were represented in the meeting, no one had women in their board of directors. The participants introduced *Xenergia* with the aim of increasing the visibility of the work of women in this 'highly masculine energy sector' and become the 'reference female driven energy project'. Attention was also paid to the cooperatives' statutes as well as the matter of language that needs to be more inclusive to inspire the participation of more women in the movement.

Mujeres con Energía was officially established as a cooperative in February 2021; the outbreak of the pandemic delayed this process. Today, apart from project development for renewable energy generation, especially for self-consumption, *Mujeres con Energía* also engages in community education around energy issues, with sessions focusing on the explanation of the details of electricity bills, the issue of energy poverty as well as the issue of equality in the energy sector. The cooperative mobilises the participation of women across the value chain of its projects: from project development to the installation of solar PVs. It is nevertheless stressed that finding female engineers for its projects is not always easy.

Looking at the EU-wide level, *REScoop.eu*, the European federation of citizen energy cooperatives, considers gender equality as a key area of its operation. In fact, it has been agreed that since the director is a legal person, two natural persons of different gender shall be appointed as a representative and its deputy, to complete the assignment on behalf of and for the account of the legal entity. REScoop.eu wishes to increase its efforts to close the still persisting gender gap. In 2020, the organisation in the context of the General Assembly of Cooperatives Europe adopted the Charter of commitment on equality between women and men in cooperatives: 'Cooperatives provide decent employment and help reduce inequalities which affect the most vulnerable people, including women. The equality between women and men and for all is at the core of the cooperative identity and as well as of the European Union' (Fig. 9.3).

Fig. 9.3 Charter for Gender Equality Cooperatives Europe ©Cooperatives Europe

The cooperatives underline that only through the inclusion, participation, cohesion and enhancement of differences, we can proceed on the path of change which is crucial in responding to the social and economic challenges that the world is facing. As Stefania Marcone, vice-president of Cooperatives Europe and Chair of the Gender Equality Working Group, commented:

> Cooperatives, as people-centred businesses, are an important tool for women's empowerment, their access to labour as well as their path to self-entrepreneurship, ensuring their voice being heard in the economy and the society. With a large number of women on the frontline of the fight against the Covid-19 pandemic, the role of women is more important than ever for Europe to build back better. By adopting this charter, the European cooperative movement commits to make a difference for millions of women cooperators of today, and tomorrow, in Europe, and in the world. (Cooperatives Europe, 2020)

In 2021 REScoop.eu also launched a Gender Power working group. This working group aims at mapping the needs and challenges that energy cooperatives face and providing the space for reflection and exchange about good practices, as well as tools and instruments that energy communities can apply to diversify their boards, member base, and reflect the principle of gender justice in their overall operation.

Mapping

During the preparation phase of the Gender Power working group, a short survey was launched. Its outcome might reflect the general conditions in the cooperative energy movement with regard to gender equality. The survey received only 12 responses (from 10 cooperatives). In terms of gender, the respondents were balanced: 50% men and 50% women. The cooperatives represented in the survey marked wide geographic coverage: Centre, North and South of Europe.

The questions of the survey were not exhaustive. Its purpose was to initiate mapping of the current situation in the field and start a conversation with the participants of the working group. The first issue we hoped to shed light on was the issue of participation. How many women are on the board of directors of the cooperatives? How many men? How many are the women and how many are the men who participate in the executive board of the cooperatives? The responses showed that for every woman in the board of directors, there were two men on average. (On one occasion there were only women and in another only men on the board). When it comes to the Executive Board, responsible for the day-to-day operation of the cooperatives, the picture was slightly better. First of all, it has to be stressed that not all the cooperatives of our sample have such a board in place. For those who do, some mark 50% or even 60% women participation in the board, while others note participation of 25%, 20% or even no participation at all in that board.

The participants of the survey argued that gender equality is reflected in their statutes (60%). Yet, this does not necessarily mean that they had a concrete strategy on how to implement it in their operation. Gender equality appeared as an above-average priority for the cooperatives that participated in the survey (2.8 out of 5). Yet, even if the issue is quite significant for them, very few initiatives volunteered to share any good practices in relation to gender equality.

During the kick-off meeting of the Gender Power working group community, energy leaders from the REScoop.eu network gathered to reflect on the current status of gender equality in the community energy movement and coordinate their efforts to help mainstream gender-just energy communities. The REScoop.eu team presented the preliminary results of a performed members' survey, which exposed that the participation of women in the board of directors of energy cooperatives is still quite low, while there seems to be a better picture regarding the participation of women in Executive boards (for those that have one).

During the meeting, the participants took a moment to lay-out their vision for the future; where 'women are deeply involved in their local energy communities and actively contribute to the development of decentralised citizen-centred renewables systems'. A future in which 'more *female faces supporting the mission and day-to-day operations of renewable energy cooperatives will become visible*', and in which the sector will go beyond the issue of representation, and be able to consider, evaluate, and '*benefit from the specific added value of women*'. An agreement was reached that it's worth striving for a future where a gender-balanced board of a renewable energy cooperative will be the reflection of its gender-balanced membership.

The Gender Power working group has the will to learn and implement tools that can help close the existing gender gap in the movement, and the REScoop.eu team is committed to strengthen and coordinate its members' activities to contribute to a gender-just energy transition.

First of all, the working group aims to bring to the fore the voices and faces of women in community energy and make visible their impact as well as the consequences of the (gender) inequality in the energy system. The group wishes to proceed with its mapping and linking exercise. The ambition is to map the (inter)national networks and initiatives working on the topic of Gender and Energy, and also connect to other relevant projects and issues, like for instance in relation to energy poverty, as well as other social and solidarity initiatives. In addition, the Gender Power working group has the ambition to prepare a prototype for an empowerment and mentoring programme. The programme should focus on communication but also on digitalisation and energy training, among other areas, for women. Furthermore, the idea is that the training will involve both girls and women and boys and men as mentors and mentees.

Discussion and Recommendations for the Way Forward

In the context of gender inequality, prevailing stereotypes and widespread injustice in our societies, the community energy initiatives presented in this contribution illustrate inspiring alternative cultures, structures and practices that have the potential to radically transform the energy system. The focus of investigation has been on the structures and strategies developed, the tools and instruments that citizen-driven, renewable energy cooperatives use to help craft a just energy transition. In fact, these community energy initiatives turn to their cooperative principles and especially their *Voluntary and open membership*, without gender, social, racial, political or religious discrimination, the *Democratic member control*, achieved through the one person, one vote principle, and their focus on *Education, training, and information* for the self-empowerment of people of all genders, so that the energy transition becomes truly inclusive and gender-just. One thing is certain: proceeding without recognising the existence of the problem will perpetuate the problem. The signs show that the awareness is there, and it keeps increasing, and so is the will for the cooperatives to improve their impact.

First of all, to attract new- and empower their existing female members, more and more cooperatives consider the importance of gender-just communication and visibility. Increasing the representation of women, as well as non-binary people in the media and cooperative social activities in general can help change perceptions of both men and women on what community energy is and who it is for and from. The energy system should undeniably take on board the needs and aspirations of everyone. And, just as 'we cannot create what we can't imagine', we cannot become what we cannot see. Role models are, thus, very important for enhancing women's empowerment.

Beyond communication, other measures that the initiatives engage in involve training, coaching, and mentoring for both men and women. Such training can help enable women to reach their full potential in the community energy field, also by shifting attitudes of men and women about what is and what is not possible. It is worth stressing that when organising workshops, those should not be organised to encourage women to be more like men, or with the mistaken assumption that women have plenty of free time. Training should be adjusted to the specific conditions that women are in and their specific needs and ambitions. Facilitation is also crucial. The initiatives recognise that effort is needed to create the space for women to speak up during meetings and other events so that people of all genders can claim their space in the energy transition.

When it comes to their hiring procedures or the selection of the members of their boards of directors, the initiatives do not adopt strict policies such as quotas but they opt for more flexible ways of reaching gender equality in their operation. Yet, as fewer opportunities occur in society for women and non-binary people, there are voices in the field that suggest that it may be worth prioritising getting them on board through different kinds of activities, even if they are different from what is traditionally considered 'competitive candidate'. Besides, quotas are a temporary corrective measure to a hidden male bias, which can be lifted when the problem cease to exist.

Furthermore, policy and advocacy is another important tool for mainstreaming gender justice. Adequate resources should be allocated to national or regional programmes and initiatives targeting the elimination of gender stereotypes, including the stereotypical gender roles in the energy transition and beyond. Inclusive and gender-responsive strategies and policies need to be adopted. Gender justice should be a cross cutting issue. Awareness-raising campaigns are central in helping change the broader conditions in which energy communities operate. All in all, engendering the energy transition requires action beyond the empowerment of women. A gender-just energy transition requires strategies and policies that recognise and pay attention to the existing social imbalances. Structural changes are necessary to transform the existing patterns of domination and (gender) discrimination in the energy field and consolidate institutions that enable an open space for women and men to flourish. Community energy initiatives illustrate how the democratic and cooperative principles that guide their operation already start to shape alternative practices and structures that have the potential to transform the energy sector contributing to a socially and environmentally just energy transition.

References

European Commission. (2020). A Union of Equality: Gender Equality Strategy. (2020–2025). https://ec.europa.eu/info/policies/justice-and-fundamental-rights/gender-equality/gender-equality-strategy_en

AEMENER. (2018). *Observatorio sobre el rol de la mujer en las empresas del sector energetico 2018*. https://aemener.es/pdf/AEMENER%20Informe%20Observatorio%202018.pdf

Allen, E., Lyons, H., & Stephens, J. C. (2019). Women's leadership in renewable transformation, energy justice and energy democracy: Redistributing power. *Energy Research & Social Science, 57*, 101233.

Angel, J. (2016). Towards energy democracy. In *Discussions and outcomes from an international workshop.*

Azocar, M. J., & Ferree, M. M. (2016). Engendering the sociology of expertise. *Sociology Compass, 10*(12), 1079–1089.

Becker, S., & Naumann, M. (2017). Energy democracy: Mapping the debate on energy alternatives. *Geography Compass, 11*(8), e12321.

Besley, T., Folke, O., Persson, T., & Rickne, J. (2017). Gender quotas and the crisis of the mediocre man: Theory and evidence from Sweden. *American Economic Review, 107*(8), 2204–2242.

Burke, M. J. (2018). Shared yet contested: Energy democracy counter-narratives. *Frontiers in Communication, 3*, 22.

Burke, M. J., & Stephens, J. C. (2017). Energy democracy: Goals and policy instruments for sociotechnical transitions. *Energy Research & Social Science, 33*, 35–48.

Cecelski, E., & Dutta, S. (2011). Mainstreaming gender in energy projects–a practical handbook. *ENERGIA.*

Clancy, J. S., Daskalova, V. I., Feenstra, M. H., & Franceschelli, N. (2017). Gender perspective on access to energy in the EU.

Cooperatives Europe. (2020, February 10). *Gender equality charter released.* https://coopseurope.coop/resources/news/gender-equality-charter-released

Directive 2013/33/EU of the European Parliament and of the COUNCIL of 26 June 2013 laying down standards for the reception of applicants for international protection (recast). Retrieved February 2, 2017, from http://eur-lex.europa.eu/legal-content/EN/TXT/PDF/?uri=CELEX:32013L0033&from=EN

Duwe, M., Velten, E. K., Evans, N., Freundt, M., Pestiaux, J., Martin, B., Vermeulen, P., & Vermeulen, P. (2019). *Planning for Net Zero: Assessing the draft national energy and climate plans as cited in WECF (2021) Why the European Green Deal needs Ecofeminism: Moving from gender-blind to gender-transformative environmental policies.*

Ely, R. J., & Thomas, D. A. (2020). Getting serious about diversity. *Harvard Business Review, 98*(6), 114–122.

Feenstra, M., & Özerol, G. (2021). Energy justice as a search light for gender-energy nexus: Towards a conceptual framework. *Renewable and Sustainable Energy Reviews, 138*, 110668.

Feldpausch-Parker, A. M., Endres, D., & Peterson, T. R. (2019). Editorial: A research agenda for energy democracy. *Frontiers in Communication, 4*(2019), 19.

Fraune, C. (2015). Gender matters: Women, renewable energy, and citizen participation in Germany. *Energy Research & Social Science, 7*, 55–65.

IEA. (2014). *Energy supply security.* Emergency Response of IEA Countries 2014. https://www.iea.org/publications/freepublications/publication/ENERGYSUPPLYSECURITY2014.pdf

IRENA. (2019). *Renewable energy: A gender perspective.* IRENA.

Jenkins, K., McCauley, D., Heffron, R., Stephan, H., & Rehner, R. (2016). Energy justice: A conceptual review. *Energy Research & Social Science, 11*, 174–182.

Johnson, O. W., Han, J. Y. C., Knight, A. L., Mortensen, S., Aung, M. T., Boyland, M., & Resurrección, B. P. (2020). Intersectionality and energy transitions: A review of gender, social equity and low-carbon energy. *Energy Research & Social Science, 70*, 101774.

Kaijser, A., & Kronsell, A. (2014). Climate change through the lens of intersectionality. *Environmental politics, 23*(3), 417–433.

Kooij, H. J., Oteman, M., Veenman, S., Sperling, K., Magnusson, D., Palm, J., & Hvelplund, F. (2018). Between grassroots and treetops: Community power and institutional dependence in the renewable energy sector in Denmark, Sweden and The Netherlands. *Energy Research & Social Science, 37*, 52–64.

Lieu, J., Sorman, A. H., Johnson, O. W., Virla, L. D., & Resurrección, B. P. (2020). Three sides to every story: Gender perspectives in energy transition pathways in Canada, Kenya and Spain. *Energy Research & Social Science, 68*, 101550.

Moodley, L., Holt, T., Leke, A., & Desvaux, G. (2016, August). *Women matter Africa.* McKinsey & Company. https://www.mckinsey.com/~/media/mckinsey/featured%20insights/Women%20matter/Women%20matter%20Africa/Women%20Matter%20Africa%20August%202016.ashx

Noland, M., Moran, T., & Kotschwar, B. (2016). *Is gender diversity profitable? Evidence from a global survey.* Peterson Institute for International Economics.

ÖGUT (Österreichische Gesellschaft für Umwelt und Technik). (2016). *Chancengleichheit von Frauen und Männern in der Energiebranche.* https://www.oegut.at/downloads/pdf/Endbericht_Chancengleichheit_Energiebranche.pdf?m= 1480502641

Pateman, C. (1985). *Women and democratic citizenship.* The Jefferson Memorial Lectures at the University of California. Unpublished manuscript.

Pearl-Martinez, R., & Stephens, J. C. (2016). Toward a gender diverse workforce in the renewable energy transition. *Sustainability: Science, Practice and Policy, 12*(1).

Proka, A. (2021). *Organising for power change: Transformative business models for the energy transition.* PhD Thesis, Erasmus University Rotterdam.

Radtke, J. (2016). *Bürgerenergie in Deutschland—Partizipation zwischen Gemeinwohl und Rendite.*

Ryan, S. E. (2014). Rethinking gender and identity in energy studies. *Energy Research & Social Science, 1*, 96–105.

Sovacool, B. K., & Dworkin, M. H. (2015). Energy justice: Conceptual insights and practical applications. *Applied Energy*.

Steinfort. (2018, September 12). *Ecofeminism: Fueling the journey to energy democracy*. https://www.tni.org/my/node/24366).

Stephens, J. C. (2019). Energy democracy: Redistributing power to the people through renewable transformation. *Environment: Science and Policy for Sustainable Development, 61*(2), 4–13.

Sweeny, S. (2014). "Working toward energy democracy." In *State of the World 2014: Governing for Sustainability* (Ed.), (pp. 215–227). Worldwatch Institute. Washington, DC: Island Press.

Szulecki, K., & Overland, I. (2020). Energy democracy as a process, an outcome and a goal: A conceptual review. *Energy Research & Social Science, 69*, 101768.

UN. (2015). The 17 Goals. https://sdgs.un.org/goals

UNFCCC, Paris Agreement, United Nations Framework Convention on Climate Change. (2015). https://unfccc.int/process-and-meetings/the-paris-agreement/the-paris-agreement

Van Veelen, B., & Van Der Horst, D. (2018). What is energy democracy? Connecting social science energy research and political theory. *Energy Research & Social Science, 46*, 19–28.

Vansintjan, D. (2015). *The energy transition to energy democracy. Power to the people*. Final results oriented report of the REScoop 20-20-20 Intelligent Energy Europe project.

Vasilakis, A., Vrettos, C., Kitsikopoulos, D., Kontolati, A., Koukoufikis, G., Margosi, M., & Palaiogiannis, F. (2020). *Mapping of energy communities in Greece*. https://www.greenpeace.org/static/planet4-greece-stateless/184045bd-mapping_of_energy_communities_v1.2.pdf

Verbong, G., & Loorbach, D. (Eds.). (2012). *Governing the energy transition: Reality, illusion or necessity?* Routledge.

WECF. (2021). *Why the European green deal needs ecofeminism: Moving from gender-blind to gender-transformative environmental policies*.

Winther, T., Ulsrud, K., & Saini, A. (2018). Solar powered electricity access: Implications for women's empowerment in rural Kenya. *Energy Research & Social Science, 44*, 61–74.

WISE (Women in Solar Energy). (2017). *Women employment in urban public sector*.

WWEA. (2021). Policy Paper Series (PP-01-21-EN) Women in Community Energy Through Openness to Diversity July 2021.

Yildiz, Ö., Rommel, J., Debor, S., Holstenkamp, L., Mey, F., Müller, J. R., Radtke, J., & Rognli, J. (2015). Renewable energy cooperatives as gatekeepers or facilitators? Recent developments in Germany and a multidisciplinary research agenda. *Energy Research and Social Science, 6*, 59–73.

PART IV

Women at the Intersection of Gender and Energy Equality

CHAPTER 10

Women in Energy Communities: An Intersectional Analysis of Their Participation

Aurore Dudka

Introduction: Towards an Energy Democracy: The Role of Energy Communities

According to the European Environment Agency (EEA), the European Union is already and will be strongly impacted by climate change, provoking an increase in the society's vulnerability. One of the main factors contributing to this vulnerability is the energy sector, provocating for instance air pollution responsible each year for 500,000 premature deaths (European Environment Agency, 2018). Changing how energy is produced and consumed has been deeply considered as an important objective point of the European Union roadmap. Especially, one core aspect is to support the development of renewable energies, intending to have in 2030, 40% of the energy coming from renewable sources and achieving the carbon neutrality by 2050. In this context, the energy market is

A. Dudka (✉)
University of Trento, Trento, Italy
e-mail: aurore.dudka@unitn.it

N. Rocha Lawton, C. Forson (eds.), *Women and the Energy Sector*,
https://doi.org/10.1007/978-3-031-43091-6_10

undergoing a profond transformation, shifting towards a decentralized energy production and consumption model. But one key problem is the fact that this reconfiguration has been mostly apprehended under a technological view. However, going towards polycentric governance, characterized by multiple decision-making centres and especially small and local initiatives in a bottom-up logic, implies not only technological transformations but also social and political changes (Bauwens, 2017). This involves reshaping the link between society and energy, an issue that, for the moment, remains understudied.

Indeed, renewable energies have opened the possibility for citizens individually or collectively to become active players in the energy market, producing and consuming their own energy. Especially, pushing by the Directive (EU) 2019/944 and 2018/2001, a new form of collective action has been flourishing: energy communities. By implying a shift in the ownership of energy previously held by private companies or national states, they have been considered as the most important institutional innovation of energetic transition (Yildiz et al., 2015). Energy communities are defined as: *projects where communities (of place or interest) exhibit a high degree of ownership and control of the energy project, as well as benefiting collectively from the outcomes* (either energy-saving or revenue-generation) (Seyfang et al., 2013). Oft, they are also framed as a form of collective action initiatives (Gregg et al., 2020), where local communities, possibily in collaboration with local state and private actors, adopt a democratic management, based on the principle of *one people one voice* to build, and manage their renewable energy installations bringing also broader social and environemental benefits for the whole population (European Parliament, 2018; Lowitzsch & Hanke, 2019).

This model has been a rising success since civil society, has been very responsive and dynamic to engage with (Caramizaru & Uihlein, 2020). To give a short overview, actually, there are more than 10.000 energy communities initiatives, in the European Union, with most of them taking the form of cooperatives (Wierling et al., 2023). While it is anticipated that by 2050, these initiatives will have the potential to reach up to 37% of European citizens (Kampman et al., 2016). However, their diffusion has been more concentrated among northern countries, some of which are already at an incumbent stage, while southern countries have lagged behind (Candelise & Ruggieri, 2020). In particular, Flanders is among the dynamic regions in the development of energy communities. These organisations already supply more than 2% of electricity, with one of the first citizens' energy cooperative, Ecopower, created in 1991. Led at the beginning by environmental activists, Ecopower has been strongly growing. Today, it has more than 67,000 members and is one of the most important energy communities in the European Union.

The business model of Ecopower consists of directly engaging citizens into energetic transition, who become direct members of the cooperative by subscribing a limited number of shares, for €250 each, fixed for six years to avoid speculation. Profits are returned to the members as dividends, paid up to a maximum of 6% (as prescribed by the Belgian law) or reinvested in the cooperative. Moreover, since 2003, Ecopower has been allowed to become an energy supplier. Regarding their renewable energy park, all their installations have been situated in Flanders, where Ecopower has developed wind turbines, hydropower, cogeneration , and PV installations. As energy is locally produced and surplus is not retained, the cooperative can offer electricity tariffs lower than the market price. For instance, Ecopower's members saved up to €500 compared to other suppliers last year during the energy crisis.

Beyond the economic and financial aspects, Ecopower has also put a strong emphasis on the democratic and fair aspects of this model, looking for an energy democracy characterized by civic ownership, popular sovereignty, and participatory governance (Szulecki, 2018; Szulecki & Overland, 2020). The aim of Ecopower is that *each cooperator has a say in the company*. And indeed, each member has equal voting rights, whatever their initial investment, in the cooperative, organized at each level democratically. Moreover, Ecopower regularly organizes meetings like the Energie café, and each year all members are invited to participate to the general assembly to approve cooperative decisions and orientations. Ecopower is also an active member of the european federation of citizen energy cooperatives: Rescoop.eu, heading in Bruxelles, which is a network of 1900 European energy cooperatives gathering 1,250,000 citizens, emphasizing the superiority of environmental and social values over the profit in the energy transition (Rescoop, 2020).

The Challenge of Inclusivity for Energy Communities

While energy communities have already been successful in implanting an energy democracy by strongly focusing on equitable governance, they still lack emphasis on the recognitional aspect of their democratic process, focusing on its inclusive character. Energy communities bring the intuitive idea of an energy citizen (Devine-Wright, 2007), with equitable rights and responsibilities to shape energetic transition. However, the risk is to fall into the local trap (Purcell, 2006; Taylor Aiken, 2015), meaning to postulate that inclusivity will appear per se in these projects and 'to assume rather than demonstrate that community projects are more democratic or just' (Van Veelen, 2018). Indeed, Ecopower, like all cooperatives belonging to Rescoop's network, subscribed to the principles of *equality, equity, and solidarity, and 'belief in the ethical values of honesty, openness, social responsibility and caring for others*' (Rescoop, 2020). Despite this, few concrete actions has been taken and in this regard, Ecopower is not an isolated case. A recent study shows that the majority of European energy communities does not adopt any kind of actions to support underrepresented groups, such as women, in these organizations, and only one-third of them talked about this issue (Hanke et al., 2021).

This lack of emphasis on the recognitional tenet of energy justice is a serious problem (Cleaver & De Koning, 2015). Without rules regarding the fact that access should be equally granted to all citizens, the representativity and thus the strength of energy democracy could be largely undermined (Jenkins, 2018; Sovacool & Dworkin, 2015; Sovacool et al., 2017). Collective actions are characterized by propriety rights and social norms that regulate peoples' inclusion (Ostrom, 1992). These rules don't apply to women and men in the same way, especially in the energy field, that can be more discriminant for women (Agarwal, 2001; Meier zu Selhausen, 2016). For example, participating in an energy community requires financial resources and, on average, in the European Union, women earn much less than men, around 20%, resulting in a gap of €560 per month (EIGE, 2020). Investing is also considered more as a male business and women are usually more risk-averse (Charness & Gneezy, 2007). Joining a community and being actively involved in it requires time, which is scarce for women, mainly if mothers (Fraune, 2015; Łapniewska, 2019).

Relevant also, is the fact that the energy field has been very discriminatory, especially for women, as it has traditionally been considered a male domain. To give a quick overview, in the fossil fuel energy sector, women represent 6% of the technical staff and 1% of the top management (UN, 2016). The situation is quite better in renewable energy sectors, where women represent on average 32% of the workforce but are still largely employed on 'feminine' tasks, e.g., administrative function or marketing (Pearl-Martinez & Stephens, 2016). Moreover, while energy communities are presented as not requiring a specific knowledge, still they belong to a technological field, where women can feel less suited due to the presence of strong gender stereotypes. Female graduate students in science, technology, engineering and mathematics (STEM) are less than 30% and only 19% in engineering (UNIDO, 2014; Wang & Degol, 2017). Moreover, energy has been considered as an expert domain where people have been largely disempowered on energetic issues, leading also to exclude 42 million people according Eurostat in 2023, mostly women, living currently in energy poverty in Europe (Clancy, 2019; Martiskainen et al., 2018).

For these reasons, energy communities are likely to produce internal and external forms of exclusion for women (Van Veelen, 2018), aligning with the conclusions of other scholars who have emphasized the risk that energy communities may foster a form of democracy dominated by well-resourced, well-meaning middle-class men (Johnson & Hall, 2014). Indeed, the development of renewable installations technologies even when adoting a bottom-up approach led to create new patterns of inequalities (Tjørring, 2016; Bosch & Schmidt, 2020; Grossmann & Creamer, 2017). It is the case for example between those having the financial resources to invest in an energy communitiy and can benefit from their economic and social outcomes like low energy pricesor empowerment on energy issues and the others who face barriers such as women to access to this opportunity (Hanke & Lowitzsch, 2020; Martiskainen et al., 2018). This is why it is important to ask, at a moment in which energy communities are presented as an alternative model able to bring more fairness in energy transition: is gender inequality produced and reproduced by energy communities?

Towards an Analysis of the Intersectionality of Gender Inequalities

An Overview of the Women's Participation in EC

In the energy world, and especially in energy communities, sex-disaggregated data is still largely missing, showing that in a world dominated by numbers, what is not counted does not count (Clancy & Roehr, 2003; Clancy et al., 2017). Joining an energy community has been mostly studied under the lens of the willingness to participate, where motivations have been paramount (Bauwens, 2016), but with little consideration of the fact that social logic and constraints can undermine the potential and the capabilities of citizens to participate (Coy et al., 2021). The focus on women's participation in these organizations is anecdotal. The attention on participants' heterogeneity put by scholars is more an accidental focus, aiming to describe rather than analyse gender inequalities. More generally, looking at women into collective action, the term gender has been used for the first time only in 2005 by Ostrom, and it is generally not considered as a variable of interest when studying these initiatives (Łapniewska, 2016; Ostrom, 2005).

It is to fill this research gap that this survey, realized in collaboration with the project ASSET Horizon 2020 has been launched in March 2021. It collects the answers of 5387 members of Ecopower, representing 17% of those who read the email message, of whom 5114 were exploitable. This study gives an overview of members' profiles in the cooperative (LaRose & Tsai, 2014). The first important results regard gender representation across members in these initiatives. As we can see, women are largely underrepresented in Ecopower. While men represent around 79% of members, women are only 21%. This gender gap confirms previous results found in other countries, e.g., Germany where 80% of the 2000 energy communities' members are men (Yildiz et al., 2015).

Towards an Intersectional Analysis of Women's Decision to Join the Cooperative

Participating in the cooperative is mostly, but not only, gender-related. The problem of representation in Ecopower regards also other characteristics, such as family income, educational level, occupation, and age. For example, in Belgium, 37.6% of inhabitants have a tertiary education level while they represent 61% of cooperative participants. Furthermore, people over 50 represent 39% of the population but 62% of Ecopower's members. High incomes are also underrepresented with the median of personal incomes in the cooperative situating between €3500 and €4000 per month. In contrast, people having a monthly personal income below €2000 represents only 11% of members. This is why, as a multidimensional phenomenon, inequalities crossing energy communities, Ecopower should be considered as a whole. Gender inequalities are complex and cannot be summarized to a binary approach, being a man or a woman. They intersect with factors such as income level, marital status, and age. To understand women's barriers in their participation in energy communities, women have to be considered not as a unique entity but in their diversity. If some studies have already warned on the underrepresentation of women in energy communities (Fraune 2015), this analysis could be only meaningful by understanding how different social markers can marginalize or privilege some women compared to others. But for the moment, this kind of approach is still strongly missing in the literature (Johnson et al., 2020; Søraa et al., 2020) since no study adopted an intersectional perspective to state how inequalities can overlap in energy communities and thus self-reinforce a cycle of privileges regarding economic, environmental, and social benefits (Lennon et al., 2020; Mundaca et al., 2018). By consequence, the next section aims to analyse this previous descriptive data, not considering women as a monolithic category but going deeper, using a quantitative approach with sex-disaggregated data, to study gender under the lens of intersectionality.

Joining the Cooperative

When looking at who is joining the cooperative, the descriptive results showed that women are largely underrepresented. Here, we aim to analyse what can foster or undermine their participation. One important dimension is that, despite the fact that women are at the core of energetic use by being mostly responsible for reproductive work at home, they tend to have less power in decision-making than men regarding energy issues and that is one of the reason why join an energy community (Clancy, 2002; Łapniewska, 2019; Winther et al., 2020).

When answering the question 'In your family who chooses to join the cooperative?' men are much more likely to take this decision when they are in a couple. Indeed, for a male, the odds of being in a couple and joining the cooperative are 3.06 times larger than the odds for a female being in a couple. When comparing singles, women are more likely to join the cooperative than men. This means that one first barrier for women to join the cooperative is their marital status. In most cases, men are more likely to become the unique member representing the whole family. This has a strong consequence since men are going to be in the household those strongly implicated in energy issues, like participating in formations or assembly, and voting. Indeed, when in couple, in 80% of the cases, men are the one choosing to represent the family when joining the cooperative, meaning that energy decision is largely conditioned by gender, being a male domain, even in energy communities promoting fairness and inclusivity.

Although, by adopting an intersectional perspective and including age, occupation, personal income, field, and level of study as control variables in the model, the probability for men to join the cooperative is weakened, as the odds ratio drop from 3.06 to 1.67. This underlines the fact that inside the couple, power relations regarding energy issues not only are a matter of gender but also depend on other factors. For example, income is a very important dimension impacting the probability to represent the family in the cooperative. The odds of being in a couple and representing the family in the energy communities are 101 times higher for people earning more than €5000 compared to those earning less than €1000 per month. Age is also positively correlated with being the one joining the cooperative. People over 60 years old are more likely to represent the family. Finally, surprisingly, having a degree in STEM compared to having a degree in humanities is not associated with highest probability to join the cooperative, despite the fact that energy being related to the STEM, they could have felt more entitled than others.

And indeed, looking more deeply to the interactions between gender and other control variables, interesting results emerge:

Occupational effect: Males are less impacted by variations in their social position, while for women, things are more complex. Women employed in higher administrative functions and unskilled jobs have a high probability to be those joining the cooperative when they are in couple, respectively 79% and 82%, while the probability is much lower for female semi-skilled workers (29%) or for those with clerical functions. As regards men, the probability is on average higher compared to women. Those in higher administrative functions have a probability of 91% of joining the cooperative, unskilled workers 84% of probability, while clerical workers and farmworkers 79%. This means that for men, regardless their

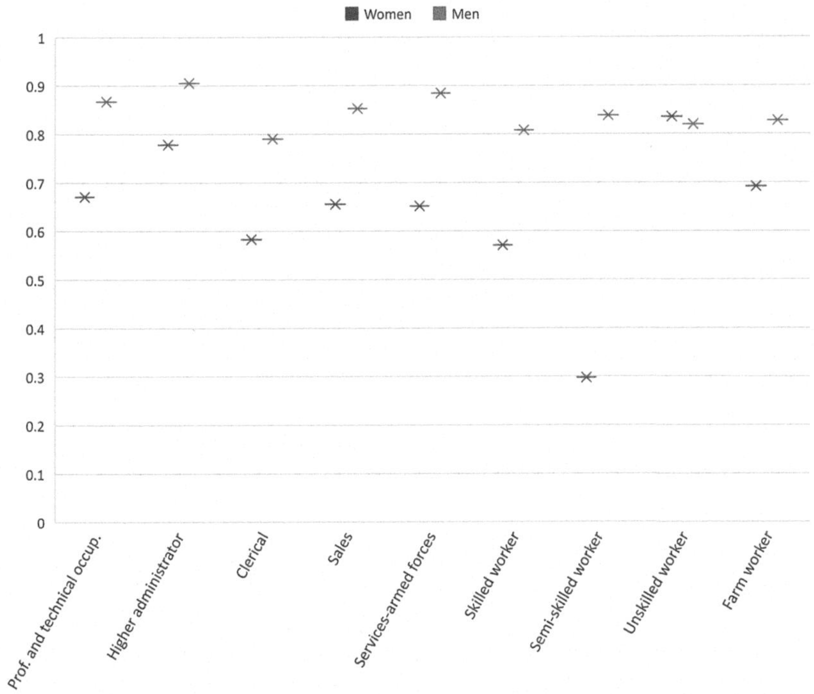

Fig. 10.1 Interaction between gender and occupation for joining the cooperative

occupation, are almost always sure to be the one representing the family in the cooperative and the main member, while for women, occupation explains in part the variation in the participation rate (Fig. 10.1).

STEM **effect**: While previously we find that having studied STEM does not affect participation in the cooperative, the interaction effect between gender and field of study reveals that STEM studies impact women's probabilities of joining the cooperative. The predictive margins show that a woman being in a couple with at least a degree in STEM has a 71.91% chance of choosing to join the cooperative, while those having at least a degree in humanities has a 63.21% chance to be the representant. As regards men, those having a degree or above in STEM have an 86.89% chance of being in a couple, while those who studied in the humanities field an 84.86%.

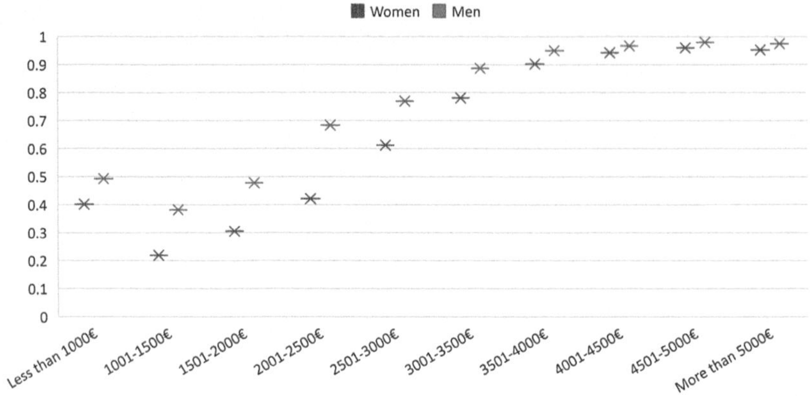

Fig. 10.2 Interaction between gender and income for joining the cooperative

Income **effect**: Women, having low incomes, are more much more likely to be single, when joining the cooperative. Moreover, if they are in a couple, they have almost no probability to be the one joining the cooperative. Indeed, the predictive margins show a threshold effect. Women earning less than €2500 are much more likely to be single to those winning more than €2500, while for a man, earning less this threshold effect is much lower and equal to €1500. This means that for women earning a minimum salary, who are largely underrepresented in the cooperative, deciding to join the cooperative is very difficult, especially when they are in couple. For higher incomes, the gender differences disappear (Fig. 10.2).

This analysis shows that, differently from men, women's decision to join the cooperative depends on their socio-economic characteristics. While, regardless of gender, poor people have difficulties in joining the cooperative, the effect is stronger for women. Moreover, the presence of poor women is residual, since the one having a monthly income inferior to €1000 per month represents only 1.84% of women joining the cooperative.

Children effect: When including children, the odds ratio of joining the cooperative for men tends to rise from 1.67 to 1.93, meaning that when women become mothers the gender roles division is more unbalanced, and men are in charge of the decisions regarding energetic issues.

Having a child reduces the possibility for women to join the cooperative even though being a mother tends to undermine women's careers and

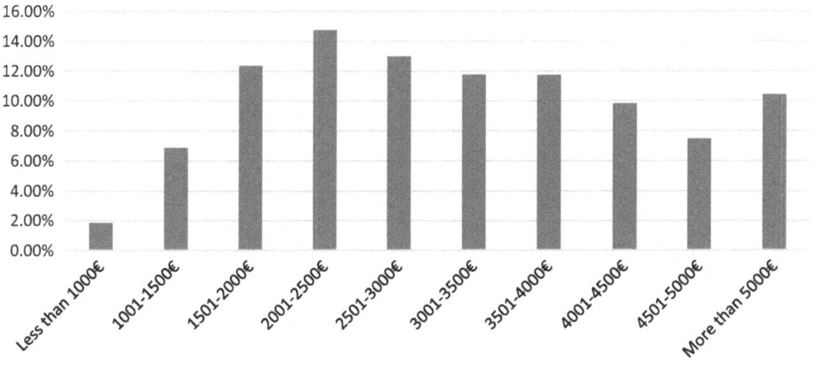

Fig. 10.3 Repartition of women by income in the cooperative

increase their charge of domestic works, putting them at the core of energetic use. Finally, some social positions are more legitimate to justify the fact that women engage with energy transition and the social construction of women representation to fit with energy transition (Fig. 10.3).

To conclude, this first analysis shows that while descriptive results show that women are largely underrepresented in energy communities, further investigation reveals that power relations and inequalities can in some cases be amplificated. In this way, inequalities are not only reproduced but even accentuated by energy communities.

Investing

A second dimension related to the decision to join the cooperative is the investment in these organizations. Indeed, joining Ecopower requires a minimum investment, consisting in buying at least one share at the cost of €250. Since investing is a gendered decision, it is interesting to see how men and women differ in their investment choices, especially if we adopt an intersectional point of view. Compared to the previous model, the seniority in the cooperative has been introduced since it has been shown that the motivations to invest by new members in Ecopower differ from the formers, being more driven by instrumental motivations (Bauwens, 2019), with the risk to be more stereotyped as a male business.

Preponderance of gender effect: In the cooperative, the majority of Ecopower's members were very prudent in their investments. About 64% of them invested the minimum amount required. But, contrary to what

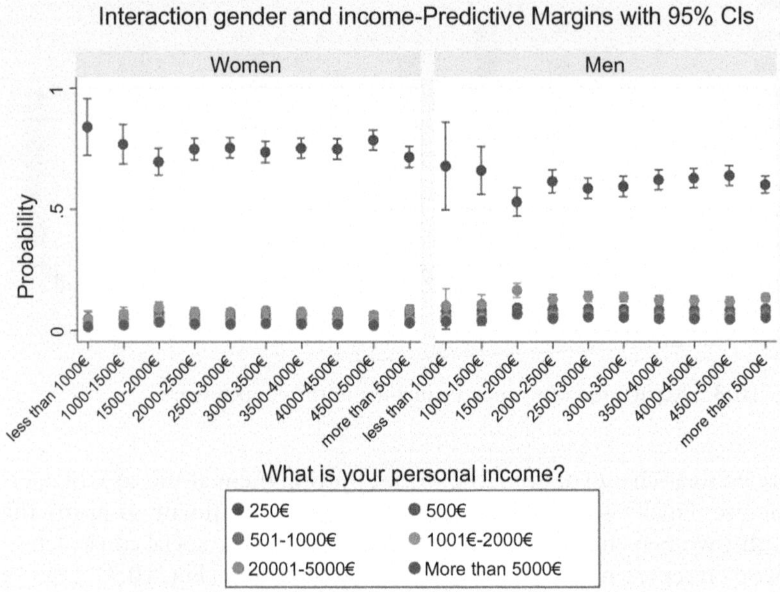

Fig. 10.4 Interaction between gender and income for investing in the cooperative

could be expected, the decision on the amount to invest does not depend on income and type of occupation. Gender is, instead, one of the main determinants. Indeed, other things equal, the odds of men of high investment versus the combined middle and low investments is 1.63 times higher than that of women. As regards gender differences, on an all-other thing equal basis, women are more likely than men to invest the minimum (72% as opposed to 62%), about as likely to invest between €500 and €5000 (17% to 21%) and less likely to invest the maximum of €5000 (12% versus 18%) (Fig. 10.4).

Age **effect**: Age is also positively correlated with the amount invested, especially for people up to 50 who tend to invest more. This is especially true for women over 60, of whom around 65% invested the minimum amount, compared to 85% of women between 18 and 30 years old. As regards men, they have around 50% of probability to invest the minimum, 79% if they are 18–30 years old.

Seniority in the cooperative: A negative relationship is depicted, meaning that previous cohorts of investors tended to invest less than those

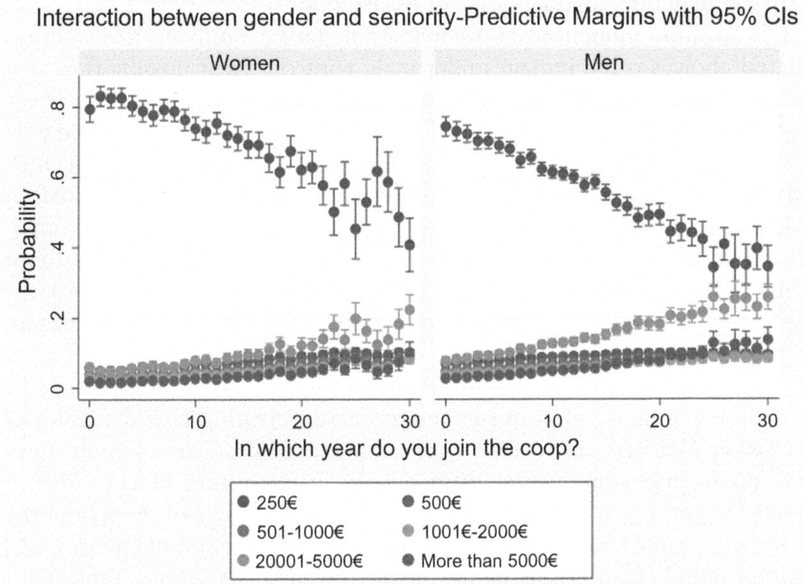

Fig. 10.5 Interaction between gender and seniority for investing in the cooperative

currently joining the cooperative. Once again, a gender difference is observed, with women more cautious to invest (Fig. 10.5).

When women invest in the cooperative, they capitalize always less than men and this is true regardless of their income, occupation, field, level of study, age, and seniority. This means that psychological barriers are related to gender. Some factors as age can reinforce this relation, especially for young women.

Conclusion

Consistent with the results of other studies (Yildiz et al., 2015; Fraune, 2015), the case of Ecopower show that in this kind of initiatives the participants are predominantly male. Since also here around 8 members out of 10 are men. However, what this study delves into is the fact that this problem is not solely related to gender but also intersectional, with many other factors intersecting with and, in some instances, significantly reinforcing gender inequalities. The decision for women to join a cooperative is strongly influenced by their marital status, signifying that energy-related choices often remain under male control. Their possibility is also determined by their economic income. Since women tend to have fewer resources and are more risk-averse, a concrete obstacle is the fact that participating in the cooperative requires financial participation, in a technical sector still considered as an incumbent and therefore a higher risk of losing their investment. Moreover, discriminatory effects against women are not only economic but also social. Some women, for example those from higher social occupation, may feel they fit in more, while others, lacking of trust or having the perception of being less legitimate might hesitate to participate in such initiatives.

This is why energy communities could led to the reinforcement of women's marginalization in the energy sector, creating new dynamics of exclusion. With collective action generating an atmosphere of conformity and dissuading some social groups from participating (Little, 2002). While, according to the IRENA, women constitute 32% of the workforce in the renewable energy industries, they make up only 20% of energy community members, and this figure drops to a mere 2% among those with fewer resources. Energy communities further reinforce traditional gender roles, as women with children, despite being the primary users of domestic energy, those who would benefit the most from these initiatives, and those with a legitimate voice to speak on these matters, face the most limited opportunities to participate.

As a consequence, although energy communities endorse the idea of democracy and justice, achieving equal participation of men and women in collective energy initiatives remains a complex challenge (Łapniewska, 2019). Indeed, energy communities are currently mainly focused, even if democratic and based on social values, on investing and managing renewable energy installations, framed mostly as a technological issue. However, this statement does not want to condemn the transformative potential of energy communities. The role of these organizations to bring more fairness is more than promising (Bauwens & Eyre, 2017; Mundaca et al., 2018) while their commitment to enhancing women's inclusion should be acknwoledged. Rather, it highlights the necessity for strong institutions within the framework of energy democracy, ensuring equal representation of all citizens.

Indeed, the fact that little attention has been put on women's issues can be easily understandable since to impose their model energy communities have been fighting against incredible obstacles to be recognized as a credible alternative to the current energy system. Energy communities must have proved their economical sustainability, their positive environmental impact, and their capacity to deal with technical problems like connecting to the electricity network, overcoming juridical issues, and fighting against disinformation and opposition especially regarding wind power (Martinez, 2020; Jobert et al., 2007; European Parliament, 2018). As the model of energy communities is currently becoming more established, it is a critical juncture, and it is essential to prioritize the implementation of robust policies to promote inclusive participation (EU, 2020).

In this sense, some countries have already been pioneers by proposing innovative measures to promote women's participation in renewable energy projects. For example, a solution to overcome the fact that marital status can hinder women's participation was adopted in Norway who included a double participation for people joining energy communities: one share for the man and one share for women (Łapniewska, 2019). But other measures can be imagined. Incentives could be stronger by imagining that one part of the benefits-like micro-donations made by the energy cooperative Enercoop in France-finances social programmes, for example, attributing shares to a specific underrepresented population such as women for free. The launch of the new platform Equity by the European Union and the task force established by Rescoop.eu offer valuable opportunities to establish a benchmark of best practices, further aiding in the development of stronger inclusivity.

However, since inequalities are multidimensional, it would be too easy to think that this kind of policy could escape from considering other dimensions. Barriers to including women in the energy world are numerous and policies have to take strong measures to empower those generally excluded from the public sphere and those who feel less legitimate to deal with an expert domain like energy. This would pass by focusing on energy communities 'process in terms of active participation but also outcomes by fostering the voice of those, who generally do not have one. Fostering a more fair and democratized engagement requires questioning deeper who are and what are the aspirations of various energy users and how they will and can interact with the energy system (Søraa et al., 2020). In this regards, projects like Horizon 2020's EmpowerMed (Empowering Women to Take Action against Energy Poverty, 2019–2023), which empowered 10,000 vulnerable individuals in managing their energy consumption, and the European Parliament's resolution of July 5, 2022, addressing women's poverty, are highly encouraging. They demonstrate the European Union's commitment to taking concrete actions in the right direction, especially in the context of intersectionality: "*to protect women living in energy poverty by providing a timely and coordinated response to address the long-term impact of the energy crisis; highlights that access to affordable utilities must be guaranteed to low-income households, and in particular older women and single mothers*" (2021/2170(INI)).

Finally, another important aspect to highlight when addressing gender inequalities in energy communities through the concept of intersectionality is that it goes beyond merely advocating for women's inclusion and extends to combating poverty and other intersecting factors of exclusion. Therefore, promoting women's inclusion is a commitment to addressing multiple social challenges simultaneously. Only developing an intersectional consciousness could bring a real change towards environmental sustainability, requiring a fair representation of all society members (Terriquez et al., 2018).

References

Agarwal, B. (2001). Participatory exclusions, community forestry, and gender: An analysis for South Asia and a conceptual framework. *World Development, 29*(10), 1623–1648. https://doi.org/10.1016/S0305-750X(01)00066-3

Bauwens, T. (2017). Designing institutions for collective energy action: The roles of renewable energy cooperatives in a polycentric low-carbon transition.

Bauwens, T. (2016). Explaining the diversity of motivations behind community renewable energy. *Energy Policy, 93*(June), 278–290. https://doi.org/10.1016/j.enpol.2016.03.017

Bauwens, T. (2019). Analyzing the determinants of the size of investments by community renewable energy members: Findings and policy implications from Flanders. *Energy Policy, 129*(June), 841–852. https://doi.org/10.1016/j.enpol.2019.02.067

Bauwens, T., & Eyre, N. (2017). Exploring the links between community-based governance and sustainable energy use: Quantitative evidence from Flanders. *Ecological Economics, 137*(July), 163–172. https://doi.org/10.1016/j.ecolecon.2017.03.006

Bosch, S., & Matthias, S. (2020). Wonderland of Technology? How Energy Landscapes Reveal Inequalities and Injustices of the German Energiewende. *Energy Research & Social Science, 70*(December): 101733. https://doi.org/10.1016/j.erss.2020.101733

Candelise, C., & Ruggieri, G. (2020). Status and evolution of the community energy sector in Italy. *Energies, 13*(8), 1888. https://doi.org/10.3390/en13081888

Caramizaru, A., & Andreas, U. (2020). Energy Communities: An Overview of Energy and Social Innovation.

Charness, & Gneezy. (2007). Strong evidence for gender differences in investment. *UC Santa Barbara: Department of Economics*. https://escholarship.org/uc/item/428481s8

Clancy, J. (2002). Blowing the smoke out of the kitchen: Core issues in household energy and gender.

Clancy, J., Feenstra, M., & Daskalova, V. (2017). Gender perspective on access to energy in the EU. https://doi.org/10.13140/RG.2.2.14577.02406.

Clancy, J. (2019). *Women, gender equality and the energy transition in the EU*. European Union.

Clancy, J., & Roehr, U. (2003). Gender and energy: Is there a northern perspective? *Energy for Sustainable Development, 7*(3), 44–49. https://doi.org/10.1016/S0973-0826(08)60364-6

Cleaver, & De Koning. (2015). Furthering critical institutionalism. *International Journal of the Commons, 9*(1), 1. https://doi.org/10.18352/ijc.605

Coy, D., Malekpour, S., Saeri, A. K., & Dargaville, R. (2021). Rethinking community empowerment in the energy transformation: A critical review of the definitions, drivers and outcomes. *Energy Research & Social Science, 72*(February), 101871. https://doi.org/10.1016/j.erss.2020.101871

Devine-Wright, P. (2007). Energy citizenship: Psychological aspects of evolution in sustainable energy technologies. In J. Murphy (Ed.), *Governing Technology for Sustainability* (pp. 63–86). Earthscan.

EIGE. (2020). *Gender equality index, 2020*. European Institute for Gender Equality edition.

EU. (2020). *Towards an inclusive energy transition in the European union: Confronting energy poverty amidst a global crisis*. LU: Publications Office. https://data.europa.eu/doi/10.2833/103649

European Parliament. (2018). Directive (EU) 2018/2001 of the European parliament and of the council of 11 December 2018 on the promotion of the use of energy from renewable sources.

Fraune, C. (2015). Gender Matters: Women, Renewable Energy, and Citizen Participation in Germany. *Energy Research & Social Science, 7*(May), 55–65. https://doi.org/10.1016/j.erss.2015.02.005

Gregg, J. S., Nyborg, S., Hansen, M., Schwanitz, V. J., Wierling, A., Zeiss, J. P., Delvaux, S., et al. (2020). Collective action and social innovation in the energy sector: A mobilization model perspective. *Energies, 13*(3), 651. https://doi.org/10.3390/en13030651

Grossmann, M., & Creamer, E. (2017). Assessing diversity and inclusivity within the transition movement: An urban case study. *Environmental Politics, 26*(1), 161–182. https://doi.org/10.1080/09644016.2016.1232522

Hanke, F., Guyet, R., & Feenstra, M. (2021). Do renewable energy communities deliver energy justice? Exploring insights from 71 European cases. *Energy Research & Social Science, 80*(October), 102244. https://doi.org/10.1016/j.erss.2021.102244

Hanke, F., & Lowitzsch, J. (2020). Empowering vulnerable consumers to join renewable energy communities—Towards an inclusive Design of the Clean Energy Package. *Energies, 13*(7), 1615. https://doi.org/10.3390/en13071615

Jenkins. (2018). Energy justice, energy democracy, and sustainability: Normative approaches to the consumer ownership of renewables. In *Energy transition: Financing consumer co-ownership in renewables*. Springer.

Jobert, A., Laborgne, P., & Mimler, S. (2007). Local acceptance of wind energy: Factors of success identified in French and German case studies. *Energy Policy, 35*(5), 2751–2760. https://doi.org/10.1016/j.enpol.2006.12.005

Johnson, O. W., Han, J. Y.-C., Knight, A.-L., Mortensen, S., Aung, M. T., Boyland, M., & Resurrección, B. P. (2020). Intersectionality and energy transitions: A review of gender, social equity and low-carbon energy. *Energy Research & Social Science, 70*(December), 101774. https://doi.org/10.1016/j.erss.2020.101774

Johnson, V., & Hall, S. (2014). Community energy and equity: The distributional implications of a transition to a decentralised electricity system. *People, Place and Policy Online, 8*(3), 149–167. https://doi.org/10.3351/ppp.0008.0003.0002

Kampman, Blommerde, & Afman. (2016). *The potential of energy citizens in the European Union*. CE Delft.

Łapniewska, Z. (2016). Reading Elinor Ostrom through a gender perspective. *Feminist Economics, 22*(4), 129–151. https://doi.org/10.1080/13545701.2016.1171376

Łapniewska, Z. (2019). Energy, equality and sustainability? European electricity cooperatives from a gender perspective. *Energy Research & Social Science, 57*(November), 101247. https://doi.org/10.1016/j.erss.2019.101247

LaRose, R., & Tsai, H.-y. S. (2014). Completion rates and non-response error in online surveys: Comparing sweepstakes and pre-paid cash incentives in studies of online behavior. *Computers in Human Behavior, 34*(May), 110–119. https://doi.org/10.1016/j.chb.2014.01.017

Lennon, B., Dunphy, N., Gaffney, C., Revez, A., Mullally, G., & O'Connor, P. (2020). Citizen or consumer? Reconsidering energy citizenship. *Journal of Environmental Policy & Planning, 22*(2), 184–197. https://doi.org/10.1080/1523908X.2019.1680277

Little, A. (2002). The politics of community: Theory & Practice. *Edinburgh University Press*. https://doi.org/10.1057/palgrave.cpt.9300066

Lowitzsch, J., & Hanke, F. (2019). Renewable energy cooperatives. In J. Lowitzsch (Ed.), *Energy transition* (pp. 139–162). Springer International Publishing. https://doi.org/10.1007/978-3-319-93518-8_7

Martinez, N. (2020). Resisting renewables: The energy epistemics of social opposition in Mexico. *Energy Research & Social Science, 70*(December), 101632. https://doi.org/10.1016/j.erss.2020.101632

Martiskainen, M., Heiskanen, E., & Speciale, G. (2018). Community energy initiatives to alleviate fuel poverty: The material politics of energy Cafés. *Local Environment, 23*(1), 20–35. https://doi.org/10.1080/13549839.2017.1382459

Meier zu Selhausen, F. (2016). What determines Women's participation in collective action? Evidence from a Western Ugandan coffee cooperative. *Feminist Economics, 22*(1), 130–157. https://doi.org/10.1080/13545701.2015.1088960

Mundaca, L., Busch, H., & Schwer, S. (2018). "Successful" low-carbon energy transitions at the community level? An energy justice perspective. *Applied Energy, 218*(May), 292–303. https://doi.org/10.1016/j.apenergy.2018.02.146

Ostrom, E. (1992). Governing the commons: The evolution of institutions for collective action. *32 Nat. Resources J., 415.*

Ostrom, E. (2005, September 21–24). Developing a method for analyzing institutional change. In *"Who should do what in environmental governance: Institutions and constraints" symposium.* Porto Conte Ricerche Centre in Tramariglio.

Pearl-Martinez, R., & Stephens, J. C. (2016). Toward a gender diverse workforce in the renewable energy transition. *Sustainability: Science, Practice and Policy, 12*(1), 8–15. https://doi.org/10.1080/15487733.2016.11908149

Purcell, M. (2006). Urban democracy and the local trap. *Urban Studies, 43*(11), 1921–1941. https://doi.org/10.1080/00420980600897826

Rescoop. (2020). *Charter REScoop.*

Seyfang, G., Park, J. J., & Smith, A. (2013). A thousand flowers blooming? An examination of community energy in the UK. *Energy Policy, 61*(October), 977–989. https://doi.org/10.1016/j.enpol.2013.06.030

Søraa, R. A., Anfinsen, M., Foulds, C., Korsnes, M., Lagesen, V., Robison, R., & Ryghaug, M. (2020). Diversifying diversity: Inclusive engagement, intersectionality, and gender identity in a European social sciences and humanities energy research project. *Energy Research & Social Science, 62*(April), 101380. https://doi.org/10.1016/j.erss.2019.101380

Sovacool, B. K., Burke, M., Baker, L., Kotikalapudi, C. K., & Wlokas, H. (2017). New Frontiers and conceptual frameworks for energy justice. *Energy Policy, 105*(June), 677–691. https://doi.org/10.1016/j.enpol.2017.03.005

Sovacool, B. K., & Dworkin, M. H. (2015). Energy justice: Conceptual insights and practical applications. *Applied Energy, 142*(March), 435–444. https://doi.org/10.1016/j.apenergy.2015.01.002

Szulecki, K. (2018). Conceptualizing energy democracy. *Environmental Politics, 27*(1), 21–41. https://doi.org/10.1080/09644016.2017.1387294

Szulecki, K., & Overland, I. (2020). Energy democracy as a process, an outcome and a goal: A conceptual review. *Energy Research & Social Science, 69*(November), 101768. https://doi.org/10.1016/j.erss.2020.101768

Taylor Aiken, G. (2015). (local-) Community for Global Challenges: Carbon conversations, transition towns and governmental elisions. *Local Environment, 20*(7), 764–781. https://doi.org/10.1080/13549839.2013.870142

Terriquez, V., Brenes, T., & Lopez, A. (2018). Intersectionality as a multipurpose collective action frame: The case of the undocumented youth movement. *Ethnicities, 18*(2), 260–276. https://doi.org/10.1177/1468796817752558

Tjørring, L. (2016). We Forgot Half of the Population! The Significance of Gender in Danish Energy Renovation Projects. *Energy Research & Social Science, 22*(December), 115–24. https://doi.org/10.1016/j.erss.2016.08.008

UN. (2016). *Gender and climate change: Gender and sustainable energy.* United Nations Development Programme.

UNIDO. (2014). Sustainable energy for all: The gender dimensions. https://www.unido.org/sites/default/files/2014-02/GUIDANCENOTE_FINAL_WEB_s_0.pdf

Van Veelen, B. (2018). Negotiating energy democracy in practice: Governance processes in community energy projects. *Environmental Politics, 27*(4), 644–665. https://doi.org/10.1080/09644016.2018.1427824

Wang, M.-T., & Degol, J. L. (2017). Gender gap in science, technology, engineering, and mathematics (STEM): Current knowledge, implications for practice, policy, and future directions. *Educational Psychology Review, 29*(1), 119–140. https://doi.org/10.1007/s10648-015-9355-x

Wierling, A., Valeria J. S., Jan P. Z., Constantin von B., Heather A. P., Ingrid K. K., Tobias K., et al. (2023). "A Europe-Wide Inventory of Citizen-Led Energy Action with Data from 29 Countries and over 10000 Initiatives." *Scientific Data 10*(1), 9. https://doi.org/10.1038/s41597-022-01902-5

Winther, T., Ulsrud, K., Matinga, M., Govindan, M., Gill, B., Saini, A., Brahmachari, D., Palit, D., & Murali, R. (2020). In the light of what we cannot see: Exploring the interconnections between gender and electricity access. *Energy Research & Social Science, 60*(February), 101334. https://doi.org/10.1016/j.erss.2019.101334

Yildiz, Ö., Rommel, J., Debor, S., Holstenkamp, L., Mey, F., Müller, J. R., Radtke, J., & Rognli, J. (2015). Renewable energy cooperatives as gatekeepers or facilitators? Recent developments in Germany and multidisciplinary research agenda. *Energy Research & Social Science, 6*(March), 59–73. https://doi.org/10.1016/j.erss.2014.12.001

CHAPTER 11

Gender and Energy Poverty in Africa: An Intersectional Approach

Verena Tandrayen-Ragoobur

Introduction

Energy supply is a key enabler of economic development as it drives productivity, creates new jobs and modern economic activities and reduces household workloads for more productive paid work (Pueyo & Maestre, 2019). Greater energy use is thus vital to economic growth and poverty alleviation. The United Nations Sustainable Development Goal (SDG) 7 emphasises on the universal access to affordable, reliable, sustainable and modern energy by 2030. SDG 7 interacts with other Sustainable Development Goals in an attempt to improve the livelihoods of people around the globe. Energy remains a necessity in many aspects of basic well-being, from heating and cooking, to education, health, transport and working.

The COVID-19 pandemic has further stressed the importance of energy access in developing countries, in particular, as a key for

V. Tandrayen-Ragoobur (✉)
Department of Economics and Statistics, University of Mauritius, Moka, Mauritius
e-mail: v.tandrayen@uom.ac.mu

N. Rocha Lawton, C. Forson (eds.), *Women and the Energy Sector*,
https://doi.org/10.1007/978-3-031-43091-6_11

appropriate health care provision and for saving lives (Ogunbiyi, 2020). SDG 7 is salient for Africa, with large number of people living in energy poverty whereby energy is too expensive, unreliable or unavailable. According to the International Energy Agency (2020), 770 million people worldwide lack access to electricity, out of which three-quarters lived in Sub-Saharan Africa (SSA). In fact, the number of people in SSA with access to electricity has steadily increased since 2010 from 33% to 46% in 2019. However, the COVID-19 pandemic is likely to pull many into poverty and reverse the progress made across different African countries, thus making access to energy more problematic. An estimated 555 million people in SSA will still be without electricity by 2030 (UN Stats, 2021). Energy poverty remains a serious concern for the African region.

Globally, the biggest victim of energy poverty and external shocks are women. Energy poverty is considered to have a woman's face (Sánchez et al., 2020). Recent studies have highlighted the linkage between gender, energy and poverty (Clancy et al., 2017; Espinoza-Delgado and Klasen, 2018). Within the African region, lack of access to energy affects women's wellbeing and living conditions and accrues in time poverty, which undermines their educational and economic opportunities (OECD, 2021a). In addition, the impact of energy poverty on educational outcomes has negative intergenerational effects. Links between mother's education and children's health are well established both theoretically and empirically. Existing work has shown that lower education rate of mothers is associated with high infant stunting levels (Abuya et al., 2012), lower immunisation levels (Özer et al., 2018), and even child mortality (Gakidou et al., 2010). Limited access to education discourages schooling in future generations, thus perpetuating the vicious cycle of unemployment and poverty (Azomahou & Yitbarek, 2016).

Though, energy poverty has a strong gender dimension, other forms of social difference including ethnicity, race, socioeconomic class, sexual orientation and age as well as other axes of social power and oppression, which women face, may further accentuate the gender energy gap. Gender is mutually constructed with other forms of social factors or inequalities and these interwoven factors exacerbate gender inequality differences (Benería & Roldan, 1987; Crenshaw, 1991). These interconnections with axes of social power and oppression mean that gender relations are constituted in varied and uneven ways (McDowell, 1999). This chapter thus contributes to the existing literature on energy poverty and gender inequality by using an intersectional approach to explain the gendered energy gap. Adopting the intersectionality model developed by Crenshaw (1991) and Großmann and Kahlheber (2017), the study analyses how various axes of inequality like education, age, health, socioeconomic class, ethnicity and decision-making processes, amongst others influence the gender energy poverty link.

Second, to our knowledge, the existing empirical work on energy poverty and gender inequality is rather scant, especially for African countries. Existing evidence has concentrated mainly on a group of developing countries or a specific country analysis. For instance, Nguyen and Su (2021) investigate the influence of energy poverty reduction on gender inequality across 51 developing nations from 2002 to 2017. Four dimensions are analysed; namely employment, health, education and socio-political economic rights by applying the two-step system generalised method of moments. Their findings show that a reduction in energy poverty creates employment opportunities for women in the industrial and services sectors. It also improves female wages and women's health conditions. Their study concludes that reducing energy poverty may generate gender equality in education and socio-economic rights. In contrast to Nguyen and Su (2021), we probe into the causes of the gender energy poverty rather than its consequences. In addition, we adopt an intersectional perspective of explaining gender energy poverty via a combination of gender and social differences measured though survey data for a group of Sub-Saharan African countries. The analysis uses the Demographic Health Survey data for 20 countries from 2010 to 2015. This study innovates by assessing the gender energy gap via the intersection of gender with other aspects of social differences, pertinent to the African context.

The structure of the chapter is as follows. The next section reviews the existing theoretical and empirical literature explaining the gendered energy poverty from an intersectional approach. Section 'Conceptualising Gender Energy Poverty from an Intersectionality Perspective' explains the conceptual framework on gender and poverty. Section 'Situational Analysis of Energy Poverty in Sub-Saharan Africa' discusses the data on access to electricity across the African region. The section 'findings' discusses the results. The section 'Data and Methodology' analyses the data and methodology used. The chapter finally concludes with important policy options on gender energy poverty for Africa.

Literature: Gender Energy Poverty from an Intersectional Lens

The concept of energy poverty has received much attention in the energy literature (Boardman, 2013; Bouzarovski & Petrova, 2015). There is, however, no universal and accepted definition of energy poverty (Bouzarovski et al., 2012; González-Eguino, 2015). Since poverty itself is a multi-dimensional and complex concept, energy poverty incorporates a mix of elements, which makes it difficult to be measured via a single indicator. Energy poverty has been conceptualised as a 'situation of inability to realise the essential capabilities due to insufficient choice in accessing affordable, reliable, adequate, quality and safe energy services in a reasonable manner' (Day et al., 2016). This is in line with Sen (1976) capability approach, whereby poverty is the absence of opportunities and choices for a basic living. Similarly, using the concept of relative poverty by Townsend (1979), Buzar (2006) considers the energy poor as those individuals living without access to the socially and materially necessary domestic energy services that enable them to have a proper living. The inability of households to meet basic domestic energy needs is the core element of the definition (Nguyen & Su, 2021; Thomson et al., 2016).

Energy poverty can be analysed into availability and affordability of energy sources. The availability and access of basic energy resources such as electricity is a central issue in developing countries (González-Eguino, 2015; Pachauri & Spreng, 2004, 2011) while in developed nations, attention is on access to modern and clean energy services and technologies. Studies have also differentiated between energy poverty and fuel poverty (Bouzarovski et al., 2012; Li et al., 2014). The former is a broader term capturing all the problems related to the inadequate energy access while fuel poverty relates to low energy affordability. The interactions between energy prices, low-income levels and energy inefficiency of households explain the low energy affordability (Primc et al., 2021). However, the two terms are often used interchangeably, which will also be the case in this study.

Differences across men and women in their access to and use of energy are at the heart of the gender-energy nexus literature (Feenstra & Özerol, 2021). There is a misconception that women and men are equal in their uses of energy (Petrova & Simcock, 2021). There are distinct gender dimensions in which the lives of women and men are affected by energy use. Though, there has been an increasing role of women in the economic development of many developing countries, there remain major structural inequalities in both the labour market and in the welfare system that perpetuate gender inequalities. Clancy and Roehr (2003) argue that similar disparities prevail in relation to energy use, and this is explained by the intersection of gender with other axes of social differences.

Energy-related deprivations have serious implications on well-being and livelihoods (Churchill & Smyth, 2021; Llorca et al., 2020; Thomson et al., 2017). Despite the emerging agreement of the critical role of gender in energy poverty literature, there remains relatively scant analysis of the energy-gender nexus from feminist development scholars (Listo, 2018). This is because from the feminist scholarship, gender is not a discrete category, but it is constructed mutually with other forms of social differences including class, ethnicity, age and sexuality, to name a few (Benería & Roldan, 1987; Crenshaw, 1991). In addition, McDowell (1999) views these interconnections with axes of social power and oppression. Thus, gender is perceived as the social, economic and political constructions of 'feminity' and 'masculinity', a fundamental axis of social power that shapes social relations in an unequal manner (Moreno-Bella et al., 2019; Robinson, 2019). This definition perceives gender as a social construct, which changes over time, space and context. It also translates into different inequalities. The gender dimension to energy poverty varies across social, economic and political situations and takes on board an intersectional perspective.

The intersectional approach to energy poverty integrates the interconnection as to how gender inequality correlates with other inequalities of ethnicity, race, class and age amongst others (Azocar & Ferree, 2016). Social inequalities and systems of power all interact with the existing gender dimension and bring in issues of overlapping inequities hence making energy poverty a highly gendered problem, with women bearing the brunt of the consequences of inadequate energy access while suffering from systemic discrimination as well as decreased access to resources and decision-making (Pachauri & Rao, 2013). The intersectionality approach, which is a strand of the social stratification theory, unravels the mechanisms of multiple discriminations that arise at the same time.

The intersections of such axes of inequalities need to be used as a theoretical underpinning to explain the gender energy deprivation.

Energy poverty results from the negative outcome of unequal social relations that are shaped by multi-dimensional aspects of gender disadvantage such as housing, basic services and rights (Robinson et al., 2018).

The 'triad' theory as founded by Boardman (1991) explains the causes of energy poverty from income as the only axis of inequality (Moore, 2012). It however fails to engage with the deeper systemic structures and inequalities that underpin domestic energy deprivation (Großmann & Kahlheber, 2017). As per Großmann and Kahlheber (2017), while the income dimension of the triad theory is an important axis of inequality, other dimensions of inequality or social characteristics of individuals or households are present in the energy poverty debate. Similarly, Brunner et al. (2011) point out that for energy poor families, energy bills are just one problem among several others. For instance, age and health (illness) can trigger energy poverty by causing increased energy needs and consumption. These families may need longer hours of heating, hence consuming more energy, which leads to higher bills and increased risk of energy poverty.

Further, households with single parents and small children are often associated with low incomes (Lawson et al., 2015) and as such may suffer from energy deprivation. Petrova and Simcock (2021) further argue that experiences of energy poverty in the home are differentiated along gender lines, stressing the agency of household members who are actively involved in the household decision-making process. They link the operation of power surrounding gender and women in the energy poverty discourse. Women and men have distinct access to resources and participate differently in the decision-making process. In particular, women's participation in decision making within the household and community is often restricted. Being involved more in unpaid work relative to income-generating activities makes women voiceless in decision-making. Men who are the main breadwinners of the family gain greater control over financial decisions among other household decisions.

The gender dimension of the causes of energy poverty appears in varied ways and women in general experience energy poverty more severely than men do. They are often disproportionately responsible for the household duties that are connected to energy. This is particularly acute in rural settings across developing nations whereby women spend considerable amount of time on tasks like collecting firewood for basic cooking, heating and lighting needs (O'Dell et al., 2019). Lack of access to energy represents a major obstacle to women's and girls' wellbeing and economic opportunities as their greater time commitment on unpaid care work due to energy poverty reduces their ability to go to school, or to be employed, which would have in turn enhanced their livelihoods. This translates into lost opportunities for education and remunerated labour (OECD, 2018), perpetuating poverty and inequality (Dutta et al., 2017). Access to energy allows for efficient products that reduce the time burdens of domestic responsibilities and create time for greater productive women's engagement in the local economy. Various studies have shown that electrification has a positive impact on female labour participation rates in developing countries. For example, Lewis (2014) show that household electrification improves school attendance for girls and in the long run women's employment opportunities. Though both men and women benefit from access to electricity, women gained the most as the use of electric appliances frees up time spent on household chores. Further, energy poverty has intergenerational effects via its influence on educational outcomes. Limited access to education discourages schooling in future generations, thus prolonging the vicious cycle (Azomahou & Yitbarek, 2016). The links between mothers' education and children's health are also well-established so that education rate of mothers tends to be highly correlated with infant stunting levels (Abuya et al., 2015) and higher levels of immunisation (Özer et al., 2018).

Energy poverty has important implications on women's and girls' health especially when they rely on polluting sources of energy for basic needs and bear the effects of air pollution. The negative consequences of the use of traditional fuels like paraffin and biomass such as burning of wood, dung and crop residues on human health have been widely reported (Cecelski et al., 2015; Venkataraman et al., 2010). They are thus exposed to challenges such as health hazards, premature deaths, time poverty and physical hazards, amongst others. Exposure to ambient air pollution has also adverse consequences on fertility, pregnancy and new-borns. Women and children are the main casualties from indoor pollution generated by inefficient cooking stoves widely used in the Global South. They are likely to suffer from stroke, pneumonia, lung diseases, asthma, cancer and other diseases (Lam et al., 2012). Energy poverty harms not only physical health but also mental health across many developing countries (Cecelski et al., 2015). Köhlin et al. (2011) emphasise that energy interventions, such as improved cooking technologies or access to electricity, can bring significant gender benefits. In essence, women tend to endure a greater proportion of energy poverty's harmful consequences (Clancy et al., 2007; Day et al., 2016) due to existing inequalities at all levels. In fact, gender inequalities can be noted at all levels in the energy system which thus create gender injustices (Clancy & Feenstra, 2019; Kooijman-Van Dijk, 2020; Winther et al., 2020).

The energy and gender literature has, however, focussed mainly on the household realm where women suffer heavily from the burdens of energy poverty (Johnson et al., 2020). The household-centred literature provides strong evidence of the connection between energy poverty and gender inequality and relates to the negative impact on women's health burdens, use of time, education, amongst others. However, rather than focussing on the care and reproductive roles of women, there is a need to include the productive positions of women as entrepreneurs and employees. Women as employers and workers face different challenges than men in their uses of and benefits from electricity because they operate in different types of productive activities in varied locations and their access to assets, finance, market infrastructure and skills differ considerably. Further, women's agency, mobility and time use tend to be restricted and as such influence their choice of the type, location and nature of paid work undertaken.

Women are thus important agents in the energy system and it is vital to acknowledge women's triple role as producer, decision-maker and consumer (Clancy & Feenstra, 2019; Winther et al., 2018). In addition, the gender-energy justice nexus has been criticised for adding a binary dimension to the energy poverty debate (Sovacool et al., 2020). Sovacool et al. (2020) postulate the need for a whole-system approach so that justice is achieved at every stage of the energy supply chain to have equitable outcomes. Women are not only unequally experiencing the effects of energy poverty but access to energy and resources is a manifestation of existing gender inequalities (Köhlin et al., 2011).

Conceptualising Gender Energy Poverty from an Intersectionality Perspective

This study adopts a holistic framework to assess gender energy poverty from an intersectional perspective. The conceptual model is as per Fig. 11.1.

Gender energy poverty worsens when age, class, ethnicity, race, sexual orientation and disability, amongst others, aggravate gender inequality. The connection between gender and energy poverty is further explained by a multiplicity of factors that exist at the micro, meso and macro levels hence exacerbating the gender energy disparity.

Within the framework, under the micro dimension, we include individual characteristics, family dynamics and housing conditions. Individual

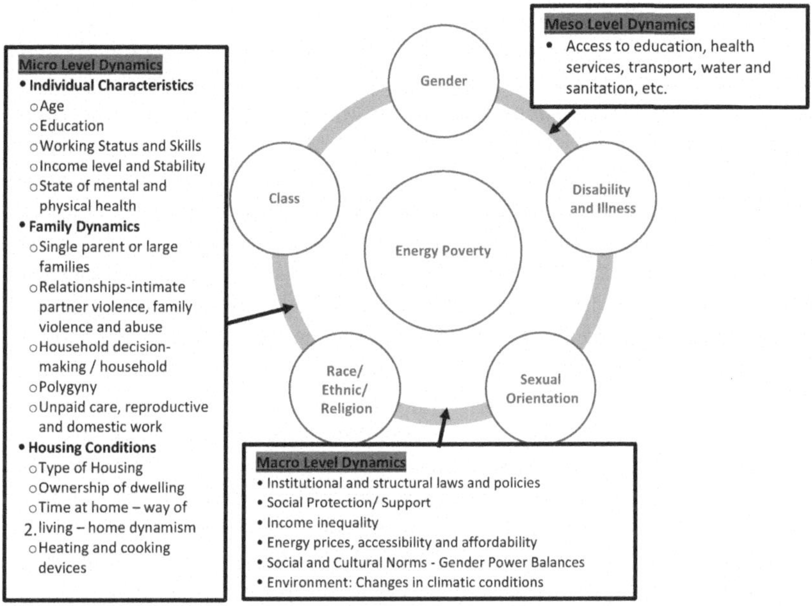

Fig. 11.1 Conceptual framework—Modelling gender energy poverty from an intersectionality perspective. (Source: Author's adaptation and extension on various models namely Großmann and Kahlheber (2017) and Gonzalez Pijuan (2018))

characteristics take into account, age, educational level of the individual and the working status (can be operating in the formal or informal sector or may be unemployed or a pensioner). Energy poverty is associated with lack of income and the inability to achieve decent living standards. Those with inadequate or unstable incomes may suffer from stigma and feeling of rejection, which creates psychological and health problems (Gonzalez Pijuan, 2018). Disability and health status are crucial psychological factors that worsen vulnerability to energy poverty among women (Robinson, 2019). Within family dynamics, the structure of the family is also analysed in terms of single parent or large families. The family structure will influence the energy needs and consumption. One parent families headed by women are more likely to be materially deprived than those headed by men (Živčič et al., 2021).

In addition, household size and number of occupants is an important factor influencing energy use and poverty as larger dwelling sizes imply higher energy bills (Antepara et al., 2020). Energy poverty is often associated with difficulties in securing a decent housing and income level. The types of housing and the heating and cooking appliances available, as well as the amount of time spent at home on unpaid work, are the important elements to ensure a decent living. Lack of access to energy supplies increases the vulnerability of individuals with inadequate housing and income, which thus carries the risk of social exclusion (Gonzalez Pijuan, 2018). Another important element within family dynamics relates to the distribution of roles in the household. Women do not have equal voices on policy, economy and the household. Women lack power in energy-related decisions, which cover both purchasing and usage decisions (Pachauri & Rao, 2013). If the man is the main breadwinner and the woman is responsible for domestic chores and caregiving, which are tied to energy use, therefore making women more vulnerable to energy accessibility and affordability. This reinforces the patriarchal division of gender role and gives women the sole responsibility for addressing energy poverty. Another feature of many households is the prevalence of domestic violence and abuse. Munien and Ahmed (2012) highlight that the underlying structures that perpetuate poverty and energy poverty are patriarchy, marginalisation, vulnerability and exclusion from participation in decision-making processes. These characteristics are more prevalent amongst women, which makes energy poverty take a female face. Likewise, Hailemariam et al. (2021) find that energy poverty increases the likelihood of experiencing physical violence.

Within the meso-level dynamics, access to education, health services, transport, water and sanitation, amongst others are important preconditions to reduce energy poverty among women. The cycle of exclusion from education and literacy can lead to early onset of reproduction, higher risk of maternal health and thus higher risks of poverty and energy deprivation. Banerjee et al. (2021) further demonstrate that lower energy poverty is associated with negative health and education outcomes. Lastly, across the macro-level dynamics, it is the legitimate right of all men and women to have access to energy and the state as the bearer to protect, promote and uphold the rights of its citizens must ensure access to basic energy services. Still across many developing economies, men and women face institutional barriers to the realisation of rights (often referred to as 'rights failures'), mainly because gender inequalities along with other social disparities are produced and reproduced by the energy system governance institutions (Danielsen, 2012). Right failures in energy systems are connected to two interrelated factors namely the lack of recognition of women's energy needs, knowledge and contribution and the unequal distribution of control over the energy sources. These result mainly from cultural and social norms that shape the roles and relationships between men and women. In essence, women tend to internalise social norms that place a low value on their worth and contribution. Such norms permeate energy system structures, institutions, including its legal, regulatory frameworks, programmes and policies (Danielsen, 2012; Moser & Norton, 2001).

The macro-level dimension also links social protection and energy access as energy deprivation is highly connected to poverty. Thus, targeted social protection systems can reach those most in need including the chronically poor and those who are economically vulnerable. There is a two-way link between lack of access to energy services and poverty. Access to energy can alleviate poverty, improve incomes and enhance living conditions and at the same time, lack of access to electricity amongst chronically poor is due to both the unavailability of energy supply and unaffordability (UN, 2015). Appropriate social support measures can alleviate the gendered energy gap, which tends to have primarily a women's face in the Global South. Women, especially single mothers, and elderly single women are the most affected by lower energy access due to lower income, psychological characteristics and behavioural patterns (Clancy et al., 2017). Higher energy prices may make access to energy difficult for many women who cannot afford the basic energy services. Lastly, the impacts of the environment and climate change are not gender neutral, as women in the developing region tend to be more exposed to natural hazards and domestic fire accidents, amongst others since they are more likely to be poor, face time poverty, drudgery in households and farm tasks. Hence, due to these pre-existing inequalities, changes in climatic conditions have more severe effects on them compared to their male counterparts.

Situational Analysis of Energy Poverty in Sub-Saharan Africa

Using data on the proportion of people with access to electricity as an indicator of energy access across different African regions over the last two decades from 2000 to 2019, it can be observed that the percentage of people gaining access to electricity has increased from 36% in 2000 to 56% in 2019. As per the World Energy Outlook (2020), much of that increase in electricity access originates from a small group of countries like Kenya, Senegal, Rwanda, Ghana and Ethiopia where access rate improved from 20% in 2013 to 85% in 2019. There has been huge investment in grid and off grid connections across these countries. For instance, countries like Kenya, Tanzania and Ethiopia accounted for around half of the five million people gaining access through new solar systems in 2018 in Africa. For Sub-Saharan Africa, in particular, the percentage of people having access to electricity has doubled from 24% in 2000 to 48% in 2019.

However, across Africa, the disparity in electricity access between urban and rural regions is very pronounced with 81% of the urban population having electricity compared to only 37% in the rural region. Such urban-rural energy gap is further noted for Sub-Saharan Africa with 76% having access to electricity in cities and towns relative to only 29% in rural areas (see Table 11.1). With the health crisis and the social and economic downturn caused by the COVID-19 pandemic, energy access may not be one of the top concerns of many African governments who have been shifting priorities to prevent the spread of the virus through stringent and restrictive measures. This has slowed energy access programmes and hindered activities in the decentralised energy access domain. The Sub-Saharan African region, home to three-quarters of the global population without access to electricity, is being particularly hit by the pandemic, and recent progress achieved by the region in energy access may be reversed due to the effects of the virus.

Data and Methodology

Data

The study uses Demographic and Health Survey (DHS) data for 20 Sub-Saharan African countries over the period 2010 to 2015. Comparable data across the different nations is collected via standard questionnaires and similar survey procedures are applied in each country. Data is available on socio demographics such as age, education level, gender and working status of the respondent as well as household characteristics namely the sex of the household head, wealth index, decision-making power, and access to electricity. Information on different forms of intimate partner violence and attitudes towards violence are also available. The DHS uses multistage

Table 11.1 Electricity access by region from 2000 to 2019

	Proportion of the population with access to electricity							*Population without electricity (million)*
	National					*Urban*	*Rural*	
	2000 (%)	*2005 (%)*	*2010 (%)*	*2015 (%)*	*2019 (%)*	*2019 (%)*	*2019 (%)*	*2019*
WORLD	**73**	**77**	**80**	**85**	**90**	**96**	**85**	**771**
Africa	**36**	**40**	**44**	**49**	**56**	**81**	**37**	**579**
Sub-Saharan Africa	**24**	**28**	**33**	**40**	**48**	**76**	**29**	**578**
North Africa	91	97	>99	>99	>99	>99	>99	<1
Central Africa	15	16	19	22	24	44	6	108
East Africa	10	17	21	31	47	79	35	165
West Africa	33	38	45	50	56	87	28	174
South Africa	77	78	84	92	94	95	92	3
Other Southern Africa	14	16	21	32	39	67	20	128

Source: IEA, World Energy Outlook (2020)

cluster sampling techniques to develop nationally representative samples. In the first sampling stage, each country is stratified into regions, from which census-based enumeration areas are designed and selected with a probability proportional to size. In the second sampling stage, households are randomly selected within each designated enumeration area. Sample weights are used to reduce sample variability especially when the latter is selected with unequal probability across countries. Hence, the cross-tabulations apply weights to produce a proper representation and control for the sample survey design.

Methodology and Survey Characteristics

Indicators of energy access are used in terms of whether the household has electricity and the type of cooking fuel used. As per SDG 7, universal access to affordable and reliable energy is important (UN, 2020). Access to electricity and clean fuels for cooking are crucial factors that ensure good living conditions for everyone (Halkos & Gkampoura, 2021). These indicators are analysed from an intersectional perspective in terms of micro-, meso- and macro-level dynamics as per the conceptual model. A profiling of women across 20 SSA countries is shown in Table 11.2.

The data reveals that women are mainly within the age group of 25 and 29 years. They are likely to be uneducated with 36.26% being unschooled

Table 11.2 Socio-demographics of women surveyed across 20 SSA countries from 2010 to 2015

Profile of respondents	%	*Profile of respondents*	%
Age Categories		*Wealth Index/ Household Income Range*	
15–19	6.77	Poorest	21.04
20–24	18.19	Poorer	21.47
25–29	23.03	Middle	20.56
30–34	19.39	Richer	19.54
35–39	15.48	Richest	17.39
40–44	10.15	*Types of Earnings*	
45–49	6.69	Not paid	28.30
Education Level		Cash only	46.19
No education	36.26	Cash and in-kind	19.94
Incomplete primary education	27.05	In-kind only	5.49
Complete primary education	12.51	*Household Characteristics*	
Incomplete secondary education	17.79	*Head of Household*	
Complete secondary education	3.74	Male	94.19
Higher education	2.65	Female	5.81
Religion		*Household's Access to Basic Necessities*	
Muslim	30.74	Household with electricity	25.00
Catholic	32.58	Toilet facilities shared with other households	41.93
Protestant	10.62	*House Ownership*	
Traditional/ Animist	4.48	Does not own house	43.11
Other religion	21.57	Own house alone	6.96
Working Status		Own house jointly	42.62
Working Women	63.82	Other arrangements	7.30
Non-working women	36.18	*Household Decision Making Process*	
Husband Working	91.29	Women can decide alone on daily household purchases	45.44
Husband Not Working	8.71	Women can decide alone on visits to family	17.40
		Women can decide alone on how to spend her earnings	46.66
		Women can decide alone on large household purchases	11.36
		Women can decide alone on her health care	16.07

Source: Author's Compilation from DHS Surveys

while 27.1% have incomplete primary education. Only 2.7% have reached higher education level. In terms of religion, 32.6% are Catholics, 30.7% are Muslims, 10.6% are Protestant, around 4.5% are Animist and the remaining ones around 21.6% are either Hindus, Buddhists, amongst many others.

We next analyse the employment status of women in the survey whereby 63.8% of women work, and around 46.2% are paid in cash only, 19.9% have earnings in terms of both cash and in-kind while around 5.5% receive their salary in kind. The data also indicates that 28.3% of women are not paid for their work. In terms of household income, 21.0% of women surveyed come from the poorest households within the lowest income bracket and around 21.5% are within the category of poorer households. This demonstrates that poverty in Africa has a woman's face, confirming the various disparities, which women encounter. In terms of household characteristics, around 94.2% of respondents state that their husband is the head of household. Around 43.1% do not own the house in which they are living whilst 42.6% own the house jointly with their partner or husband. Women's lack of homeownership often translates into lack of decision-making power over electricity (Clark, 2021). In essence, it is highlighted in the survey that only 16.1% of women can decide alone on their health care. Sole decisions on how to spend their earnings can be made by around 47% of women surveyed while 45.4% of women can decide on their daily household purchases. However, the percentage is much lower when it concerns large household purchases and visits to the family, where only 11.4% and 17.4% of women are able to take these respective decisions on their own.

Findings

Micro-Level Dynamism

From the conceptual framework, the study first probes into the micro-level dynamism in explaining gendered energy poverty from an intersectional approach. The micro-level perspective is split into individual characteristics, family dynamics and household characteristics. Table 11.3 first analyses the individual features and family structures of African women. Energy use changes over the human lifespan and increases with age (Powell, 2019). Across the different age groups, women aged between

Table 11.3 Micro-level dynamics: Individual characteristics and family dynamics of African women

	No electricity (%)	*Has electricity (%)*		*No electricity (%)*	*Has electricity (%)*
Individual Characteristics			**Family Dynamics**		
Age			Family Size		
15–19	82.46	17.54	2 members	63.62	36.38
20–24	76.69	23.31	3–4 members	70.97	29.03
25–29	74.15	25.85	5–7 members	76.45	23.55
30–34	72.98	**27.02**	8–10 members	78.86	21.14
35–39	74.85	25.15	11–13 members	78.98	21.02
40–44	74.91	25.09	>13 members	76.32	23.68
45–49	73.69	26.31	**Family Relationships / Violence and Abuse**		
Education			Subject to Intimate Partner Violence	75.35	24.65
No education	88.23	11.77	Not subject to IPV	74.92	25.08
Incomplete primary	83.43	16.57	**Patriarchal Nature of the Household**		
Complete primary	74.43	25.57	Woman decides alone on **daily household purchases**	71.59	28.41
Incomplete secondary	52.3	47.7	Woman and husband decide on daily household purchases	74.16	25.84
Complete secondary	37.32	62.68	Husband decides alone on daily household purchases	86.51	13.49
Higher Education	14.01	85.99	Woman decides alone on **family visits**	71.98	28.02
Working Status			Woman and husband decide on visits to family	72.00	28.00
Non-working women	71.96	28.04	Husband decides alone on family visits	80.27	19.73

(*continued*)

Table 11.3 (continued)

	No electricity (%)	*Has electricity (%)*		*No electricity (%)*	*Has electricity (%)*
Working Women	76.73	23.27	Woman decides alone on **how to spend her earnings**	66.53	33.47
Type of Earnings from Work			Woman and husband decide on how to spend her earnings	67.21	32.79
Not paid	91.11	8.89	Husband decides alone on how to spend her earnings	84.69	15.31
Cash only	62.12	37.88	Woman decides alone **on large household purchases**	66.64	33.36
Cash and in-kind	87.95	12.05	Woman and husband decide on large household purchases	70.21	29.79
In-kind only	92.34	7.66	Husband decides alone on large household purchases	81.34	18.66
Wealth Index			Woman decides alone **on her health**	68.40	31.60
Poorest	96.75	3.25	Woman and husband decide on her health	72.57	27.43
Poorer	92.76	7.24	Husbands decides alone on her health	79.39	20.61
Middle	87.48	12.52	Woman decides alone **on food to be cooked each day**	67.32	32.68
Richer	67.68	32.32	Woman and husband decide on food to be cooked each day	61.23	38.77

(*continued*)

Table 11.3 (continued)

	No electricity (%)	*Has electricity (%)*		*No electricity (%)*	*Has electricity (%)*
Richest	20.2	79.8	Husbands decides alone on food to be cooked each day	71.09	28.91
State of Mental and Physical Health			**Polygamy**		
Do not experience any emotional violence	74.94	25.06	Husband has one wife	72.82	27.18
Experienced any emotional violence	75.71	24.29	Husband has two wives	85.02	14.98
			Husband has more than two wives	89.09	10.91
			Unpaid care, reproductive and domestic work		
			Time to Collect Water		
			Water available on Premises	34.14	65.86
			Takes less than 10 minutes	77.3	22.7
			Takes between 10 to 20 minutes	85.6	14.4
			Takes between 21 to 40 minutes	89.25	10.75
			Takes between more than 40 minutes	89.54	10.46

Source: Author's Compilation from DHS Surveys

30 and 34 years are more likely to have access to electricity relative to other age groups. Since the survey covers only women aged between 15 and 49 years, we are unable to capture older women from the data. However, it can be noted that as we move to higher age brackets, more women can access electricity while the younger ones seem to be disadvantaged. Likewise, the statistics reveal that across the group of African women who have completed higher education, around 14% do not have

electricity compared to 88.2% of uneducated having no access to electricity. A similar high percentage is observed for women with incomplete primary education where 83.4% has no electricity. More educated women appear to suffer less from energy poverty relative to poorly educated ones. This is in line with studies of Acharya and Sadath (2019) and Donev et al. (2018). When taking into account, the working status of women, around 23% of working women have access to electricity compared to 28% of non-working ones. The literature addressing the connection between energy access and women's labour force participation has so far produced mixed results (Clark, 2021). The macroeconomic conditions may matter in explaining these inconclusive findings as the market may have little capacity to absorb additional labour force, even if energy access lead to more opportunities for women to enter the labour market (Rathi & Vermaak, 2018). We further probe into the types of earnings and those women who are not paid or receive payment in kind for their job, are likely to be most energy poor, with only 8.9% and 7.7%, respectively, having access to electricity. The data also differentiates across different categories of households using the wealth index. Women from poorest and poorer households are most affected with the lowest percentage having electricity. There is an important energy gap across the richest households where 79.8% of women have electricity compared to only 3.3% and 7.2% of women from poorest and poorer families, respectively.

Next, as per the conceptual framework, we investigate into the family dynamics to explain gendered energy poverty in Africa. The analysis probes into family size, family relationships, the patriarchal nature of the household, polygamy (an important feature of African societies) and unpaid care and domestic work. A higher percentage of women in larger families do not have electricity compared to those women in smaller families. The figure declines to only 21% of women accessing electricity in households with 11 to 13 members. The percentage (23.7%) is slightly higher for those within family size of 13 or more members. A non-linear relationship may be possible because as family size increases, more young people are able to work and are in a better position to access energy. Family relationships is analysed in terms of the occurrence of intimate partner violence where those women who are victims of abuse from their partner or husband are less likely to access electricity (24.7%) compared to those not facing any abuse (25.1%). Though the difference seems to be small, it is often postulated that electrification can reduce gender-based violence in Sub-Saharan Africa (Sievert, 2015). Sievert's theory of change posits that improved electricity access may change women's preferences for independence versus being in a violent relationship, hence leading to societal changes on the perceptions and attitudes towards intimate partner violence (Sievert, 2015).

The patriarchal nature of the household is then discussed in terms of the decision-making process on daily and large household purchases, on food to be cooked each day, visits to the family and who decides on how to spend the woman's earnings as well as decision on her health care. Across all six indicators, there is evidence that in households where man has more power in the decision-making role, there is a larger percentage of these households without access to electricity compared to those families where such decisions are either shared or decided by the woman alone. This patriarchal attitude in decisions accentuate existing gender inequalities while women's power matters across various family responsibilities linked to consumption, health, education, children's welfare and reproductive outcomes amongst others (Annan et al., 2019). The intra-household decision-making in responsibilities, which are energy intensive and where women are not consulted, reinforced the gendered poverty gap. Polygyny (the practice of having more than one wife) is common in many religious and cultural communities in Africa and is strongly correlated with a wide array of violence against women and children (McDermott & Cowden, 2014; Tandrayen-Ragoobur, 2020). The data shows that those households where the husband has two wives or more than two wives are likely to be the ones without electricity. Access to electricity appears to be lower at 14.98% for those women whose husband has two wives and it is worse at 10.9% for those women whose husband has more than two wives. Anjorin and Yaya (2021) found similar results across 32 countries within Sub-Saharan Africa where only a few women living in polygynous families had access to electricity. The last dimension of family dynamics relates to unpaid care and domestic work. Women are more likely to be responsible for chores like washing clothes, retrieving water, cooking and cleaning, all of which are energy-intensive (Dutta et al., 2017). Traditional gender roles lead to different patterns of energy use between men and women, implying that women will gain disproportionately from electrification. Time to collect water is used as a proxy where it is noted that those women spending more time in domestic work, are less likely to access electricity. In fact, electrification can alleviate women's time poverty and direct this time to productive work in the labour market. These tasks like water collection over long distances and the time required to complete them are referred to as 'drudgery' (Winther et al., 2018) and hinder women's empowerment and access to energy.

Still, within the micro dynamism of gendered energy poverty, the housing conditions of the African women are gauged (see Table 11.4). We differentiate the type of housing by the main floor material. Among women living in houses with earth flooring, only 5.3% has electricity compared to 22.97% in houses with wood flooring and 53.9% in houses with concrete flooring. A clear association can be highlighted between poverty and energy deprivation. Ownership of the house and ownership of land are used as indicators to highlight the women's access to assets and properties. Widespread patriarchal practices prevent women's access to and control over land and properties, which undermine women's empowerment. Land access and rights are critical in empowering women economically, socially and politically and help in addressing practical and strategic needs (Nadasen, 2012) and one of these needs is energy access and use. The data shows that it is mainly those with joint ownership over land and houses, who are more likely to face difficulties in accessing electricity. Among those women with joint ownership of land or house, only 18.3% and 18.7% have electricity, respectively. Type of cooking fuel is also used as a variable to analyse the gendered energy gap across Africa, those women who use charcoal, wood, straw, agricultural crops or animal dungs (that is the poorest segments of the population), are less likely to access electricity.

Meso-Level Dynamism

Within the meso component, access to education, health services, transport, water and sanitation are evaluated. The education aspect focusses on those women who cannot read and as seen above many either have never attended schools or have only completed basic primary schooling. From Table 11.5, only 9.7% of those women who cannot read in the survey have access to electricity. There is a high two-way causation between education level and energy access across women. Those with less education may find it more difficult to access electricity due to their inability to have enough earnings from a decent job that can allow for energy access. Uneducated women also tend to be deprived of their rights at home and do not

Table 11.4 Micro-level dynamics: Housing conditions of African women surveyed

a	*No electricity (%)*	*Has electricity (%)*		*No electricity (%)*	*Has electricity (%)*
Housing Conditions					
Type of Housing-Main Floor Material			**Heating and Cooking Devices**		
Concrete flooring	46.15	53.85	**Type of Cooking Fuel**		
Wood flooring	77.03	22.97	Electricity	0.55	99.45
Earth flooring	94.73	5.27	Liquefied Petroleum Gas	6.71	93.29
Ownership of Dwelling			Natural gas	12.22	87.78
Woman owns house alone	71.25	28.75	Biogas	17.65	82.35
Woman and husband own house jointly	81.31	18.69	Kerosene	17.43	82.57
Woman does not own house	66.73	33.27	Coal, lignite	22.57	77.43
Ownership of Land			Charcoal	57.11	42.89
Woman owns land alone	74.87	25.13	Wood	88.49	11.51
Woman and husband own land jointly	81.74	18.26	Straw/shrubs/grass	90.75	9.25
Woman does not own land	68.05	31.95	Agricultural crop	83.33	16.67
			Animal dung	100	0

Source: Author's Compilation from DHS Surveys

participate in the household decision making process. The gender gap in education is thus reflected in the gendered energy disparity. Similarly, household electrification has a positive effect on educational attainment. The heavy burden of household drudgery due to lack of energy access falls very often on young girls. Better household access to electricity will reduce household chores and therefore, girls can spend more time towards studying and remain in school for a longer period. Increased educational attainment may lead to higher wages and better labour outcomes in the long

Table 11.5 Meso-level dynamics of African women surveyed

	Women (%)	*No electricity (%)*	*Has electricity (%)*
Education			
Women cannot read at all	33.75	90.27	9.73
Health Access			
Did not visit a health facility in the last 12 months	40.24	76.85	23.15
Visited a health facility in the last 12 months	59.73	73.9	26.1
Water Access			
Main Sources of Drinking Water			
Piped into dwelling	5.39	9.27	90.73
Piped to yard/plot	8.42	21.95	78.05
Public tap/standpipe	15.05	60.21	39.79
Tube well or borehole	19.86	90.21	9.79
Protected well	8.77	84.11	15.89
Unprotected well	14.33	93.30	6.70
Protected spring	4.49	86.14	13.86
Unprotected spring	7.29	94.66	5.34
River/dam/lake/ponds/stream/canal	10.35	95.10	4.90

	Women (%)	*No electricity (%)*	*Has electricity (%)*
Sanitation			
Toilet Facilities			
Toilet facilities shared with other households	41.93	65.56	34.44
Toilet facilities NOT shared with other households	57.35	69.80	30.20
Transport			
Household has a bicycle	41.70	83.61	16.39
Household has no bicycle	57.77	68.81	31.19
Household has a motor cycle/ scooter	15.84	66.22	33.78
Household has no motor cycle/ scooter	83.62	76.68	23.32
Household has a car/truck	4.79	19.07	80.93
Household has no car/truck	94.66	77.84	22.16

Source: Author's Compilation from DHS Surveys

run and as such a better standard of living. In addition, health access is proxied by a visit to the health facility in the last year, where 59.7% of women went to a health centre/ hospital. Among, those who could access a health facility, 26.1% have electricity compared to 23.2% for those who did not go to the hospital or health centre in the last 12 months. The disparity in access to health services is also translated into the gendered energy disparity across African women in the sample.

Next, we measure access to water by the main sources of drinking water available to women in Africa. The data shows that 19.9% of women get their water from tube well or borehole, followed by 15.1% from piped tap or standpipe, 14.3% from unprotected well and 10.3% from river, dam or lakes. Those who access water from wells, boreholes, rivers and dams are more disadvantaged and faced not only greater time poverty but also are more likely to be economically vulnerable. This disadvantage is further reflected in their poor energy access. For instance amongst those who access drinking water from rivers, dams or lakes, only 4.9% have electricity followed by 5.3% for those getting drinking water from unprotected spring and only 9.8% of those who use tube wells and boreholes have electricity. These methods of accessing drinking water have major health implications, as the water often have chemicals, bacteria, parasites and viruses that may make people sick. Women's health and that of their families and children are likely to be more at risk with inappropriate sources of drinking water. Sanitation facilities is subsequently analysed and proxied via the sharing or non-sharing of toilet facilities with other households. Women living in households with shared toilets are more vulnerable to poverty and lack basic services needed for a decent living. The data indicates that among those women who are sharing toilet facilities with other households, a higher percentage (34.4%) do not have electricity compared to 30.2% for those women having their own toilet amenities. The last indicator within the meso dimension is access to transport. Having a bicycle, motorcycle or car may not be an ideal measure; however, due to unavailability of data, it is used to differentiate across households having these transport facilities and their access to energy. There is evidence that those people who have a car or truck or a motorcycle tend to be better off so that the electrification rate is much higher at 80.9% and 33.8%, respectively.

Macro-Level Dynamism

One major challenge encountered to measure the macro-level dynamism is that the study uses primarily household level data, which provides mainly micro- and meso-level information. To offset partially this limitation, we use data from the OECD Development Centre's Social Institutions and Gender Index (SIGI) (OECD, 2021b), which measures discrimination against women in social institutions. The SIGI is a multidimensional index and takes into account laws, social norms and practices. The four dimensions of the SIGI are discrimination in the family, restricted physical integrity, restricted access to productive and financial resources and restricted civil liberties. These four dimensions are used as well as the overall SIGI index to assess differences in energy access across women in Africa. The composite index of discrimination in the family is made of information on child marriage, household responsibilities, divorce and inheritance. The restricted physical integrity index comprises of data on violence against women, female genital mutilation, missing women and the reproductive economy. Restricted access to productive and financial resources is computed with data on secure access to land assets and non-land assets, access to formal financial services and workplace rights. Lastly, the restricted civil liberties index covers data on citizenship rights, political voice, freedom of movement and access to justice. The SIGI and its sub-indices range from zero for no discrimination to one for very high discrimination. Based on the conceptual framework, other variables like energy prices, accessibility and affordability, social protection and environmental changes should have been included. However, microdata on these variables are not available for all countries under study.

Table 11.6 denotes the difference in electricity access among African women across the five indicators. Electricity access appears to be much lower among those women living in countries where the SIGI index is on the high side. A similar picture is noted across all sub-indices where energy access is very poor in the categories of high restricted resources, restricted liberties, restricted physical integrity and high family discrimination compared to access to electricity across women in households with lower restrictions and discrimination. The findings are in line with Tully (2006) where the elimination of discrimination against women at all levels is argued as a pre-condition to ensure adequate living conditions for women in relation to access to housing, sanitation, electricity, water supply,

Table 11.6 Macro-level dynamics of African women surveyed

	Women (%)	*No electricity (%)*	*Has electricity (%)*		*Women (%)*	*No electricity (%)*	*Has electricity (%)*
SIGI				**Restricted Physical Integrity**			
Low SIGI	36.87	65.90	34.10	Low Restricted Physical Integrity	–	–	–
Medium SIGI	54.10	73.70	26.30	Medium Restricted Physical Integrity	15.85	66.24	33.76
High SIGI	9.03	77.42	22.58	High Restricted Physical Integrity	84.15	76.65	23.35
Restricted Liberties				**Discrimination in the Family**			
Low Restricted Liberties	53.85	69.97	30.03	Low Discrimination in the Family	2.13	44.88	55.12
Medium Restricted Liberties	32.06	73.21	26.79	Medium Discrimination in the Family	31.42	74.19	25.81
High Restricted Liberties	14.09	77.38	22.62	High Discrimination in the Family	66.46	76.34	23.66
Restricted Resources							
Low Restricted Resources					–	–	–
Medium Restricted Resources					13.95	69.36	30.64
High Restricted Resources					86.05	75.92	24.08

Source: Author's Compilation from DHS Surveys

transport and communications. These discriminations and inequalities remain embedded in cultural ideals and social relationships (Kabeer, 2005), hence accentuating the gendered energy deprivation.

Conclusion

The chapter adopts an intersectional perspective of the gender-energy poverty nexus across 20 Sub-Saharan African countries over the period 2010 to 2015. Using household survey data, the study probes into how the intersection of gender with other axes of social differences can lead to gendered energy poverty. We extend the framework of Crenshaw (1991) and Großmann and Kahlheber (2017), and group the different axes of inequality into micro-, meso- and macro-level dynamism that influence the gender-energy poverty link. The data reveals that women who are vulnerable to poverty, have low education levels, have difficult access to health and transport services, spend more time in care and domestic work as well as those women victims of intimate partner violence and live in a patriarchal relationship are more likely to be energy poor. Similarly, women who do not participate in the various decisions within their household and who face discrimination in the family and restrictions in terms of access to resources, restricted physical integrity and restricted liberties are energy-deprived. There is evidence that intertwined factors and social differences exacerbate the existing gender inequalities and thereafter access to energy. Electrification is considered a crucial factor in alleviating women's time poverty, creating opportunities for women to enter the labour force or young girls to focus on schools and reducing exposure to intimate partner violence. Social and gender norms seem to represent a major challenge in the various inequalities and discriminations which women face. Changing norms and perceptions is a long-term process and in the meantime, women are deprived of many essential services including energy access, which can give them a better life and reduce the many risks and challenges they are currently facing. It is very important when analysing gendered energy poverty to take an intersectional gender approach by combining the social, economic, political and environment factors that have a greater impact on women and hence increase their vulnerability to energy poverty. Integrating a gender perspective into policies, programmes and projects will allow the development of relevant interventions that will address gender inequalities and discrimination and as such help in meeting the energy needs of both men and women.

One limitation of the analysis is the time dimension of the data. More recent data on a large sample of African countries could be used when more up to date Demographic Health Surveys are made available. Latest and comparable data will also help in adopting a comparative analysis of gender energy poverty across specific countries. This will provide an in-depth analysis on how the intersection of gender with other axes of social differences can lead to gendered energy poverty at the national level. It will better portray how these different dynamics are influenced by national geographic placement. Further, the analysis has adopted a quantitative approach, but a qualitative methodology may also be used together with the quantitative work, in order to assess many dimensions which are country specific and which are unmeasurable by either the survey data or country-level data.

References

Abuya, B. A., Ciera, J., & Kimani-Murage, E. (2012). Effect of mother's education on child's nutritional status in the slums of Nairobi. *BMC Pediatrics, 121*, 1–10.

Abuya, B. A., Mutisya, M., & Ngware, M. (2015). Association between mothers' education and grade six children numeracy and literacy in Kenya. *Education, 3–13*(43(6)), 653–665.

Acharya, R. H., & Sadath, A. C. (2019). Energy poverty and economic development: Household-level evidence from India. *Energy and Buildings, 183*, 785–791.

Anjorin, S., & Yaya, S. (2021). Anaemia among under-five children: Is maternal marriage at 18th birthday and above protective? Evidence from 15 countries in Sub-Saharan Africa. *Maternal & Child Nutrition, e13226.*

Annan, J., Donald, A., Goldstein, M., Gonzalez Martinez, P., & Koolwal, G. (2019). *Taking power*. World Bank.

Antepara, I., Papada, L., Gouveia, J. P., Katsoulakos, N., & Kaliampakos, D. (2020). Improving energy poverty measurement in Southern European regions through equalisation of modelled energy costs. *Sustainability, 12*(14), 5721.

Azocar, M., & Ferree, M. (2016). Engendering the sociology of expertise. *Sociology Compass., 10*, 1079–1089.

Azomahou, T. T., & Yitbarek, E. A. (2016). Intergenerational education mobility in Africa: Has progress been inclusive? *World Bank Policy Research Working Paper, 7843.*

Banerjee, R., Mishra, V., & Maruta, A. A. (2021). Energy poverty, health and education outcomes: Evidence from the developing world. *Energy Economics, 101*, 105447.

Benería, L., & Roldan, M. (1987). *The crossroads of class and gender: Industrial homework, subcontracting, and household dynamics in Mexico City*. University of Chicago Press.

Boardman, B. (1991). *Fuel poverty: From cold homes to affordable warmth*. Pinter Pub Limited.

Boardman, B. (2013). *Fixing fuel poverty: Challenges and solutions*. Routledge.

Bouzarovski, S., & Petrova, S. (2015). A global perspective on domestic energy deprivation: Overcoming the energy poverty–fuel poverty binary. *Energy Research & Social Science, 10*, 31–40.

Bouzarovski, S., Petrova, S., & Sarlamanov, R. (2012). Energy poverty policies in the EU: A critical perspective. *Energy Policy, 49*, 76–82.

Brunner, K. M., Spitzer, M., & Christanell, A. (2011). NELA-nachhaltiger Energieverbrauch und Lebensstile in armen und armutsgefährdeten Haushalten.

Buzar, S. (2006). Estimating the extent of domestic energy deprivation through household expenditure surveys. *CEA Journal of Economics, 1*(2).

Cecelski, E., Dunkerley, J., & Ramsay, W. (2015). *Household energy and the poor in the Third World*. Routledge.

Churchill, S. A., & Smyth, R. (2021). Energy poverty and health: Panel data evidence from Australia. *Energy Economics, 97*, 105219.

Clancy, J., & Feenstra, M. (2019). *Women, gender equality and the energy transition in the EU*. Publications Office of the European Union.

Clancy, J., & Roehr, U. (2003). Gender and energy: Is there a Northern perspective? *Energy for Sustainable Development, 7*(3), 44–49.

Clancy, J., Ummar, F., Shakya, I., & Kelkar, G. (2007). Appropriate gender-analysis tools for unpacking the gender-energy-poverty nexus. *Gender & Development, 15*(2), 241–257.

Clancy, J. S., Daskalova, V. I., Feenstra, M. H., & Franceschelli, N. (2017). Gender perspective on access to energy in the EU.

Clark, L. (2021). Powering households and empowering women: The gendered effects of electrification in Sub-Saharan Africa. Powering households and empowering women: The gendered effects of electrification in Sub-Saharan Africa. Retrieved October 10, 2021, from https://jpia.princeton.edu/news/powering-households-and-empowering-women-gendered-effects-electrification-sub-saharan-africa

Crenshaw, K. (1991). Mapping the margins: Intersectionality, identity, politics, and violence against women of colour. *Stanford Law Review, 43*(6), 1241–1299.

Danielsen, K. (2012). Gender equality, women's rights and access to energy services. A paper presented to the Ministry of Foreign Affairs of Denmark.

Day, R., Walker, G., & Simcock, N. (2016). Conceptualising energy use and energy poverty using a capabilities framework. *Energy Policy, 93*, 255–264.

Donev, J., Lyndon, G., Hanania, J., & Pomerantz, C. (2018). Natural vs anthopogenic climate change. Energy Education. Retrieved October 2, 2021, from https://energyeducation.ca/encyclopedia/Natural_vs_anthropogenic_climate_change

Dutta, S., Kooijman, A., & Cecelski, E. (2017). *Energy access and gender*. International Bank for Reconstruction and Development. The World Bank, Washington, DC.

Espinoza-Delgado, J., & Klasen, S. (2018). Gender and multidimensional poverty in Nicaragua: An individual based approach. *World Development, 110*, 466–491.

Feenstra, M., & Özerol, G. (2021). Energy justice as a search light for gender-energy nexus: Towards a conceptual framework. *Renewable and Sustainable Energy Reviews, 138*, 110668.

Gakidou, E., Cowling, K., Lozano, R., & Murray, C. J. (2010). Increased educational attainment and its effect on child mortality in 175 countries between 1970 and 2009: A systematic analysis. *The lancet, 3769745*, 959–974.

Gonzalez Pijuan, I. (2018). Gender inequality and energy poverty: A forgotten risk factor.

González-Eguino, M. (2015). Energy poverty: An overview. *Renewable and Sustainable Energy Reviews, 47*, 377–385.

Großmann, K., & Kahlheber, A. (2017). Energy poverty in an intersectional perspective: On multiple deprivation, discriminatory systems, and the effects of policies. In *Energy poverty and vulnerability* (pp. 12–32). Routledge.

Hailemariam, A., Sakutukwa, T., & Yew, S. L. (2021). The impact of energy poverty on physical violence. *Energy Economics, 105336.*

Halkos, G. E., & Gkampoura, E. C. (2021). Evaluating the effect of economic crisis on energy poverty in Europe. *Renewable and Sustainable Energy Reviews, 144*, 110–981.

IEA. (2020). *World Energy Outlook 2020.* IEA. https://www.iea.org/reports/world-energy-outlook-2020

Johnson, O. W., Han, J. Y. C., Knight, A. L., Mortensen, S., Aung, M. T., Boyland, M., & Resurrección, B. P. (2020). Intersectionality and energy transitions: A review of gender, social equity and low-carbon energy. *Energy Research & Social Science, 70*, 101774.

Kabeer, N. (2005). Gender equality and women's empowerment: A critical analysis of the third millennium development goal 1. *Gender & Development, 13*(1), 13–24.

Köhlin, G., Sills, E. O., Pattanayak, S. K., & Wilfong, C. (2011). Energy, gender and development: What are the linkages? Where is the evidence?

Kooijman-van Dijk, A. (2020). ENERGIA's gender and energy research programme: Findings and experience from research for policy. *IDS Bulletin, 51*(1).

Lam, N. L., Smith, K. R., Gauthier, A., & Bates, M. N. (2012). Kerosene: A review of household uses and their hazards in low-and middle-income countries. *Journal of Toxicology and Environmental Health, Part B, 15*(6), 396–432.

Lawson, R., Williams, J., & Wooliscroft, B. (2015). Contrasting approaches to fuel poverty in New Zealand. *Energy Policy, 81*, 38–42.

Lewis, J. (2014). Short-run and long-run effects of household electrification. In Economic History Workshop, Queens University, Kingston, Ontario, Canada.

Li, K., Lloyd, B., Liang, X. J., & Wei, Y. M. (2014). Energy poor or fuel poor: What are the differences? *Energy Policy, 68*, 476–481.

Listo, R. (2018). Gender myths in energy poverty literature: A critical discourse analysis. *Energy Research & Social Science, 38*, 9–18.

Llorca, M., Rodriguez-Alvarez, A., & Jamasb, T. (2020). Objective vs. subjective fuel poverty and self-assessed health. *Energy Economics, 87*, 104736.

McDowell, L. (1999). *Gender, identity and place: Understanding feminist geographies*. University of Minnesota Press.

McDermott, R., & Cowden, J. (2014). Polygyny and violence against women. *Emory LJ, 64*, 1767.

Moore, R. (2012). Definitions of fuel poverty: Implications for policy. *Energy Policy, 49*, 19–26.

Moreno-Bella, E., Willis, G. B., & Moya, M. (2019). Economic inequality and masculinity–femininity: The prevailing perceived traits in higher unequal contexts are masculine. *Frontiers in Psychology, 10*, 1590.

Moser, C., & Norton, A. (2001). To claim our right: Livelihood security, human rights and sustainable development.

Munien, S., & Ahmed, F. (2012). A gendered perspective on energy poverty and livelihoods advancing the Millennium Development Goals in developing countries. *Agenda, 26*(1), 112–123.

Nadasen, N. (2012). Rural women's access to land in sub-Saharan Africa and implications for meeting the Millennium Development Goals. *Agenda, 26*(1), 41–53.

Nguyen, C. P., & Su, T. D. (2021). The 'vicious cycle 'of energy poverty and productivity: Insights from 45 developing countries. *Environmental Science and Pollution Research*, 1–18.

OECD. (2018). *Education at a Glance 2018: OECD Indicators*. OECD Publishing. https://doi.org/10.1787/eag-2018-en

OECD. (2021a). *Going for Growth 2021: Shaping a Vibrant Recovery*. OECD Publishing.

OECD. (2021b). Social institutions and gender index. Retrieved September 30, 2021, from https://www.genderindex.org/sigi/

Ogunbiyi, D. (2020). OPINION: Power in a pandemic-why energy access matters during coronavirus.

Özer, M., Fidrmuc, J., & Eryurt, M. A. (2018). Maternal education and childhood immunization in Turkey. *Health Economics, 27*(8), 1218–1229.

Pachauri, S., & Rao, N. D. (2013). Gender impacts and determinants of energy poverty: Are we asking the right questions? *Current Opinion in Environmental Sustainability, 5*(2), 205–215.

Pachauri, S., & Spreng, D. (2004). Energy use and energy access in relation to poverty. *Economic and Political Weekly*, 271–278.

Pachauri, S., & Spreng, D. (2011). Measuring and monitoring energy poverty. *Energy Policy, 39*(12), 7497–7504.

Petrova, S., & Simcock, N. (2021). Gender and energy: Domestic inequities reconsidered. *Social & Cultural Geography, 22*(6), 849–867.

Primc, K., Dominko, M., & Slabe-Erker, R. (2021). 30 years of energy and fuel poverty research: A retrospective analysis and future trends. *Journal of Cleaner Production, 127003.*

Pueyo, A., & Maestre, M. (2019). Linking energy access, gender and poverty: A review of the literature on productive uses of energy. *Energy Research & Social Science, 53,* 170–181.

Rathi, S. S., & Vermaak, C. (2018). Rural electrification, gender and the labor market: A cross-country study of India and South Africa. *World Development, 109,* 346–359.

Robinson, C. (2019). Energy poverty and gender in England: A spatial perspective. *Geoforum, 104,* 222–233.

Robinson, C., Yan, D., Bouzarovski, S., & Zhang, Y. (2018). Energy poverty and thermal comfort in northern urban China: A household-scale typology of infrastructural inequalities. *Energy and Buildings, 177,* 363–374.

Sánchez, C. S. G., Fernández, A. S., & Peiró, M. N. (2020). Feminisation of energy poverty in the city of Madrid. Energy and Buildings, 223, 110157.

Sen, A. (1976). Poverty: An ordinal approach to measurement. *Econometrica: Journal of the Econometric Society,* 219–231.

Sievert, M. (2015). *Rural electrification and domestic violence in Sub-Saharan Africa* (p. 570). *Ruhr Economic Papers.*

Sovacool, B. K., Hess, D. J., Amir, S., Geels, F. W., Hirsh, R., Medina, L. R., et al. (2020). Sociotechnical agendas: Reviewing future directions for energy and climate research. *Energy Research & Social Science, 70,* 101617.

Tandrayen-Ragoobur, V. (2020). Intimate partner violence and women's labour force participation in Sub Saharan Africa. *Community, Work & Family, 23*(1), 19–43.

Thomson, H., Snell, C., & Bouzarovski, S. (2017). Health, well-being and energy poverty in Europe: A comparative study of 32 European countries. *International Journal of Environmental Research and Public Health, 14*(6), 584.

Thomson, H., Snell, C. J., & Liddell, C. (2016). Fuel poverty in the European Union: A concept in need of definition? *People, Place & Policy Online,* 5–24.

Townsend, P. (1979). *Poverty in the United Kingdom: A survey of household resources and standards of living.* University of California Press.

Tully, S. (2006). The human right to access electricity. *The Electricity Journal, 19*(3), 30–39.

UN Stats (2021). The Sustainable Development Goals Report, 2021. United Nations Statistics Division, Development Data and Outreach Branch New York, NY 10017.

Venkataraman, C., Sagar, A. D., Habib, G., Lam, N., & Smith, K. R. (2010). The Indian national initiative for advanced biomass cookstoves: The benefits of clean combustion. *Energy for Sustainable Development, 14*(2), 63–72.

Winther, T., Ulsrud, K., & Saini, A. (2018). Solar powered electricity access: Implications for women's empowerment in rural Kenya. *Energy Research & Social Science, 44*, 61–74.

Winther, U., Skontorp Hognes, E., Jafarzadeh, S., & Ziegler, F. (2020). *Greenhouse gas emissions of Norwegian seafood products in 2017.* SINTEF Ocean.

WWAP (United Nations World Water Assessment Programme). (2015). *The United Nations World Water Development Report 2015: Water for a Sustainable World*. UNESCO.

Živčič, L., González-Pijuan, I., Guiteras Blaya, M., Mazreku, V., & Tirado-Herrero, S. (2021). Combatting energy poverty with gender just policies. In *Why the European Green Deal needs Ecofeminism – Moving from gender-blind to gender-transformative environmental policies.* EEB and WECF.

CHAPTER 12

Participants or Recipients? Negotiating Gender and Energy as Empowerment in the Displaced Setting

Alison Halford

Introduction

In the first months of 2022, more than 100 million individuals were displaced worldwide (UNHCR, 2022a). According to the United Nations High Commission for Refugees (UNHCR), this increase means forcibly displaced people represent over 1% of the global population or is equivalent to the 14th most populous country in the world. In countries such as Ukraine, Ethiopia, Burkina Faso, Myanmar, Nigeria, Afghanistan, and the Democratic Republic of the Congo, war and political instability have seen people forcibly displaced internally, which sees them leave their homes but still remain within their own national borders. While over 30 million people become refugees, after fleeing their country, often to neighbouring countries, with some finding themselves housed in refugee camps

A. Halford (✉)
Centre for Computational Science and Mathematical Modelling, Coventry University, Coventry, UK
e-mail: ad4480@coventry.ac.uk

N. Rocha Lawton, C. Forson (eds.), *Women and the Energy Sector*,
https://doi.org/10.1007/978-3-031-43091-6_12

(UNHCR, 2021). With conflicts becoming more protracted, resulting in continuing waves of violence or human rights violations, these camps are rarely temporary structures, with a duration of stay between 5–20 years for those refugees (World Bank, 2019).

An estimated seven million displaced people in camps have access to electricity for less than four hours a day and heavily depend on traditional biomass and kerosene for household cooking (Lahn & Grafham, 2015). Forcibly displaced women and young children are more vulnerable to health implications arising from inefficient cookstoves and poor ventilation caused by continuing use of fossil-fuelled and wood-burning indoor fires and cookstoves (Sovacool et al., 2013; World Health Organization, 2016). If the situation is temporary, the distribution of free energy products will meet the basic needs of displaced people by preventing starvation and providing basic lighting for safety, but as humanitarian situations become protracted, these mechanisms are not suitable or sufficient to meet the continuing energy needs of refugees (Campbell & Knox Clarke, 2018).

To improve life outcomes for refugees and address the increasing demands for energy in camps, the United Nation's Clean Energy Challenge (2019) is working towards all refugee settlements having access to modern, safe, and reliable energy by 2030. New technologies using renewables, such as solar microgrids, clean energy cookstoves, and solar lighting, can be instrumental in a transition to safe and adequate energy that can respond to the complexity of the humanitarian setting (Nixon et al., 2021). Yet, camp-based refugees' deployment and uptake of energy systems using new technologies are beset with implementation difficulties, high upfront costs, and concerns around long-term sustainability (Lehne et al., 2016). Therefore, new frameworks and design of energy models are needed to ensure households have reliable, affordable energy sources, communities have street lighting and power for basic public services like health and water pumping, and humanitarian operations have sufficient low-carbon electricity.

Central to these innovative energy systems should be the adoption of a gender perspective that involves men and women in new practices around energy (Dutta et al., 2017; Winther et al., 2020). One way is to design energy solutions that shift ideas around women as passive users of energy to actively encourage women to become participants, suppliers, designers, policymakers, and stakeholders in energy solutions in the displaced setting (Bradley, 2019). Another way is to question how new forms of energy services, like solar-based technologies, are gendered in the humanitarian setting and whether new technologies, such as advanced cookstoves, are reinforcing or combating gender inequality (Clancy, Skutsch, & Batchelor, 2003.

This chapter will explore to what extent improved energy access using renewables benefits men and women equally in the following three ways. First, it will start with an overview of discourse around energy, gender, and the humanitarian setting to give insights into how camp-based women and young girl refugees are framed in the literature. Second, drawing upon the work of the Humanitarian Engineering and Energy for Displacement (HEED) project with Congolese refugees in Rwanda and internally displaced people in Nepal, it will discuss how women and young girls can engage in meaningful participation in the design, implementation, and maintenance of renewable energy systems. Finally, the chapter will conclude on gendered negotiations, if any, that occur for women and young girls to shift from recipients to participants of energy systems in the humanitarian setting. By bringing to the fore the ways the improved access to energy can be instrumental in combating gender inequality, this can aid decision-making in energy planning, procurement, and delivery for greater gender equity for all people living in refugee camps.

Energy, Gender, and the Humanitarian Setting

In aiming for safe, modern, affordable, and sustainable energy for all, Sustainable Development Goal 7 (SDG 7) encompasses how improved access to energy can increase economic development, build resilient infrastructures, promote gender empowerment, and increase health and well-being outcomes. Humanitarian engineering has a critical opportunity to meet SDG 7 when designing and delivering energy projects and address the gendered impacts of poor access to energy. For example, women and young girls have suffered more from health implications arising from inefficient cookstoves and poor ventilation by the continuing use of fossil-fuelled and wood-burning indoor fires and cookstoves (Ritchie & Roser, 2019). While approaches to gender-based violence (GBV) and energy should seek more complex narratives, with energy interventions, such as street lighting, seen as one action amongst many to combat all forms of violence against women in displaced camps (Listo, 2018).

Energy poverty contributes to food insecurity in the refugee settlements as the scarcity of fuelwood or energy sources for cooking can see camp-based women refugees do most of the cooking, having to miss meals, reduce portion sizes, or exchange food for cooking fuel (Pueyo & Maestre, 2019). Similarly, carrying firewood bundles weighing 20 kg and over has seen some refugee women and young girls at much greater risk of short- and long-term physical injury (Rogers, Sovacool, & Clarke, 2013). Having to shoulder the responsibility of providing and conserving energy, women and young girls are required to spend more time collecting water and fuel, which reproduces unequal gendered tasks in the home (Malakar & Day, 2020). The additional burden of these activities impacts women's and young girls' ability to secure and improve livelihoods, access information, and participate in decision-making at the household and community levels (Bradley & Liakos, 2019). As a result, women are more inclined to become involved with less energy-intensive enterprises that limit their ability to access the full benefits of more productive use of energy interventions (Pueyo & Maestre, 2019; Rafa et al., 2022).

Humanitarian energy design and deployment protocols would benefit from understanding how the socio-cultural context can reinforce the gendered division of labour in the domestic and public spaces (Rewald, 2017). Socio-technical energy design frameworks are a useful tool when working to address gender inequality and access to energy (Sovacool et al., 2020). Conceptualising design frameworks through a socio-technical lens will amplify the voices of women and young girls as the method ensures end users are acknowledged in how technology is applied, integrated or adapted (Rohracher, 2003). There is also the potential to refute 'techno saviourism' that oversimplifies complex socio-political challenges as locating technology within everyday practices, policies, and values will evidence to what extent these interventions benefit all groups within society.

However, humanitarian responses to energy needs often have been reactive, resulting in arbitrary solutions that seek the distribution of no cost to refugee products and services and little or no discussion around longevity (Thomas, Rosenberg-Jansen, & Jenks, 2021). This approach has seen funding priorities prefer approaches and models for energy programmes and planning to adopt a technological-led framework, which may fail to meet the needs of women and girls, or address broader gender issues (Barbieri, Riva, & Colombo, 2017). Arguably, this has resulted in projects focusing more on women and young girls as the main firewood collectors and positioning their energy needs and aspirations around domestic chores. This includes overstating the role of technological solutions, such as alternative fuels or cookstoves, in resolving violence against women (Abdelnour & Saeed, 2014).

Energy interventions that reduce insecurity and sexual and gender-based violence by increasing protection and safety for women are critical in the displaced setting. The literature on gender, refugees, and energy spanning almost two decades continually re-emphasises that women and girls are disproportionality subject to all forms of gender-based violence (GBV) when collecting firewood (Global Alliance for Clean Cookstoves, 2016). There are also increased risks due to poor public lighting resulting in reduced privacy and security in camps and displacement settings (Corbyn & Vianello, 2018). Women and young girls living in refugee camps have been subject to or fearful of acts of sexual violence when visiting the latrines or meeting friends and family at night (Aubone & Hernandez, 2013). In a context where people are experiencing heightened trauma and poverty, along with the destabilising of social structures, violence against women and girls can become normalised in the refugee setting (Jensen, 2019). This impacts their psychological and physical well-being and places considerable restrictions on women and young girls' freedom of movement.

At times, the lack of understanding of the embedded nature of GBV in societal structures and norms has seen engineering projects generate simplistic equations that less wood collecting for cooking will automatically reduce rape, sexual assault and physical abuse (Listo, 2018). Energy-saving cooking stoves can reduce wood usage, but operational roadblocks and economic demands still force refugee women and girls to collect wood to sell, which continues to make them vulnerable to acts of violence (Malakar & Day, 2020). Samer Abdelnour and Akbai Saeed (2014) noted in their study of refugee women in Darfur that suggesting 'stoves reduces rape' in the refugee context causes an 'immensely complex crisis' to be seen as 'manageable problems'. Focusing solely on tangible energy solutions, such as improving cookstoves, as the only means of challenging GBV reinforces the representation of refugee women and girls as passive rather than agential and could limit engagement with women-led community-based responses.

The issue of GBV is complex as it reflects the power imbalance between genders, embedded cultural inequalities, and normative values that reinforce gender hierarchy. Violence against women should not be seen as isolated incidents, but rather as a recurring pattern of behavior that can manifest in various ways. Involving women and young girls in co-designing energy interventions could indeed develop substantive dialogues on which energy interventions are best at challenging

GBV. Moreover, there is a move away from essentialising women and young girls to a greater discussion on how refugee women and young girls challenge gender stereotypes in significant ways through improved access to energy (Irena, 2019).

The Case of HEED

From 2017 to 2021, the Humanitarian Engineering and Energy for Displacement (HEED) project worked with Congolese refugees living for protracted periods in three refugee camps in Rwanda (Nyabiheke, Gihembe and Kigeme) and internally displaced persons (IDPs) forced to leave their homes as a result of the 2015 earthquake in Nepal. The HEED project aimed to understand the energy needs of forcibly displaced people to increase access to safe, sustainable, and affordable energy services. HEED explored the potential of alternative energy ownership models informed by sensor data and using renewable technologies.

Rwanda

One of the smallest countries in Africa, Rwanda is a landlocked country in Central Africa that borders Tanzania, Burundi, the Democratic Republic of the Congo (DRC), and Uganda. With a population of just under 13 million, Rwanda host around 127,112 refugees, mainly from Burundi and the Democratic Republic of the Congo (DRC), of which 90% live in six camps (UNHCR, 2021). About 75% of refugees are women and children (UNHCR, 2021). In 2019, the Rwandan government committed to including energy and environmental preservation as a mechanism for refugee integration. HEED worked in three of the six refugee camps in Rwanda: Gihembe, Kigeme, and Nyabiheke.

Gihembe Refugee Camp, Gicumbi District, Northern Province: One of the smaller but longest-established refugee camps in Rwanda, it was created in 1997 in Gicumbi District in Northern Province. In December 2021, the camp was closed, and refugees relocated to Mahama camp, Kirehe District, Eastern Province (UNHCR, 2022b).

Kigeme Refugee Camp, Gatsibo District, Eastern Province, Rwanda: The opening of Kigeme in Nyamagabe District in Southern Province in 2012 was in response to a sudden influx of refugees from the DRC as they fled violence between government forces and rebel militias. In 2021 Kigeme Camp was home to around 15,000 refugees (UNHCR, 2022c).

Nyabiheke Refugee Camp, Nyamagabe District, Southern Province, Rwanda: Since the early 2000s, refugees from the DRC have been arriving at Nyabiheke Camp, Gatsibo District, Eastern Province, which currently hosts around 14,000 refugees (UNHCR, 2022d).

Most refugees in the camps reside in metal-roofed mudbrick dwellings with no connection to the main electricity grid (Sandwell et al., 2020). Traditional mud (clay) cookstoves or three-stone stoves are the primary heating and cooking methods, and a regular lighting source was either mobile phones or candles. UNHCR provide access to cooking energy through cash assistance programs. (Halford et al., 2022). Most cooking occurs inside the homes or in a separate building with no ventilation, with most households' fuel stack' (Gaura et al., 2020). For the past four years, until it ceased operating, the cookstove company Inyenyeri in partnership with the UNHCR, provided free fuel-efficient stoves that used sustainable biomass fuel pellets purchased from the suppliers (Irena, 2017).

Nepal

Nepal is located at the junction of the Indian and Eurasian tectonic plates, making it especially prone to earthquakes. In 2015, the Gorkha earthquake and subsequent landslides affected 31 of Nepal's 75 districts, saw 8856 people dead, and around 750,000 private houses damaged or destroyed. According to the Internal Displacement Monitoring Centre (2022), over 29,000 people are still internally displaced in Nepal. Many families displaced by the 2015 earthquake in Rasuwa were moved to various parts of the district by the National Reconstruction Authority (NRA) in the Uttargaya Rural Municipality and Gosaikunda Rural Municipality. According to the District Project Implementation Unit of the NRA, the government provided grants to families to buy land to rebuild their houses, but it has been suggested that there is still a gap in housing provision for internally displaced people (Gihemere, 2020).

HEED worked with an internally displaced community formed after the 2015 earthquakes in Khalte, Rasuwa, district, Nepal.

Khalte, Uttargaya Municipality, Rasuwa district, Nepal: The camp was an informal settlement and hosted around 220 households, with 90% of the camp population identified as Tamang. Tamangs are one of the largest indigenous groups in Nepal, with their own distinct language, culture, and social structures. The majority of the camp residents lived in one or two-room tin houses and were connected to the national grid for general household energy needs. The camp was managed by a Camp Management Committee, which are authorised to negotiate and make decisions on behalf of the camp.

Methodology

HEED project had a four-stage methodology. The first stage of the project was data collection using sensors and surveys. In Nepal, the sensors in footfall monitors and individual energy appliance monitors documented electricity usage, costs, and sufficiency in grid-connected sub-metered scenarios. Over 1,000 survey responses were collected on existing energy provision and refugees' energy needs and aspirations (Bhargava et al., 2022).

Informed by the quantitative and qualitative research data gathered from stage one, the next stage of the project HEED delivered 'Design for Displacement (D4D)' workshops. These workshops brought together policymakers in the UK, Rwanda, and Nepal with social entrepreneurs, academics, manufacturers, distributors, and maintenance services in the off-grid energy and ICT sector, to develop energy designs. In addition, refugees and displaced people were involved in the design and planning process through 12 'Energy for End users (E4E)' workshops in four displacement settings (three in Rwanda and one in Nepal). These workshops targeted a number of key user groups, including young people, women, social entrepreneurs, and local business members.

The third stage of the project was the piloting of community co-designed solar interventions. In Khalte, Nepal, the interventions were seven advanced solar streetlights. In Rwanda, HEED installed a stand-alone solar system for a community hall and 40 solar mobile lanterns in Nyabiheke; a PV-battery micro-grid for two nursery buildings and a playground in Kigeme; 8 solar streetlights and four advanced solar streetlights in Gihembe. The insights are summarized in Table 12.1.

Table 12.1 Overview of data collected, interventions, and handover of interventions

Location	*Data collected*	*Intervention 1*	*Intervention 2*	*Handover*
Gihembe Refugee Camp, Gicumbi District, Northern Province, Rwanda	Surveys Workshops Footfall Sensor data	Four advanced solar streetlights provide both night-time lighting and ground-level socket	Eight regular solar streetlights	Alight (NGO) Managed by community leaders
Nyabiheke Refugee Camp, Nyamagabe District, Southern Province, Rwanda	Surveys Workshops sensors on solar mobile lanterns,	One Standalone Solar System supplying electricity for a community hall	40 mobile solar lanterns	Alight (NGO) Managed by the community leaders
Kigeme Refugee Camp, Gatsibo District, Eastern Province, Rwanda	Sensors on cook stoves Surveys workshops	PV-battery micro-grid to supply electricity to a playground and two nursery buildings	Sensors-based cooking behaviour study with 20 households	World Vision Managed by the community leaders
Internally Displaced Camp, Khalte, Uttargaya Municipality, Rasuwa District, Nepal	Surveys Workshops Footfall Sensor data	Seven advanced solar street lights	Electrical appliances use monitored in 20 households	Community Leaders

Engaging Women and Young Girls in Meaningful Participation in the Design, Implementation, and Maintenance of Energy Systems

A key strategic focus of the HEED Project was understanding gendered barriers to energy access and energy-driven services. To identify how gender informs decision-making around energy, the team began by asking in what ways gendered divisions of labour and power shape differential energy access, usage, and needs. Many energy programmes, surveys, and research studies do include sessions with all-female groups to encourage participation in the research and listen to women's energy aspirations and needs (Sandwell et al., 2020). Typically, however, discussions tend to frame women as the primary acquirers and users of energy in the home and as sole beneficences when switching to cleaner fuels for cooking and other uses in the domestic space (Bradley & Liakos, 2019). In doing so, involving only women in discussions around household energy can reinforce traditional gender roles within the household and within societies (Wilhite, 2017) and assume that improvements in the domestic space are sufficient to address gender inequality in access to energy.

There is also the critical and often overlooked factor of how male heads of households can hold decision-making power over finances that could impact how women decide what fuel, cookstoves, and cooking utensils are to be used (Clancy & Dutta, 2005). For example, during the D4D workshops in Nepal, there were clear gender differences when participants were asked, 'if interventions resulted in less costly electricity, how would you use savings?' Men indicated the savings would be used to purchase additional appliances such as a fridge and a heater. Responses from women were more centred on immediate family needs: spending the savings on children's school expenses, more food or investing or expanding their businesses. Therefore, programmes and policies must explicitly consider the gender dynamics that inform not only the division of labour in the home but also how unequal gendered roles limit their household decision-making. Instead, there needs to be an account of how energy access impacts both women *and* men to identify more gender-responsive solutions (Winther et al., 2017).

Recognising gender differences in energy needs, use, and priorities is critical because energy interventions that improve access to energy for communities are not necessarily the interventions with the greatest positive impact on women (Govindan & Murali, 2020). This can be seen during the feedback sessions on the impact of the electrification of the community hall in Nyabiheke refugee camp, Rwanda, that allowed for a TV to be installed and used, where the overriding consensus was that men benefited more from this intervention than women:

> The boys do receive more benefits after the lights were introduced. Household activities are assigned to girls, and the boys fetch water. After fetching water, the boys can come to watch TV in the Hall while the girls are still doing housework at home. *(Male, 35 years old, Congolese, Nyabiheke Camp)*

> That is true. Women are busy at home more than men with household activities. Men have more free time, so they benefit more from the

> Community Hall lights. Men have more freedom to move around. The men might watch some TV as well while picking up their phone. Women can't so easily come to pick up the phone during the day. *(Female, 22 years old, Congolese, Nyabiheke Camp)*

Domestic responsibilities not only mean women and young girls have less freedom on how to spend their time during the day but also impact how they spend the evenings. When HEED installed lighting in the playground at Gihembe camp to encourage more engagement with sport, the intervention unequally privileged men as they could maximise the facilities both day and night:

> Different people, including young males, females and even adults, use the playground for sporting activities. In particular, it is an important place where the youth play when they return from school. During the day, both the males and females use it, but mainly the males use it at night. It is difficult for girls to come out at night because parents expect them to be home to help with domestic work. So, the females respect our culture by engaging in domestic activities. *(Male, 19 years old, Congolese, Gihembe Camp)*

Improving access to energy will not necessarily improve women's lives unless there are mechanisms to challenge longstanding traditions that limit autonomy and self-determination, both at the household level but also in society in general. This means, as Clancy et al. (2012) state, that it is not the energy technology that benefits women but the consequence of the technology on the women's lives that informs the outcomes. Hence, it is somewhat simplistic to think that involving women in a discussion about improving access to energy, for example, the use and type of cookstoves, will automatically dismantle entrenched inequality or bring about female empowerment.

Significant changes that promote egalitarian practices require more project and policy emphasis on incorporating meaningful roles for women in planning, deploying and implementing energy programmes for domestic and public spaces. Increasing the visibility of women in energy decision-making could address negative gendered-specific impacts that impede opportunities to engage in leisure and education and develop entrepreneurial enterprises.

Participants or Recipients? Empowering Women and Young Girls Through Improved Access to Energy

Humanitarian crises destabilise family and social networks, which can amplify feelings of anomie and anxiety. One way to build emotional resilience is to better understand gender differences and inequalities that are barriers to autonomy and agency for women. Identifying formal and informal structures that reproduce gender roles, which limit self-determination and reinforce gendered privilege, is central to the empowerment process. Another way to support well-being is through collective action driven by women's voices to address women's and young girls heightened vulnerability to gender-based violence (WRC, 2015). There is also the less recognised but equally important role that community-managed energy resources can play in facilitating new forms of social cohesion to promote greater self-sufficiency, equity and security for displaced women and girls.

At the local level, community-based organisations can be instrumental in empowering women by giving them access and the ability to exercise decision-making over these resources, whether material or cultural capital. If rules around the use and management of community resources are fair, just, and equal, decisions about energy resources will positively influence women's lives at the household level and beyond. However, when local institutions, such as councils or committees, do not represent the needs and priorities of all community members, the management of resources will replicate intracommunity inequality and prevent participation by women and other groups (e.g. people with a disability or lack of literacy) (Theis, Bryan, & Ringler, 2019).

During the E4E workshops, HEED encouraged women to identify what their energy needs and aspirations were in the home and their communities. The intention was to empower them as agents of change in the safety and integrity of the proposed intervention and to optimise the benefits they could gain from improved access to energy. For instance, as previously mentioned, when HEED installed the Standalone Solar System supplying electricity for a community hall in Nyabiheke camp, the hall was used to charge phones and watch television. Then gradually, the hall was used for large gatherings, such as marriages and funerals by the community. That women organised these activities at the hall was not unexpected, as many women at the workshops in Nepal and Rwanda expressed their unhappiness at a lack of a safe space in the camp for family celebrations. Accordingly, the HEED team responded to their request and allowed for sufficient sockets for a sound system and additional lights when sizing the system. However, as women became more confident in accessing community-managed energy, other activities emerged that sought to deal with individual and community healing and trauma recovery:

> We use the hall to remember the death of our family members who died in Congo. We used to remember death mostly during the day, but now that we have light in the hall at night, we can extend the remembrance to evenings. We watch films of events related to the war and teach our children where they come from. *(Female, 29 years old, Congolese, Nyabiheke camp)*

> We hold a remembrance of families that lost their lives during the war. It helps us to remember the death and also teaches our children who were born in Rwanda their origin. Women gather here, men too. The men are not excluded. The women started this evening but then invited men because they discussed family issues; this is applicable to everyone. *(Female, 33 years old, Congolese, Nyabiheke camp)*

For women to feel empowered, they must have the ability to make choices and can act upon them in a range of settings and across all societal structures (de Groot, Mohlakoana, Knox, & Bressers, 2017). Given the opportunity, for some women refugees, becoming engaged in managing community energy resources created spaces for other women, children, and men that supported pathways to greater wellbeing and dignity.

Conclusion

Focusing metrics on how improved energy access for women can meet either essential physical and health needs or strengthen human security fails to acknowledge the potentiality for women to mobilise energy resources as an act of empowerment in other ways. Introducing new technologies that address the lack of energy will not automatically improve the outlook for women in displaced camps unless there is an interrogation of gendering practices. If women are excluded from consultation when discussing energy needs, it can lead to solutions that do not meaningfully consider or understand the realities of energy-poor displaced women or the technologies most suited to address their needs. Most of all, it requires engineers, policymakers, humanitarian organisations, energy suppliers, and practitioners to engage with refugees to challenge fully and ultimately transform energy practices in ways that reduce rather than increase inequality for displaced populations.

Acknowledgements The author would like to acknowledge the financial support of the Engineering and Physical Science Research Council (EPSRC) for funding the Humanitarian Engineering and Energy for Displacement (HEED) project as part of the Global Challenges Research Fund (EP/P029531/1). The author would like to acknowledge and thank project delivery partners Practical Action and Scene Connect for their significant role in co-ordinating in-camp activities and providing technical inputs and tools. We would also like to recognise the support of Ministry of Disaster Management and Refugees (MIDIMAR) and United Nations High Commissioner for Refugees (UNHCR) and the contributions of the Global Plan of Action, Chatham House, and the Renewable Energy for Refugees (RE4R) Project (a partnership between Practical Action and UNHCR, supported by the IKEA Foundation).

References

Abdelnour, S., & Saeed, A. (2014). Technologizing humanitarian space: Darfur advocacy and the rape-stove panacea. *International Political Sociology, 8*(2), 145–163.

Aubone, A., & Hernandez, J. (2013). Assessing refugee camp characteristics and the occurrence of sexual violence: A preliminary analysis of the Dadaab complex. *Refugee Survey Quarterly, 32*(4), 22–40.

Barbieri, J., Riva, F., & Colombo, E. (2017). Cooking in refugee camps and informal settlements: A review of available technologies and impacts on the socio-economic and environmental perspective. *Sustainable Energy Technologies and Assessments, 22*, 194–207.

Bhargava, K., Verba, V., Nixon, J., Gaura, E., Brusey, J., & Halford, A. (2022, July 10). *EPSRC-funded humanitarian engineering and energy for displacement (HEED) datasets (v1.0) (Data set)*. Zenodo (Online). https://doi.org/10.5281/zenodo.5792260.

Bradley, T., & Liakos, K. (2019). Vulnerability of women and girls in refugee settings: Considerations for energy programming. In O. Grafham (Ed.), *Energy access and forced migration Routledge* (pp. 122–139). Routledge.

Bradley, T. L. (2019). Vulnerability of women and girls in refugee settings: Considerations for energy. In *Energy access and forced migration* (pp. 122–141). Routledge.

Clancy, J., & Dutta, S. (2005). Women and productive uses of energy: Some light on a shadowy area. *UNDP Meeting on Productive Uses of Renewable Energy* (Vol. 9, p. 11).

Clancy, J., Skutsch, M., & Batchelor, S. (2003). *The gender-energy-poverty nexus: Finding the energy to address gender concerns in development*. DFID project CNTR998521.

Clancy, J., Winther, T., Matinga, M., & Oparaocha, S. (2012). *Gender equity in access to and benefits from modern energy and improved energy technologies: World development report background paper*. ETC/ENERGIA in association Nord/Sør-konsulentene.

Corbyn, D., & Vianello, M. (2018). *Prices, products and priorities: Meeting refugees' energy needs in Burkina Faso and Kenya*. Chatham House the Royal Institute of International Affairs.

de Groot, J., Mohlakoana, N., Knox, A., & Bressers, H. (2017). Fuelling women's empowerment? An exploration of the linkages between gender, entrepreneurship and access to energy in the informal food sector. *Energy Research & Social Science, 28*, 86–97.

Dutta, S., Kooijman, A., & Cecelski, E. (2017). *Energy access and gender. Getting the right balance*. The World Bank.

Gaura, E., Bhargava, K., Verba, V., & Brusey, J. (2020). *Retrieved from EPSRC HEED data repository: Stove use monitoring system (v1.0)* (Data set). (Online). https://doi.org/10.5281/zenodo.3947000.

Gihemere, B. (2020, September 19). *Around 750 earthquake victims still living in landslide-prone settlements in Rasuwa*. Retrieved july 2022, from The Kathmandu Times: https://kathmandupost.com/national/2020/09/19/around-750-earthquake-victims-still-living-in-landslide-prone-settlements-in-rasuwa-1600481156

Global Alliance for Clean Cookstoves. (2016). *Gender-based violence in humanitarian settings: Cookstoves and fuels*. Global Alliance for Clean Cookstoves.

Govindan, M., & Murali, R. (2020). Gender inclusion in the energy policies: What are we missing? *Energy Review, 2*(7), 2–3. https://drive.google.com/file/d/1JCXrXxeFCxWgOHag-WLBULs8OLOTXiJL/view

Halford, A., Gaura, E., Bhargava, K., Verba, N., Brusey, J., & Nixon, J. (2022). Off the boil? The challenges of monitoring cooking behaviour in refugee settlements. *Energy Research & Social Science, 90*, 102603.

Internal Displacement Monitoring Centre. (2022). *Global report on internal displacment 2022*. Norwegian Refugee Council. Retrieved August 1, 2022, from https://www.internal-displacement.org/global-report/grid2022/

Irena. (2017). *Irena renewable energy statistics (online)*. Retrieved July, 2022, from https://www.irena.org/publications/2017/Jul/Renewable-Energy-Statistics-2017

IRENA. (2019). *Renewable energy: A gender perspective.* IRENA.
Jensen, M. (2019). Gender-based violence in refugee camps: Understanding and addressing the role of gender in the experiences of refugees. *Inquiries Journal (Online), 11*, 11. Retrieved July 18, 2022, from http://www.inquiriesjournal.com/a?id=1757
Lahn, G., & Grafham, O. (2015). *Heat, light and power for refugees saving lives, reducing costs.* Chatham House Report for the Moving Energy Initiative.
Lehne, J., Blyth, W., Lahn, G., Bazilian, M., & Grafham, O. (2016). Energy services for refugees and displaced people. *Energy Strategy Reviews, 13–14*, 134–146.
Listo, R. (2018). Preventing violence against women and girls in refugee and displaced person camps: Is energy access the solution? *Energy Research & Social Science, 44*, 172–177.
Malakar, Y., & Day, R. (2020). Differences in firewood users' and LPG users' perceived relationships between cooking fuels and women's multidimensional well-being in rural India. *Nature Energy, 5*, 1022–1031.
Nixon, J. D., Bhargava, K., Halford, A., & Gaura, E. (2021). Analysis of standalone solar streetlights for improved energy access in displaced settlements. *Renewable Energy, 177*, 895–914.
Pueyo, A., & Maestre, M. (2019). Linking energy access, gender and poverty: A review of the literature on productive uses of energy. *Energy Research & Social Science, 53*, 170–181.
Rafa, N., Van To, T. T., Gupta, M., & Uddin, S. M. (2022). The pursuit of energy in refugee contexts: Discrimination, displacement, and humanitarian energy access for the Rohingya refugees displaced to Bangladesh. *Energy Research & Social Science, 83.*
Rewald, R. (2017). *Energy and women and girls: Analyzing the needs, uses, and impacts of energy on women and girls in the developing world.* Oxfam Research Backgrounder series. https://www.oxfamamerica.org/explore/research-publications/energy-women-girls
Ritchie, H., & Roser, M. (2019). *Indoor Air Pollution OurWorldInData.org.* Retrieved April 20, 2021, from https://ourworldindata.org/indoor-air-pollution
Rogers, C., Sovacool, B. K., & Clarke, S. (2013). Sweet nectar of the Gaia: Lessons from Ethiopia's "project Gaia". *Energy for Sustainable Development, 17*(3), 245–251.
Rohracher, H. (2003). The role of users in the social shaping of environmental Technologises. *Innovation: The European Journal of Social Science Research, 16*(2), 177–192.
Sandwell, P., Tunge, T., Okello, A., Muhorakeye, L., Sangwa, F., Waters, L., Kayumba, T., & Rosenberg-Jansen, S. (2020). *Ensuring refugee camps in Rwanda have access to sustainable energy.* Practical Action.

Sovacool, B. K., Clarke, S., Johnson, K., Crafton, M., Eidsness, J., & Zoppo, D. (2013). The energy-enterprise-gender nexus: Lessons from the multifunctional platform (MFP) in Mali. *Renewable Energy, 50*, 115–125.

Sovacool, B. K., Hess, D. J., Amir, S., Geels, F. W., Hirsh, R., Medina, L. R., Miller, C., Palavicino, C. A., Phadke, R., Ryghaug, M., Schot, J., & Turnhei, B. (2020). Sociotechnical agendas: Reviewing future directions for energy and climate research. *Energy Research & Social Science, 70*, 101617. https://doi.org/10.1016/j.erss.2020.101617

Theis, S., Bryan, E., & Ringler, C. (2019). Addressing gender and social dynamics to strengthen resilience for all. In A. Quisumbing, R. Meinzen-Dick, & J. Njuki (Eds.), *2019 annual trends and outlook report: Gender equality in rural Africa: From commitments to outcomes* (pp. 126–139).

Thomas, P. J., Rosenberg-Jansen, S., & Jenks, A. (2021). Moving beyond informal action: Sustainable energy and the humanitarian response system. *Journal of International Humanitarian Action, 6*(1), 1–20.

UNHCR. (2021, December 31). *Refugee statistics*. Retrieved July 2022, from https://www.unrefugees.org/refugee-facts/statistics/

UNHCR. (2022a, May 23). *UNHCR: Ukraine, other conflicts push forcibly displaced total over 100 million for first time*. UNHCR Refugee Agency News. Retrieved July 10, 2022, from https://www.unrefugees.org/news/unhcr-ukraine-other-conflicts-push-forcibly-displaced-total-over-100-million-for-first-time/

UNHCR. (2022b, July 10). *Gihembe refugee camp*. Statistics. Retrieved Febuary, 2021, from https://reliefweb.int/report/rwanda/unhcr-rwanda-gihembe-refugee-camp-profile-08-feb-2021

UNHCR. (2022c, July 10). *Rwanda population statistics*. https://reporting.unhcr.org/document/1721

UNHCR. (2022d, July 12). *UNHCR rwanda*. Nyabiheke Refugee Camp Profile. https://reliefweb.int/report/rwanda/unhcr-rwanda-nyabiheke-refugee-camp-profile-31-march-2021

United Nations. (2019). *UN clean energy challenge*. Retrieved September 6, 2021, from https://www.unhcr.org/uk/clean-energy-challenge.html

United Nations High Commission for Refugees (UNHCR). (2021). *Rwanda*. (UNHCR). Retrieved April 21, 2021, from https://reporting.unhcr.org/rwanda

Wilhite, H. (2017). *Gender implications of energy use and energy access*. EEG State-of-Knowledge Paper Series. Oxford Policy Management.

Winther, T., Matinga, M. N., Ulsrud, K., & Standal, K. (2017). Women's empowerment through electricity access: Scoping study and proposal for a framework of analysis. *Journal of Development Effectiveness, 9*(3), 389–417.

Winther, T., Ulsrud, K., Matinga, M., Govindan, M., Gill, B., Saini, A., Brahmachari, D., Palit, D., & Murali, R. (2020). In the light of what we cannot see: Exploring the interconnections between gender and electricity access. *Energy Research & Social Science, 60*, 101334.

World Bank. (2019, December 9). *2019 update: How long do refugees stay in exile? To find out, beware of averages.* (X. Devictor, Editor). World Bank. Retrieved July 19, 2022, from https://blogs.worldbank.org/dev4peace/2019-update-how-long-do-refugees-stay-exile-find-out-beware-averages

World Health Organization. (2016). *Burning opportunity: Clean household energy for health, sustainable development, and wellbeing of women and children.* WHO.

CHAPTER 13

Complex Intersections: Reflections on Women, Gender Inequality and Energy

Natalia Rocha Lawton and Cynthia Forson

Introduction

The theme of women and energy has two dimensions, both closely related to the role of energy and its social implications in the private and public spheres of life. Firstly, the impact of contextual factors such as environmental structures on social relations at the nexus of energy-human interactions cannot be underestimated (Braudel, 1998; Harris, 2004). Secondly, this understanding is essential in unpacking the relationship between gender and energy (Braudel, 1998; Harris, 2004) and, subsequently, gender inequality in the energy sector.

From a materialist perspective, the process of gendering occurs in three different areas of human life: production, reproduction and consumption

N. Rocha Lawton (✉)
Coventry University, Coventry, UK
e-mail: natalia.rochalawton@coventry.ac.uk

C. Forson
Lancaster University, Lancaster, UK
e-mail: c.forson@lancaster.ac.uk

N. Rocha Lawton, C. Forson (eds.), *Women and the Energy Sector*, https://doi.org/10.1007/978-3-031-43091-6_13

(Bradley, 2007). Gender equality has been taken for granted in the energy studies literature (DiMuzio & Ovadia, 2016) and has been secondary in the understanding of energy capitalism in society. Gender relations and inequality in access to the production, reproduction and consumption of energy (Ryan, 2014; Sovacool, 2014a) are subsumed under energy capitalism. And the mechanisms of the ways gender serves as a resource for capital (Acker, 2012). Thus, an analysis of gendered relationships in energy capitalism is an important contribution to developing a better understanding of inequality and its consequences on men's and women's experiences (Acker, 2006; Clancy et al., 2020) in their relationship with energy production, distribution and consumption.

Gender Inequality in Energy Studies

There is a growing interest in energy studies to understand the significant elements that constitute energy inequality within the energy sector. Nevertheless, most of the literature has overlooked the relevance of the connection between the different analytical factors that constitute the nature of the complex intersections. Some of the literature has taken energy as a single dimension to explain inequality. As such there is a gap in the literature with regard to, for example, the gendered dimensions of the march towards renewable energy as well as a lack of a critical view on the 'gender-blind' decision-making in the renewable energy process (Johnson et al., 2020).

Some energy scholars, however, have been highlighting the important role that gender has played in developing energy projects; the way women have been affected by the use of hazardous indoor energy sources; the unequal access to energy sources; and the institutional barriers that limit the participation of women in energy planning, production and consumption (Clancy, 2016). Further, there are aspects of energy security, equality and justice in energy access and well-being (Sovacool, 2014b) that are central to energy conversations in both developed and developing regions of the world. These conversations cover the underrepresentation of women in the energy sector and the way the use, production and reproduction of energy have impacted men and women differently.

The chapters in this book have focused on different forms of energy production, distribution and consumption and contribute to the understanding of the complex nature of the link between gender inequality and socio-economic structures in different contexts. Also, the discussion of contextual aspects was fundamental to addressing gender and social inequalities in energy transitions (Newell & Mulvaney, 2013). In this chapter, we reflect on some of the themes that have emerged from the conversations that we hope this book has started.

Multiple Standpoints on Knowledge Construction on Gender Inequality in the Energy Sector

This book represents a great challenge in integrating distinct points of view about the consequences of diverse forms of energy production and consumption in the lives of men and women in the Global North and Global South. The 11 substantive chapters based their analysis on countries from different geographical regions ranging from North and South America; South Asia, Western Asia and Southeast Asia; and sub-Saharan African countries to Western and South Europe, thus demonstrating that energy injustice is not only, or indeed, primarily, a Global South issue. Energy patriarchy is marbled in energy systems across the globe and reflects the doggedness of patriarchy across all human systems.

Further, the chapters' contributions were based on diverse disciplines and fields of knowledge, including political economy, humanitarian engineering, sociology, geography, environmental and energy studies, political science and political economy, public policy, sustainability and business and organisational studies. This shows the extent of the interest in integrating energy studies and gender studies from practitioners, policymakers and academics alike. The diversity of project disciplines which contributed to this book highlights the importance of multiple approaches to understanding the complexities of gendered energy systems in society.

The book also went beyond the technical and economic aspects of energy systems and drew on social sciences perspectives and presented a plurality of theoretical and methodological approaches to gender inequality analysis situated in the energy context. The contributions placed gender on the centre stage of energy inequality analysis, although from rather different perspectives. The analyses have been developed within a framework of energy transition and are associated with diverse concepts and theories including ecofeminism, energy patriarchy, intersectionality and environmentalism, among others.

Finally, the chapters in this book go beyond describing the challenges and constraints of women in diverse settings. They take a solutions-based view in the way they outlined ways out of unequal conditions structured on complex intersecting social dimensions and anticipate solutions situated in context. For example, **Ovadia** in Chap. 2 proposed changes in the local regulations to include gender equality in the emerging oil and gas and mining sectors in Tanzania, while **Palit et al.** highlighted the role of policies in promoting gender empowerment in the energy industry in India and Nepal in Chap. 3. Further, **Pan et al.**, in Chap. 7, have suggested that support for women's workforce participation to build gender balance in the solar sector is fundamental to promoting gender equality, inclusion and retention of women in solar energy in Brazil. They emphasise the importance of this to making the sector more attractive to women in energy diversification and transition.

Eco-feminist Perspectives on Inequality in the Energy Sector/Systems

There is a recognition in the literature and in this book of the relevance of feminist perspectives on energy in understanding unsustainable energy cultures and advancing the design of just energy systems (Bell et al., 2020). Energy consumption and well-being are interconnected and reinforce the notions of energy poverty, energy vulnerability and energy precariousness (Day et al., 2016). **Tandrayen-Ragoobur** in Chap. 11 emphasised that social stratification analysis provides a relevant theoretical base to explain the overlapping dimensions of gender and energy poverty. The author provided a holistic intersectional framework to give a detailed account of the complexities of gender, women's vulnerability and energy poverty.

Feminist perspectives in the literature have provided a different angle for the analysis of energy systems and their implications in society. Bell et al. (2020) argue that 'feminist theory offers expertise in the study of power more broadly, which is widely applicable to the full spectrum of energy research' (p. 1). Regarding gender in energy studies, women and their relationship with natural resources are relevant in explaining how context and energy inequality are inextricably linked. Traditionally, it has been common in the literature to focus on the 'self', where the human is dominant in the relationship with natural resources, in this case, with energy systems. Ecofeminism, a sociopolitical theoretical perspective that criticises the domination of nature and patriarchal structures that perpetuate women's social oppression, pushes back against this domination approach, recognising the importance of protecting the environment, which includes the 'community'.

The main purpose of ecofeminism is to promote women's participation through agency in collective organisations (Warren, 2000). Ecofeminism's political perspective is based on an intrinsically pluralistic and inclusive vision supported by a contemporary democratic view and ethical values. Thus, this theoretical approach has been relevant in the analysis of energy communities. It opens an important discussion on the role of energy communities to promote energy democracy and justice and gender equality in energy systems.

Proka (Chap. 9) presents energy communities in energy transition as 'alternative cultures, structures and practices that have the potential to radically transform the energy system' (p. 171). In this way, energy cooperatives have a relevant role to play in contesting the traditional energy capitalist production system that perpetuates gender inequality. As Proka argues, 'Community energy initiatives illustrate how the democratic and cooperative principles that guide their operation already start to shape alternative practices and structures that have the potential to transform the energy sector contributing to a socially and environmentally *just* energy transition' (p. 173). **Freenstra and Kottari** in Chap. 6 argued that women in the Dutch energy transition provide, firstly, a picture of the role of women in the technical task, economic challenge and social impact of the energy transition and, secondly, the need for the energy sector to use the untapped labour potential of women to accelerate the energy transition.

Dudka (Chap. 10) developed a critical discussion of the development of energy communities and taking an intersectional approach concluded that the gender binary perspective of energy communities 'have led to the reinforcement of women's marginalization in the energy sector' (p. 223). She argues that women have been traditionally excluded and 'as a consequence, this lack of emphasis on the active role of women in the energy communities could lead to the production and reproduction of new gender inequalities in the energy world' (p. 223). Dudka posited that a highly male-dominated environment remains in energy communities and therefore the traditional unbalanced gender power relation prevails in these communities. Clearly, despite its advantages, energy communities do not resolve women's marginalisation and inequality in the energy sector and in some cases reinforce it or indeed create new forms of gender inequality, unhinging the argument that collective participation in energy communities automatically creates equal conditions for different social groups to participate.

In Chap. 8, taking a practitioner approach, **Habersbrunner and Kuschan** put forward a process perspective that provides possible solutions to the gendered nature of energy communities. They suggested that decentralisation, democratisation and the creation of inclusive institutions are central to achieving energy transition. Their analysis addressed and acknowledged gendered aspects of energy communities in terms of decision-making, production and energy consumption in European countries, Turkey and Uganda.

Our book has contributed to the growing body of literature on gender equality in energy communities (Łapniewska, 2019), which is directly linked with justice and energy democracy where citizens have an active role in the energy transition. The argument is built on the assumption that community energy initiatives are based on voluntary membership and on an inclusive citizen-driven environment. Distributed energy and energy communities provide a new narrative that challenges the capitalist production system based on dependence on fossil fuel energy. However, the different cases presented in this book also found the reproduction of complex inequalities. While acknowledging the potential of energy communities to challenge patriarchal energy hierarchies, the chapters also emphasised the existence of complex unequal conditions for women that are reproduced in the energy communities while proposing improving policies in different settings such as the Netherlands, Greece, Belgium and Turkey where citizens are key actors in the energy communities.

Environmental Uncertainty, Social Context and Energy

The intersection of energy, gender inequality and environment is directly linked with the perception of risk, uncertainty and different uses of energy. The analysis of some of those intersections as well as its implications for energy consumption and production has been highlighted in recent research (Amorim et al., 2017; Anfinsen & Heidenreich, 2017; Carlsson-Kanyama & Lindén, 2010; Rocha Lawton, 2019). The idea that the human being is the owner of his/her destiny breaks down in the face of the uncontrollability of environmental uncertainty as a very real agent with its own dynamics and times.

Pandemics such as COVID-19 have been seen as a factor of uncertainty and disruption, and gender inequality has been exacerbated by the pandemic in terms of energy demand and consumption (Jiang et al., 2021) and energy communities. However, as important as pandemics and environmental disasters are, gender inequality is a persistent condition that transcends the temporality of a pandemic or an environmental crisis. **Mahecha-Groot et al.** in Chap. 4 analyse the energy policies in the coal sub-sector in Colombia to address the different ways women and men cope with the consumption of energy systems and the impact of climate change on energy. They argued that policies in energy transition should acknowledge gender differences and be more aware of gendered social relationships where women are seen as agents of change that have an active voice in policymaking.

Lammers et al. in Chap. 5 explored these relationships and posited that the disproportionate impact of the climate crisis on women has slowed down progress on gender equality. Reliability of access to energy systems, they maintain, is fundamental to the facilitation of community resilience to climate change and, as a result, has benefits for women's well-being by contributing to a reduction in gender inequality. The uncertainty of environmental changes has catalysed the search for technological solutions that will provide cheap, resilient and sustainable solutions. Nevertheless, these innovations develop within particular social (gendered) contexts, with specific needs and requirements and in which women's experiences at the energy transition and climate change nexus are complex and multifaceted. **Feenstra and Kottari** (Chap. 6) took an intersectional perspective on gender, energy transition and climate change and argued for a recognition of the disparate relationships that different groups of women have with energy systems in the context of climate and environmental change. As such a diverse, inclusive and contextually contingent human capital agenda is important to meet the technological, economical and societal challenges of the energy transition and climate change.

Given that the unpredictability of global crises and the transformation of energy systems affect everybody, it is crucial to recognise the relationship among energy systems and the way the unpredictability of climate change affects social relationships and, therefore, gender dynamics. The identification of relevant interactions, components and relationships between climate change, energy systems and gendered human activity in this volume enables the exploration of ways in which gender equality can be facilitated through women's participation as active producers, distributors and consumers of energy systems in the context of changing environmental conditions.

Gender and Sustainable Consumption

Gender plays an important role with regard to sustainable consumption between men and women. The way to make consumption more sustainable is to understand that it is influenced by gender. Recognising the gender differences in motivations and behaviours as well as the social conditions that influence men and women is fundamental to planning future policies and practices related to sustainable consumption as well as addressing and improving issues of gender equality (Bloodhart & Swim, 2020, p. 1). Hence, a gender perspective is essential to understanding that in the context of energy transition, gender influences the adoption of a more sustainable energy system and with more equitable implications for women and men as consumers and producers (**Freenstra and Kottari**, Chap. 6).

There is a tendency to see a single gender dimension in the studies of energy and sustainability (Arachchi & Managi, 2021). Also, the binary perception has been common in the policies of sustainability that promotes a homogeneous view of women in different contexts. Bolsø et al. (2018) argue that it is important to problematise dualisms in order to unearth gender biases and ways of thinking that shape decision-making on all levels yet mask complexities of social practice and obscure alternative ways of doing.

The homogeneous perception of women in the literature and practice has reinforced stereotypical views and norms that shape the way women and men think about themselves, respond to its necessity and choose to act (Bloodhart & Swim, 2020, p. 1). **Mahecha-Groot et al.** (Chap. 4) proposed a move beyond gender stereotypes in order to involve women in climate policy and provide adequate mechanisms and budgets for climate change-related gender considerations to be sustainable, effectively included and implemented (Lau et al., 2021).

Gender Representation in the Construction of New Policies

It is common in the literature and in practice to establish assumptions about what women's needs are in energy consumption, instead of letting women discover and communicate their interests and priorities. This is clear in **Halford's** Chap. 12, where the author discussed the positioning of displaced women as passive users of energy and conceptualised them in terms of activated and empowered engagement in energy solutions. Listo (2018) pointed out that the lack, or under consideration, of gender has contributed to gender inequality in the energy sector. Listo (ibid.) argued that academics and practitioners on energy issues have to give special attention to gender perspectives to contribute to 'women's empowerment in research and practice' (p. 16).

The ideas behind gender representation have very important consequences in the way policies have been created and applied. This has created stereotypes about women's energy needs. **Mahecha-Groot et al.** (Chap. 4) have emphasised that women's representation as victims in the policy discourse represents an obstacle to achieve fairer energy systems and the inclusion of gendered power relations. **Ovadia** (Chap. 2) and **Palit et al.** (Chap. 3) also emphasised the relevance of women's equality and empowerment by developing policies based on the existing specific social conditions of women in Tanzania, India and Nepal respecting 'prevalent social norms and traditions to support women's participation' (**Palit et al.**, Chap. 3). This will contribute women to participating actively in the energy transition process.

The authors in this book have explored systematically and structurally issues such as gender empowerment, participation and women's rights in energy policies. Historically women have been ignored in these spaces (energy access, participation in the decision-making in the household). **Ovadia** (Chap. 2) emphasised that the government of Tanzania must follow an all-inclusive approach to gender equality in policies of training and education in the emerging oil, gas and mining sectors. He also suggested that large oil, gas and mining firms have a relevant role to play in putting into practice the gender equality principles in their recruitment process, training and procurement practices to promote women's economic empowerment.

Complex Intersections: Inequality Regimes in the Energy Sector

The different contributions in this book have showed the relevance of understanding the complex relationships of the different social interactions driven by specific social formations and determined in historical context (Gottfried, 2008, p. 4). Acker (2006) has argued that the literature on the production of inequality in organisations have focused on class or gender or race but 'rarely attempting to study them as complex, mutually reinforcing or contradicting processes' (p. 442). The chapters of this book have shown an understanding of the inequality process of gender and the interaction with different complex social aspects.

Inequality in organisations is defined by Acker (ibid.) as 'systematic disparities between participants in power and control over goals, resources, and outcomes; workplace decisions such as how to organise work; opportunities for promotion and interesting work; security in employment and benefits; pay and other monetary rewards; respect; and pleasures in work and work relations' (Acker, ibid., p. 443).

Gender inequality in energy consumption 'has received limited analytical attention' (Wu et al., 2014) except for some cases in the literature of energy (Jacobson et al., 2005; Rosas-Flores & Galvez, 2010).

Acker's theoretical construction on inequality regimes provides an analytical framework to explore the complex forms of inequalities in different contexts (Healy et al., 2018, p. 1758). The global and local nature of energy provides more layers of inequality and changes in the way inequality restructures, resurges and mutually reinforces. Acker (2006) has defined inequality regimes as the mechanisms linked with inequality in society, politics, history and culture. Also, Acker has outlined the characteristics of the base of inequality regimes that create and recreate the legitimacy of inequalities (Acker, ibid.).

If inequality in the energy sector is recognised without finding solutions and it is not faced, there is a risk of perpetuating the same condition. Also, these dynamics of inequality will be reproduced in the new renewable energy projects. **Proka** in Chap. 9 emphasised this argument in the case of European energy communities. Also, Johnson et al. (2020) have already highlighted that '...renewable energy projects alone cannot achieve gender and social equity, as energy interventions do not automatically tackle the structural dynamics embedded within socio-cultural and socio-economic contexts. If existing power asymmetries related to access and resource distribution are not addressed early on, the same structural inequalities will simply be replicated and transferred over into new energy regimes' (Johnson et al., 2020, p. 1).

Furthermore, a relevant feature of the sociopolitical condition of the energy transition is that gender and social inequality are rooted in the energy systems. As Johnson argues, 'Inequalities can constitute and persist in low-carbon energy systems; they may not be any fairer, inclusive or just than the conventional systems they displace' (Johnson et al., 2020, p. 1). As **Proka** (Chap. 9) has highlighted, regardless of the progress achieved in the renewable energy force, the gender gap still exists.

Energy Capitalism Development and Patriarchy

There is an increasing body of literature that aims to address the relationship between gender and energy systems and define the character of this connection. A variety of themes have been explored in the literature—the negative effects of energy systems on women, the oppressive environment that women have to confront and the overwhelming effects of the relationship between women and energy capitalists (Ngarava et al., 2022; Pueyo & Maestre, 2019).

Capitalism development is internally linked with the energy sector development, especially fossil fuel energy. There is a historical association between capitalist accumulation and fossil fuel energy (Altvater, 2007). The aspirations for social justice and equality have their origin in a post-capitalist system. The permanent conflict between capitalist and post-capitalist system exists because they are grounded on different assumptions and set of values, goals and social aspirations. As Altvater has argued, a 'local and regional solidarity economy is a precondition for the transition to an economy based on renewable energy because the pressure of global competition is lower (in the renewable energy space) than in the formal, open sectors, which always have to increase their local competitiveness in global competition (thus, remaining necessarily dependent on the use of fossil fuels)' Altvater, ibid., p. 31).

Renewable energy is sustained in a different logic where individuals want to be more independent, where new values of energy democracy and energy communities' participation have a fundamental role. Thus, a society based on the use of renewable energy is only feasible by promoting a solidarity economy. In this way, Altvater has highlighted the relevance of identifying the political subjects that will pursue the transition to solidarity and sustainable forms of economic activities (Altvater, 2007, p. 31). This is an issue that needs to be explored further in the literature.

There is a fundamental relationship between capitalism and patriarchy. Patriarchy has been defined in the gender studies literature as 'a system of social structures, and practices in which men dominate, oppress and *exploit women*' (Walby, 1989, p. 200). Patriarchy is a historical and social system present in all societies and based on both structural and ideological arrangements that privilege men as a dominant group over women (Hunnicutt, 2009). Capitalism is built on the patriarchal system. Bieler and Morton (2021) have argued that historically, the capitalist system emerged from an existing patriarchal system, and capitalism contributed to maintaining patriarchal relationships, across capitalist patriarchy. Thus, the central argument is that patriarchy is a constituent part of the capitalist system. 'Capitalism cannot function without patriarchy' (Bieler and Morton, ibid., p. 1760). Bieler and Morton said that 'gender oppression is a central constitutive underpinning in the making of capitalism' (Bieler and Morton, ibid., p. 1759). This means that gender inequality is a permanent feature in the process of capital accumulation.

This book has sought to understand the foundations of gender inequality by exploring the term 'patriarchy' in the analysis of capitalist structures that have been the basis of the development of the energy sector. The chapters argue that 'patriarchy' has played a crucial role in the perpetuation of energy inequality. The connection between the development of the energy sector and gender inequality sustained by patriarchal capitalist systems is identified and addressed. **Palit et al.** (Chap. 3), **Mahecha Groot et al.** (Chap. 4) and **Tandrayen-Ragoobur** (Chap. 11) posit that the heart of women's energy poverty lies at the nexus of patriarchy and capitalism, where different public and private roles are assigned to men and women, excluding women from energy decisions, while they bear the brunt of the poor decision-making around energy.

Intersectional Analysis of Inequality in the Energy Sector

Ryder (2018) has argued that 'as energy decisions and policies continue to increasingly shape the extent to which the world is impacted by climate change, we must think precisely about the complexity of identity and who is involved in energy decisions; who benefits from, and who is burdened by particular sets of energy decisions and the impacts of climate change' (Ryder, 2018, p. 266). The study of energy capitalism and gender inequality have shown the significance of complex intersecting hegemonic social and contextual factors that surround energy systems. *Intersectionality* is an analytical tool that contributes to the study of overlapping or intersecting social identities and related systems of oppression, domination or discrimination (Crenshaw, 1991). As a concept, it addresses the relationship between complex underlying forces that intersect and mutually shape one another (gender, age, class, social status, ethnicity and other social dimensions) and the resulting experiences of social actors at the nexus of two or more of these dimensions. Ashley Bohrer (2019) shows how gender, race and sexuality infiltrate capitalist inequalities and oppression.

As a power relations analytical tool, an intersectional approach seeks to understand how social relations among individuals, groups and societies at particular intersections of social experience enable or constrain oppression and inequality. Collins and Bilge (2020) said that 'intersectionality investigates how intersecting power relations influence social relations across diverse societies as well as individual experiences in everyday life. Intersectionality is a way of understanding and explaining complexity in the world, in people, and human experiences' (p. 11). Bohrer says that 'shared theoretical and political ground between Marxism and intersectionality will contribute to shaping the path for an intersectional theory of capitalism, one that would offer a sustained, nuanced critique of the logic and structure of capitalism through an analysis of race, gender, class, sexuality, imperialism and colonisation' (Bohrer, 2018, p. 64).

As such this book embeds an intersectional approach into the analysis of capitalism. The chapters employed an intersectional analysis of capitalism across the study of gender, age, race, class, colonisation and imperialism, among other factors (Bohrer, 2018, p. 64), to provide a theoretical framework to examine the effects of energy systems in society. Bohrer argues that 'the analysis of the relationality of power is the central insight of an intersectional critique of capitalism' (Bohrer, 2018, p. 68). This is the case in **Halford's** Chap. 12 who considers women at the intersection of gender and displacement and how wider patriarchal structures feature in humanitarian energy design and reinforce the gendered division of labour. She argues for a consideration of women in their location and sociocultural context in the development of humanitarian energy systems.

The analytical perspective of intersectionality in this book has helped to understand the sources of gender inequality and identify structural forms of power relation and domination from patriarchal relationships to capitalist forms of production to uncover the intersections of inequality. This has been an important contribution in this book because in the context of energy studies, intersectionality perspectives have been used very rarely (Søraa et al., 2020). **Tandrayen-Ragoobur** in Chap. 11 analyses ways in which other dimensions of inequality play such as class, age, race/ethnicity, sexual orientation, disability and illness intersect with gender to produce differential energy outcomes for diverse groups of women in sub-Saharan African countries.

An analysis based on intersectionality contributes to identifying the social relations that intersect to construct oppression, in the case of women in the energy communities. **Dudka** (Chap. 10) based her analysis on an intersectional approach to discuss and examine the different ways in which gender can intersect with other stands of inequality such as income, education, class, marital status and parenthood to deepen disadvantaged conditions for women in participating in democratic organisations such as energy communities.

Conclusion

Energy is a fundamental resource in terms of economic and human development, and there are multiple factors associated with the gendering of its production, distribution and consumption. The contributions and conceptions of the inequality women experience in this book enable a recognition of the existence of dominant complex systems that interact in different social, political, economic, environmental, disciplinary and knowledge frameworks. They also facilitate the identification of preconceptions based on homogenising views of the role of women in energy systems. Those assumptions are based on a hegemonic perception of the role or function of energy and specific constructions about the role of men and women in the private and public spheres. These hegemonic aspects that surround energy production and consumption stop women from finding equal conditions to participate in the construction of sustainable energy systems.

There are similarities and differences between sources of gender inequality and the ways this is reproduced in different geographical contexts, institutional structures and energy systems. The discussion on gender inequality in energy systems and the ways to deconstruct it has been addressed in their own contexts too. Without an understanding of these gendered complexities, there is a risk of creating more structures of oppression on women by imposing new domineering systems built for women, but not by women, to disentangle gender inequality and make invisible the relevance of cultural arrangements in different communities where men and women must take decisions on sustainable energy production and consumption.

The discussions here build on conversations that have begun on the importance of women in building a sustainable energy sector that contributes to the reduction of gendered inequalities in energy systems. This constitutes a further step in drawing up a framework to deconstruct the complex gender relationships behind gender inequality in the energy sector. It is important that this multidisciplinary, multidimensional and multi-contextual analysis is built on to provide a comprehensive understanding of the process of energy gendering and ways to mitigate its effect in order to build truly contextualised sustainable energy systems.

References

Acker, J. (2006). Inequality regimes gender, class, and race in organizations. *Gender and Society, 20*(4), 441–464. https://www.jstor.org/stable/27640904

Acker, J. (2012). Gendered organizations and intersectionality: Problems and possibilities. *Equality, Diversity and Inclusion: An International Journal, 31*(3), 214–224.

Altvater, E. (2007). The capitalist energy system and the crisis of the global financial markets: The impact on labour. *Labour, Capital and Society, 40*(1/2). Special issue Globalization, Labour and Development, pp. 18–34. https://www.jstor.org/stable/43460537

Amorim, R., López, J., Molina-Moreno, V., & Peña-García, A. (2017). Use of natural light vs. cold LED lighting in installations for the recovery of victims of gender violence: Impact on energy consumption and victim's recovery. *Sustainability, 9*(4).

Anfinsen, M., & Heidenreich, S. (2017). *Energy & Gender: A social sciences and humanities cross-cutting theme report*. SHAPE ENERGY.

Arachchi, J. I., & Managi, S. (2021). Preferences for energy sustainability: Different effects of gender on knowledge and importance. *Renewable and Sustainable Energy Reviews, 1*, 141. https://doi.org/10.1016/j.rser.2021.110767

Bell, S. E., Daggett, C., & Labuski, C. (2020). Toward feminist energy systems: Why adding women and solar panels in not enough. *Energy Research & Social Science, 68*, 2020. https://doi.org/10.1016/j.erss.2020.101557

Bieler, A., & Morton, A. D. (2021). Is capitalism structurally indifferent to gender? Routes to value theory of reproductive labour. *A Economy and Space, 53*(7), 1749–1769. https://doi.org/10.1177/0308518X211031572. https://journals.sagepub.com/home/epnSAGE

Bloodhart, B., & Swim, J. K. (2020). Sustainability and consumption: What's gender got to do with it? *Journal of Social Issues, 76*(1), 101–113. https://doi.org/10.1111/josi.12370

Bohrer, A. J. (2018). Intersectionality and marxism: A critical historiography. *Historical Materialism, 26*(2), 46–74.

Bohrer, A. J. (2019). *Marxism and intersectionality. Race, gender, class and sexuality under contemporary capitalism*. Columbia University Press.

Bolsø, A., Phillips, M., & Sabelis, I. (2018). Gendering environmental sustainability and organization: Introduction. *Gender, Work and Organization, 25*(3). https://doi.org/10.1111/gwao.12218

Bradley, H. (2007). *Gender*. Policy Press.

Braudel, F. (1998). *The wheels of commerce. Civilization and capitalism*. University of California Press.

Carlsson-Kanyama, A., & Lindén, A. L. (2010). Energy efficiency in residences – Challenges for women and men in the North. *Energy Policy, 35*(4), 2163–2172.

Clancy, J. (2016). Mainstreaming gender in energy sector practice and policy lessons from the Energy International Network, Department for International Development, UK Report: *Gender Energy Research Programme*, 30 September 2016.

Clancy, J., Özerol, G., Mohlakoana, N., Feenstra, M., & Cueva, L. S. (Eds.). (2020). *Engendering the energy transition*. Palgrave Macmillan.

Collins, P. H., & Bilge, S. (2020). *Intersectionality*. John Wiley & Sons. ebook.

Crenshaw, K. (1991). Mapping the margins: Intersectionality, identity politics, and violence against women of colour. *Stanford Law Review, 43*(6), 1241–1299. (59 pages). https://www.jstor.org/stable/1229039

Day, R., Walker, G., & Simcok, N. (2016). Conceptualising energy use and energy poverty using a capabilities framework. *Energy Policy, 93*(2016), 255–264. http://www.elsevier.com/locate/enpol

DiMuzio, T., & Ovadia, J. S. (Eds.). (2016). *Energy, capitalism and world order. Towards a new agenda in International Political Economy*. London Palgrave Macmillan.

Gottfried, H. (2008). Reflections on intersectionality: Gender, class, race and nation. *Journal of Gender Studies, 11*, 23–40.

Harris, O. (2004). Braudel: Historical time and the horror of discontinuity. *History Workshop Journal, 57*(1), 161–174. https://doi.org/10.1093/hwj/57.1.161

Healy, G., Tatli, A., Ipek, G., Özturk, M., Seirstad, C., & Wright, T. (2018). In the steps of Joan Acker – A journey in researching inequality regimes and intersectional inequalities. *Gender Work and Organization, 26*(12). https://doi.org/10.1111/gwao.12252

Hunnicutt, G. (2009). Varieties of patriarchy and violence against women: Resurrecting patriarchy as theoretical tool. *Violence Against Women, 15*(5), 553–573. https://doi.org/10.1177/1077801208331246

Jacobson, A., Milman, A. D., & Kammen, D. M. (2005). Letting the (energy) Gini out of the metrics of energy distribution and equity. *Energy Policy, 33*, 1825–1832.

Jiang, P., Van Fa, Y., & Klemeš, J. (2021, March 1). Impacts of COVID-19 on energy demand and consumption: Challenges, lessons and emerging opportunities, *Applied Energy, 285.* https://doi.org/10.1016/j.apenergy.2021.116441

Johnson, W. O., Han, J. Y., Knight, A., Mortensen, S., Aung, M. T., Boyland, M., & Resurrección, B. (2020, December). Intersectionality and energy transitions: A review of gender, social equity and low-carbon energy, *Energy Research & Social Science, 70.* https://doi.org/10.1016/j.erss.2020.101774

Łapniewska, Z. (2019). Energy, equality and sustainability? European electricity cooperatives from gender perspective. *Energy Research & Social Science, 57*(November), 101247. https://doi.org/10.1016/j.erss.2019.101247

Lau, J., Kleiber, D. L., Lawless, S., & Cohen, P. J. (2021). Gender equality in climate policy and practice hindered by assumptions. *Nature Climate Change, 11*(3), 186–192. https://doi.org/10.1038/s41558-021-00999-7

Listo, R. (2018). Gender myths in energy poverty literature – A Critical discourse Analysis. *Energy Research & Social Science, 38*(2018), 9–18. https://doi.org/10.1016/j.erss.2018.01.010

Newell, P. D., & Mulvaney. (2013). The political economy of the 'just transition'. *Geographical Journal, 179*(2013), 132–140. https://doi.org/10.1111/geoj.12008

Ngarava, S., Zhou, L., Ningi, T., Charu, M., & Mdiya, L. (2022). Gender and ethnic disparities in energy poverty: The case of South Africa. *Elsevier, Energy Policy, 161*, 112755. https://doi.org/10.1016/j.enpol.2021.112755

Pueyo, A., & Maestre, M. (2019). Linking energy access, gender and poverty: A review of the literature on productive uses of energy. *Energy Research & Social Science, 53*, 170–181. https://doi.org/10.1016/j.erss.2019.02.019

Rocha Lawton, N. (2019). Enhancing risk and safety management in large scale renewable energy generation. In G. Arruda (Ed.), *Renewable energy for the Arctic. New perspectives* (Routledge Explorations in Energy Studies) (pp. 175–190). Routledge.

Rosas-Flores, J. A., & Galvez, D. M. (2010). What goes up: Recent Trends in Mexican residential energy use. *Energy, 35*, 2596–2602. Retrieved August 2, 2018, from http://sspp.proquest.com

Ryan, S. (2014). Rethinking gender and identity in energy studies. *Energy Research & Social Science, 1*, 96–105.

Ryder, S. S. (2018). Developing an intersectional-informed, multi-sited, critical policy ethnography to examine power and procedural justice in multiscalar energy and climate change decision making processes. *Energy Research & Social Sciences, 45*, 266–275.

Søraa, R., Anfinsen, M., Foulds, C., Korsnes, M. L., Vivian, R., & Ryghaug, M. (2020, April). Diversifying diversity: Inclusive engagement, intersectionality, and gender identity in a European Social Sciences and Humanities Energy research project. *Energy Research & Social Science*, 62. https://doi.org/10.1016/j.erss.2019.101380

Sovacool, B. (2014a). What are we doing here? Analysing fifteen years of energy scholarship and proposing a social science research agenda. *Energy Research & Social Science*, 1–29.

Sovacool, B. (2014b). Deciphering energy justice and injustice. In B. K. Sovacool, R. V. Sidortsov, & B. R. Jones (Eds.), *Energy security, equality and justice.* Routledge.

Walby, S. (1989). Theorising patriarchy. *Sociology, 23*(2), 213–234. https://doi.org/10.1177/0038038589023002004

Warren, K. J. (2000). *Ecofeminist philosophy: A western perspective on what it is and why it matters.* New York: Rowman & Littlefield Publishers.

Wu, Q., Maslyuk, S., & Clulow, V. (2014). Energy consumption inequality and human development. In Z. Morvaj (Ed.), *Energy efficiency – A bridge to low carbon economy* (pp. 101–116). In Tech.

Index[1]

A
Access, 1–7, 9, 10, 17, 22, 23, 25, 27–29, 37–40, 42, 43, 46, 47, 49–51, 53, 54, 56, 57, 66, 78, 91–117, 126–128, 138, 139, 142, 152–155, 158, 168, 180, 184, 185, 187–189, 200–204, 211–213, 215, 216, 223, 232, 246, 263, 264, 266–270, 272–275, 280–284, 286, 287, 289, 298–300, 302, 303, 305–309, 316, 321, 323, 324
Adaptation, 68, 69, 76, 81, 93, 94, 97, 102, 108, 109, 111–116, 131, 215, 271
Administrators, 26
Adult literacy, 53
Affirmative action, 22, 45, 55
Affordable energy, 2, 4, 92, 94, 106, 108–110, 128, 298, 302
Afghanistan, 116, 297
African countries, 4, 9, 46, 264, 265, 274, 288, 289, 317, 328
African Development Bank, 16, 24
African Mining Vision (AMV), 16, 31
African Natural Resources Centre (ANRC), 24, 28
African socialism, 20
African Union (AU), 16
African women, 9, 277–280, 283–288
Africa, Sub-Saharan (SSA), 39, 264, 273–275, 282
Age, 8, 48, 98, 129, 129n3, 154, 156, 162, 184, 189, 191, 197, 204, 218, 249, 250, 254, 255, 264, 265, 267, 268, 270, 271, 274, 276, 277, 280, 327, 328
Agency, 5, 6, 10, 23, 25, 26, 31, 41–43, 46, 50, 81, 96, 98, 102–103, 108–110, 204, 213, 217, 268, 270, 308, 319

[1] Note: Page numbers followed by 'n' refer to notes.

N. Rocha Lawton, C. Forson (eds.), *Women and the Energy Sector*, https://doi.org/10.1007/978-3-031-43091-6

Agriculture, 25, 48, 71, 103, 130, 136, 201
Alternative Energy Promotion Centre (AEPC), 51
Alternative Energy Promotion Council (AEPC), 46
Amma Samuha, 52
Armed conflict, 67
Artisanal mining, 19, 22
Asia, 4, 39
Attributes, 65, 129, 163, 193
Autonomy, 39, 66, 83, 100, 102, 103, 105–106, 155, 180–182, 222, 307, 308

B
Baglung-Bhuji Khola Community, 51–53
Bangladesh cyclone, 101
Belgium, 9, 227, 245, 320
Bhuji Khola, 52
Bihar Rural Livelihoods Promotion Society (BRLPS), 47
Biomass fuel, 303
Black, 64, 67, 70, 75, 78, 152, 154, 155, 158, 162, 164
Black Women Platform, 78
Bottom-up approaches, 54, 179
Bowang, 51
Brazil, 7, 150–166, 168, 318
Brazilian Women's Network in Solar Energy, 151
Business, 3, 27, 31, 67, 126, 135, 164, 181, 182, 186, 193, 194, 199, 201, 202, 212, 214, 222, 232, 245, 246, 253, 256, 304, 306, 317

C
Cambio climático, 81
Capacity building, 77–79, 109, 226
Capital, 7, 27, 28, 42, 54, 55, 99, 127, 129, 134, 135, 139, 141–143, 200, 308, 316, 321, 326
Capitalism, energy, 2, 316, 325–327
Carbon, 67, 125, 126, 133, 167, 216
Care, 9, 48, 51, 55, 100, 101, 106, 138, 141, 155, 162, 163, 186, 211, 213, 229, 230, 264, 270, 277, 281, 282, 289
Career, 29, 128n2, 135, 141, 142, 153, 154, 156, 157, 164, 185, 195, 252
Care work, 17, 99–100, 103, 108, 109, 186, 268
Categories of difference, 75
Centre for Agricultural Mechanisation Rural Technology (CAMARTEC), 26
Cerrejón, 70
Cerro Largo, 70
Cesar, 64, 66, 70, 74, 75, 84
Change agents, 38, 39
Chronic illnesses, 17
Citizen driven collectives, 7
Civil society, 19, 51, 67, 68, 134, 135, 190, 193, 197, 244
Civil works, 51, 52
Class, 8, 9, 73, 75, 129, 153, 162, 197, 218, 264, 265, 267, 270, 324, 327, 328
Clean energies, 71, 92, 103, 131, 136, 183, 188, 191, 266, 298
Climate change, 5–7, 63–85, 92–107, 109–112, 114, 116, 117, 125–128, 131, 133, 135, 140, 143, 167, 168, 177, 201, 212, 213, 243, 273, 321, 322, 326, 327
Climate Change Considerations for Land Use Planning (CCOT), 76
Climate Change Law, 68

Climate crisis, 6, 65, 67, 68, 71, 74, 80–84, 91, 97, 321
Climate-gender nexus, 83
Climate mitigation, 64, 67
Climate-neutral, 67
Climate policy, 6, 66–75, 79, 81, 83, 84, 129, 134, 322
Climate resilience, 6, 94, 96, 103–113, 115, 117
Climate-resilient, 6, 91–117
Climate variability, 81, 82, 102, 103
Coal policy, 6, 65, 67–71
Coal sub-sector, 5, 64, 70, 321
Coffee, 66, 224
Collaboration, 84, 128n2, 135, 142, 219, 244, 248
Colombia, 63–70, 73, 74, 76–78, 80, 83, 84
Colombia, coal sub-sector, 321
Colombia energy sector, 83
Colonial administration, 25
Colonial economy, 25
Colonialism, 22, 217
Communication, 37, 38, 142, 167, 193, 198–204, 225, 227, 229, 230, 234, 235, 288
Community actions, 8, 51, 178
Community development, 25
Community Rural Electrification Entity (CREE), 53, 54
Community Rural Electrification Programme (CREP), 53–54
Company board, 57
Company directors, 26, 162
Complexity, 3, 4, 9, 10, 27, 298, 317, 318, 322, 326, 327, 329
Comprehensive Sectoral Climate Change Management Plans (PIGCCS), 68, 69
Comprehensive Territorial Climate Change Management Plans (PIGCCT), 68, 75, 77
Conditional factors, 43, 55, 57
Congo, 297, 309
Congolese refugees, 10, 299, 302
Consideraciones de Cambio Climático para el Ordenamiento Territorial, 76
Cooking, 78, 99, 103, 106–109, 139, 140, 188, 263, 268, 269, 272, 275, 282, 283, 298, 300, 301, 303, 306
Cooling, 104, 107, 111, 113, 128
Corporate managers, 26
Corpus-based, 65, 72, 73
Corpus-informed, 65
Corruption, 66
COVID-19, 21, 230, 232, 263, 264, 274, 321
Crisis climática, 81
Cultural norms, 22, 103
Culture, 8, 29, 30, 65, 135, 151, 160, 162, 204, 214, 216, 217, 219, 223, 235, 304, 307, 318, 319, 324
Customary law, 42

D

Damini Pathak, 49
Data generation, 57
Decentralised renewable energy (DRE), 51
Decision-making power, women, 8, 39, 40, 277, 306
Demand, 50, 55, 63, 69, 81, 82, 84, 111, 127, 130, 131, 138, 151, 153, 157, 162, 164, 205, 212, 298, 301, 321
Demand-based delivery, 57
Democratic energy systems, 126, 206, 216
Democratic Republic of Congo (DRC), 29, 302, 303

Development, 2, 4, 5, 8, 16–18, 20–28, 30, 31, 37, 46, 50, 52, 56, 65, 68, 71, 73, 76, 78, 96, 102, 103, 114, 115, 126, 132, 134, 139, 140, 142, 151, 152, 155, 157, 159, 160, 163, 165, 166, 168, 177–206, 211, 218, 223, 224, 226, 229, 231, 234, 243, 245, 247, 263, 267, 289, 299, 319, 325–328
Development Vision 2025, 20
Dhaulagiri Community Resource Development Centre (DCRDC), 51, 52
Diamond mining, 22
Direct control, 20
Direct employment, 16, 17, 20, 24, 26, 28–30
Disadvantage, 9, 189, 267, 286
Disaster management, 76
Disaster risk management, 68, 96
Discourse analysis, 65, 72, 73
Discrimination, 18, 20, 22, 29, 39, 49, 72, 101, 151, 155, 159, 168, 198, 199, 219, 222, 235, 236, 267, 287–289, 327
Dislocation, 16
Displacement, 67, 70, 98, 301, 304, 327
Dispositions, 41
Distributional impacts, 71
Diversity, 3, 7, 24, 94, 97–98, 127, 128, 130, 133, 135, 138–143, 166, 169, 180n1, 187, 214, 217, 219, 227, 229, 249, 317
Domestic, 9, 20, 28, 53, 54, 79, 139, 140, 155, 159, 218, 227, 253, 266, 268, 269, 272, 273, 281, 282, 289, 300, 306, 307
Domination, 25, 157, 196, 236, 318, 327, 328
Donors, 23, 24, 26, 31
Drudgery, 37, 38, 42, 48, 273, 283, 284
Dutch energy transition, 7, 134, 135, 140, 319
Dynamisms, 9, 277–285, 287–289

E

Earnings, 25, 55, 106, 192n4, 250, 252, 256, 277, 281–283
Ecofeminism, 71, 317–319
Economic empowerment, 15–31, 37–57, 323
Economic opportunities, 5, 7, 10, 22, 92, 152, 155–157, 168, 264, 268
Economic transformation, 31
Economic value, 70
Economic well-being, 6
Education, 7, 9, 17, 25, 26, 28, 38, 42, 48, 53, 77, 79, 92, 98, 102, 103, 108–110, 116, 128n2, 132, 136, 141, 142, 152–154, 156, 161, 162, 168, 182, 185, 189, 197, 200, 213, 222, 226, 231, 263–265, 268–270, 272, 274, 277, 280–284, 289, 307, 323, 328
Education and Skills for Productive Jobs Programme (ESPJ), 25
Egalitarian, 7, 8, 165, 197, 307
El Corozo, 70
El Descanso, 70
El Hatillo, 70
Elderly, 30, 76, 101, 155, 218, 273
Electrical equipment, 49
Electrical systems, 41
Electricity, 5, 38–40, 42–50, 52–54, 56, 57, 69, 103, 105–109, 111, 112, 128, 136, 150, 166, 179, 183, 188, 191, 194, 195, 198, 200, 211, 216, 227–229, 231, 245, 257, 264–266, 269, 270, 273–275, 277, 280–284, 286, 287, 298, 303, 304, 306, 308

Electricity bills, 49, 188, 231
Electricity distribution, 48, 50
Electricity-empowerment framework, 5, 42, 45
Electricity helpers, 49
Electricity policies, 39, 43, 45–47, 54, 56
Electricity sector, 40, 46, 48, 49, 56, 69
Electricity supply, 53, 57, 107, 108, 228, 245
Electricity supply chain, 56
Electricity-supported development, 46
Electric poles, 49
Electric power, 69
Electrification, 5, 37–57, 104–106, 108, 111, 115–117, 269, 282–284, 286, 289
Embedded agency, 41
Emissions, 67–69, 78, 79, 84, 125, 126, 131, 133, 134, 167, 177, 179, 187, 201
Employment and Labour Relations Act 2004, 20
Empowerment dimensions, 43
Empowerment Funds Department, 23
Empowerment, girls, 110
Empowerment policies, 16, 22, 26
Energy access, 2, 6, 10, 51, 91–117, 128, 212, 263, 266, 267, 273–275, 281, 283, 284, 286, 287, 289, 299, 305, 306, 309, 316, 323
Energy alliances, 8
The Energy and Resources Institute (TERI), 47
Energy capitalism, 2, 316, 325–327
Energy communities, 3, 4, 7–10, 126, 177–206, 211–236, 243–250, 253, 256, 257, 319–321, 324, 325, 328
Energy consumption, 2, 130, 138, 186–188, 205, 318, 320, 323, 324
Energy cooperatives, 8, 181–183, 191, 195, 196, 198–201, 215, 221, 222, 231, 232, 234, 235, 245, 319
Energy demand, 63, 111, 321
Energy democracy, 8, 9, 191, 206, 212, 214–218, 221, 245, 246, 256, 319, 320, 325
Energy discourse, 39
Energy disparity, 270, 284, 286
Energy diversification, 7, 151, 167, 318
Energy entrepreneurs, 38, 39
Energy equality, 2, 4, 5
Energy field, 8, 151, 203, 213, 220, 235, 236, 246, 247, 255
Energy inequality, 2, 6, 8, 316–318, 326
Energy interventions, 10, 269, 299–301, 306, 324
Energy justice, 9, 42, 71, 191, 212, 213, 222, 246
Energy labour market, 6, 7, 132
Energy landscape, 1, 132, 184
Energy market, 7, 8, 126, 137–138, 179, 182, 183, 192, 243, 244
Energy-patriarchy, 48, 55
Energy planning, 6, 39, 46, 110, 114, 115, 117, 135, 299, 316
Energy policies, 2, 4–7, 39, 41, 43, 55, 71, 73, 74, 83, 84, 130, 141, 178, 179, 185, 186, 189, 202, 205, 321, 323
Energy poverty, 38, 126, 130, 138, 139, 187, 188, 202, 204, 231, 234, 247, 263–290, 299, 318, 326
Energy production, 8, 66, 67, 126, 131, 178, 179, 189, 203, 205, 316, 328, 329
Energy relationships, 2, 4
Energy research, 41, 189, 318

Energy sector, 1–5, 7, 9, 10, 25, 63–85, 111, 126–128, 130–133, 135–138, 140–143, 149–151, 156, 157, 159, 160, 162, 164–166, 168, 182, 185–189, 194, 196, 197, 212–214, 216, 220, 221, 228, 231, 236, 247, 256, 316–320, 323–329
Energy service delivery, 39, 46
Energy supply chain, 5, 45, 49, 56, 130, 270
Energy systems, 1, 2, 4, 6, 8, 38, 48, 55, 57, 72, 83, 93–96, 110–117, 125, 126, 128, 130, 134, 135, 137, 177–179, 182, 184, 188, 191, 201, 205, 206, 211–214, 216, 218, 221, 222, 226, 234, 235, 257, 269, 270, 272, 273, 298, 299, 305–307, 317–319, 321–323, 325, 327–329
Energy transition, 4–8, 22, 57, 64, 67, 70, 71, 75, 83, 84, 115, 125–143, 149–169, 177–179, 183, 184, 186, 187, 189–191, 199–203, 205, 206, 211–236, 247, 253, 317, 319–323, 325
Energy use, 41, 130, 139, 140, 177, 263, 266, 267, 271, 272, 277, 282
Engendering energy, 83, 129–130, 136, 202, 211–236
Engineering, 3, 18, 26, 112, 132, 153, 164, 247, 299, 301, 317
Engineers, 38, 50, 116, 199, 211, 231, 310
Enhancing Employability through Vocational Training (EEVT), 23
Enterprise development, 27, 31
Entrepreneurial support, 31
Entrepreneurship, 5, 23, 38, 54
Entry, 7, 142, 143, 164
Environment, 6, 8, 10, 29, 39, 41, 73, 112, 113, 136, 157–161, 165, 185, 186, 188, 203, 212, 229, 273, 289, 319, 320, 325
Environmental and energy studies, 3, 317
Environmental damage, 64, 66, 80
Environmental degradation, 16, 66, 131
Equal access, 25, 28, 29, 46, 213
Equal pay, 29, 156, 227
Equinor, 19
Equitable communities, 6
Estrategia Nacional para la Reducción de Emisiones por Deforestación y Degradación de Bosques, 75
Ethnicity, 8, 73, 98, 129, 189, 191, 204, 218, 264, 265, 267, 270, 327, 328
European Commission, 129, 134, 138–139, 180, 181, 183, 187, 191, 214
Evolution, 68, 127
Exclusion, 22, 23, 29, 84, 127, 206, 247, 272
Expatriate labour, 20
Extraction, 16–18, 20, 22, 24, 30, 67, 70, 127, 185, 188
Extraction plants, 67
Extractive firms, 31
Extractive industries, 5, 15–31

F

Female refugees, 9
Female representation, 24, 164, 225
Female work, 29
Feminisation, 16, 19, 27
Feminist perspective, 65, 71, 318
Feminist standpoint, 72
Flexible working, 38, 163

Formal employment, 16, 25, 26, 29, 155
Fossil fuel, 67, 71, 126, 133, 142, 184, 185, 189, 205, 211, 247, 320, 325
Fuerzas Armadas Revolucionarias de Colombia (FARC), 67

G

Gas sector, 19–21, 28
Geita, 23
Gender-aware, 27, 72, 76
Gender-based discrimination, 18
Gender blind, 8, 54–55, 72, 316
Gender blindness, 5, 187, 189
Gender champions, 24
Gender dimensions, 8, 15, 16, 24, 56, 129, 185, 190, 193, 197, 199, 202, 264, 266–268, 322
Gender-disaggregated data, 18, 39, 57, 77, 78, 133, 143, 193
Gender divide, 26
Gendered energy, 2, 3, 6, 10, 265, 273, 277, 281, 283, 284, 286, 288, 289, 317
Gendered impacts, 18, 55, 71, 72, 299
Gendered meanings, 18
Gendered organisation, 43
Gendered vulnerability, 92, 96–103, 106, 117
Gender energy gap, 9, 264, 265
Gender-energy nexus, 37, 129, 130, 192, 266, 267
Gender equality, 2, 5, 6, 9, 16–20, 25, 29–31, 37, 49, 55, 64, 65, 72, 75, 78–84, 92–94, 103–110, 116, 117, 128, 133, 152, 154–156, 167, 168, 185, 195, 199, 203, 204, 213, 214, 221, 226, 231, 233, 234, 236, 265, 316, 318–323
Gender Equality and Social Inclusion (GESI) guidelines, 46
Gender gap, 8, 99, 132, 196, 213, 231, 234, 248, 284, 325
Gender-inclusive, 45, 46, 55, 143
Gender inequality, 1–4, 6, 8–10, 31, 54, 65, 72, 80, 92, 110, 116, 117, 143, 151, 156, 166, 185, 187, 189, 194–202, 218, 234, 235, 247–256, 264–265, 267, 269, 270, 272, 282, 289, 299, 300, 306
Gender in Tanzania's extractive industries, 17–20
Gender-integrating, 45
Gender justice, 7, 9, 214, 215, 217–219, 221, 223, 226, 228, 229, 232, 236
Gender mainstreaming, 7, 17, 24, 39, 45, 48, 54, 72, 75, 82, 186, 202
Gender-negative, 72
Gender-neutral, 5, 17, 45, 55, 189, 221, 273
Gender norms, 39, 41, 42, 157, 289
Gender oppression, 74, 326
Gender outcomes, 55
Gender perspectives, 6, 20, 28, 69–71, 73–79, 115, 130, 139, 140, 178, 179, 186, 187, 192, 196, 198, 289, 298, 322, 323
Gender policy, 2, 5, 6, 73, 74, 79–84
Gender policy corpus (GPC), 81
Gender relations, 2, 56, 65, 129, 186n2, 194, 197, 264, 316
Gender representation, 65, 248, 323
Gender-responsive, 45, 68, 83, 201, 204
Gender-responsive climate policy (GRCP), 65, 83
Gender-responsive policy, 5, 75

Gender roles, 18, 72, 78, 83, 97, 102, 127, 132, 142, 187, 192, 217, 236, 252, 272, 282, 306, 308
Gender-specific barriers, 39
Geology, 18
Germany, 8, 183, 186, 187, 189, 194, 195, 198–202, 216, 220, 223, 225, 248, 255
Global Gender and Climate Alliance, 100, 101
Global North, 3, 4, 127, 317
Global South, 3, 4, 94, 110, 269, 273, 317
Governance, 41, 66, 126, 134, 178, 179, 181, 191, 195, 244–246, 272
Gram Oorja, 50
Greece, 39, 188, 220, 226, 320
Greenhouse gases (GHG), 68, 84, 133, 134, 177, 179, 187, 201
Green jobs, 71, 131
Grid, 38, 43, 53, 178, 204, 216, 274, 303, 304
Grid connectivity, 53
Gross domestic product (GDP), 28, 66

H

Habitus, 41
Health, 6–9, 25, 29, 38, 42, 66, 77, 92, 96, 100–101, 103, 107–108, 110, 116, 117, 152, 153, 157–163, 168, 185, 188, 211, 213, 263–265, 268–272, 274, 277, 282, 283, 286, 289, 298, 299, 309
Heating, domestic, 227
Hegemonic discourses, 73, 74
Heterogeneous, 2, 98
Higher education, 26, 48, 153, 154, 277, 280
Hindu Succession Act, 1956, 56
Holistic, 6, 17, 50, 114, 117, 204, 270, 318
Hombre, 74, 85
Household, 38, 43, 46–48, 50–52, 55, 57, 77, 97, 98, 100, 103, 105–107, 109, 117, 130, 138–140, 152, 162, 183, 186, 188, 191, 192, 201, 202, 205, 221, 250, 263, 266, 268–275, 277, 281, 282, 284, 286–289, 298, 300, 303, 304, 306–308, 323
House wiring, 54
Human capital, 7, 127, 129, 134, 135, 139, 141–143, 321
Human development, 4, 37, 103, 211, 328
Humanitarian crises, 9, 308
Humanitarian Engineering and Energy for Displacement (HEED), 299, 302–304, 307–309
Human resources, 42, 159, 178
Human rights, 64, 67, 103, 157, 213, 298
Hurricane Katrina, USA, 98
Hydrocarbons, 69

I

Identity, 2, 41, 75, 81, 85, 98, 129, 158, 219, 231, 326, 327
Implementation, 5, 21, 40, 43, 47, 50, 53, 56, 68, 71, 77, 78, 108, 114, 115, 130, 132, 135, 140, 178, 183, 198–199, 201, 205, 206, 218, 226, 229, 298, 299, 305–307
Inclusion, 2, 7, 10, 37, 40, 43, 46, 54, 56, 71, 77–79, 127, 143, 151–155, 163, 191, 199, 232, 246, 318, 323

Inclusivity, 7, 54, 191, 246–247, 250, 257
Income, 9, 17, 37, 39, 47, 54, 98–100, 105, 129, 138, 152, 155, 184, 187, 191, 201, 249, 250, 252–255, 268, 271–273, 277, 328
Income generation, 53, 155
India, 5, 39, 40, 43, 45–54, 56, 318, 323
Indian subcontinent, 37–57
Indigeneity, 75
Indigenous, 64, 67, 70, 78, 155, 304
Individualistic organisation, 8
Indore, 50
Industrial growth, 27
Industrial mining, 27
Industrial policy, 20
Industry, 5, 15–31, 83, 127, 130–133, 135, 136, 150, 157, 161, 166, 192, 211–213, 220, 304, 318
Industry and Agriculture (TCCIA), 26
Infrastructure, 8, 49, 53, 95, 107, 110, 111, 114, 117, 126, 178, 184, 185, 187–188, 199, 211, 213, 270, 299
Inheritance rights, 56
Installation, 38, 48, 51, 112, 151, 166, 231, 244, 247, 256
Institutionalisation, 22, 189
Integrated Energy Policy 2006, 45
Integrated Labour Force Survey (ILFS) 2014, 26
Intergovernmental Forum on Mining, Minerals, Metals and Sustainable Development, 17, 18
International agencies, 26
International donors, 23
International firms, 20
International Union for Conservation of Nature (IUCN), 46
Intersectional approach/analysis/perspective/methodologies, 4, 6, 9, 83, 129, 189, 192, 203, 205, 217, 218, 243–258, 263–290, 319, 321, 326–328
Intersectionality, 3, 10, 75, 184, 248–255, 265, 267, 270–273, 317, 327, 328

J

JEEViKA, 47–48
Job creation, 25, 26, 216
Juridical law, 42
Just energy transitions, 178, 179, 214, 235, 236, 319
Justice, 20, 41, 203, 212, 217, 218, 256, 270, 287, 316, 319, 320

K

Katari Municipality, 53
Kenya, 39, 43, 45, 100, 274

L

Labour market, 6, 7, 27, 126–129, 131–136, 139, 141, 151, 155, 162, 187, 267, 281, 283
La Guajira, 64, 70, 74, 77, 84
La Loma-Pribbenow, 70
Land adjudication, 24
La Niña, 68
Large-scale, 16, 28, 101, 116, 166
Large-scale mining, 16, 19, 27, 29, 31
Latin America, 2–4, 63, 66
Law 1931, 68, 69
Legislation, 17, 20, 28, 56, 158, 183
Legislators, 26
Lighting solutions, 48
Lima Work Programme, 79
Line persons, 50

Livelihood, 9, 16, 17, 19, 27, 37, 41, 50, 54, 66, 92, 94, 97, 99, 100, 103–106, 116, 263, 267, 268, 300
Live wires, 49
Living conditions, 66, 70, 264, 273, 275, 287
Local companies, 20, 21
Local content, 5, 15–31
Local content plan/policy, 5, 18, 21
Local employment, 27
Local participation, 22, 24, 30
Low-carbon transitions, 71

M
Macro, 270
Madhya Pradesh West Power Distribution Company (MPWPDC), 50
Maharashtra, 48–50
Mahavitaran, 48–50
Male-centeredness, 8
Management, 8, 19, 27, 38, 43, 52, 54, 68, 76–79, 81, 83, 96, 102, 114, 126, 127, 135, 142, 164, 165, 183, 192, 193, 198, 199, 201, 213, 220, 227, 230, 231, 244, 247, 308
Managers, 26, 52, 139, 140, 202
Marital status, 9, 20, 129, 249, 250, 257, 328
Marriage, 48, 49, 287, 308
Masculine workplace policies, 29
Masculinity, 10, 75, 160, 189, 267
Medium-sized enterprises (MSMEs), 16, 17, 20, 25, 27, 28, 30, 31, 183
Mentoring, 51, 142, 165, 226, 231, 234, 235
Mentorship, 53
Meso, 191, 270, 283, 286
Metallurgy, 18
Meter readers, 50
Micro, 16, 27, 28, 191, 270, 283
Micro hydro, 51
Micro Hydro Power Plant, 51–53
Mine closures, 71
Mineral beneficiation value chains, 28
Mineral Policy, 15
Mines, 16, 18, 19, 27, 29, 64, 67, 70, 71, 84
Mining, 5, 15–24, 26–31, 64–66, 69, 70, 73, 74, 80–81, 83, 131, 138, 318, 323
Mining Act 2010, 15, 17, 21
Mining (Local Content) Regulations, 2018, 17, 21
Mining sector, 5, 17, 18, 20, 24, 28–30, 69, 80, 81, 318, 323
Ministry of Education, 25
Ministry of Energy and Minerals, 26
Ministry of Energy and Ministry of Water Resources and Irrigation, 51
Ministry of Health, Community Development, Gender, Elders and Children, 23
Ministry of Industry, 26
Ministry of Mines and Energy, 69
Miscellaneous Amendments Act 2017, 21
Mitigation, 64, 67–69, 91, 97, 111, 201
Mobile phones, 48, 110, 303
Modern energy, 2, 3, 37, 54, 263
Mtwara, 23
Mujeres (women), 1, 16, 38, 64, 85, 92, 127, 149, 185, 213, 246, 264, 298, 316
Multidimensionality, 9
Multidisciplinary, 3, 10, 130, 329
Multi-level, 6, 7
Multiple discriminations, 9, 267

Mumbai, 48
Mwanza, 23
My Business, 23

N
Namibia, 25, 102–103
National action, 18
National Climate Change Policy, 69
National Climate Policy, 68, 71
National Economic Empowerment Act 2004, 22, 23
National Economic Empowerment Council (NEEC), 23, 23n1, 26
National Economic Empowerment Policy, 23
National Employment Policy 2008, 20
National energy policy, 178
National Energy Strategy 2013, 46
Nationally Determined Contributions (NDC), 68, 69, 77, 79
National Plan for Disaster Risk Management (PNGRD), 76
National Policy on Climate Change 2017, 68
National Rural and Renewable Energy Programme 2012, 46
National Skills Development Strategy (NSDS), 25
National Strategy for the Reduction of Emissions from Deforestation and Forest Degradation, 75
National System for Disaster Risk Management, 76
National Women and Gender Development Policy and National Strategy for Gender Development, 2000, 20
Natural resource, 16, 20, 66, 67, 92, 96, 106–107, 212, 213, 318
Natural resource-based development, 31
Natural Resource Governance Institute (NRGI), 18
NEEC's Empowerment and Facilitation Department, 23
Nepal, 5, 10, 39, 40, 43, 45–54, 107–109, 299, 302–304, 306, 308, 318, 323
Nepalese, 45, 52
Netherlands, 127, 129, 133–136, 138–140, 187, 223, 320
Neutrality of gender, 2
Norms, 5, 22, 39, 41, 42, 48, 51, 57, 65, 75, 101, 103, 108, 110, 152, 155, 157, 189, 190, 219, 246, 273, 287, 289, 301, 322, 323

O
Oil, 15–19, 21–24, 28–31, 95, 150, 183, 213, 220, 318, 323
Open participation, 7, 8, 181, 197
Oppression, 8, 72, 74, 75, 155, 198, 264, 267, 318, 326–329
Organisational studies, 3, 317
Orientation, 8, 68, 81, 85, 158, 245, 264, 270, 328
Ownership, 27, 40, 43, 48, 51, 52, 56, 99, 106, 110, 114, 126, 179, 181, 182, 194–196, 206, 216, 220, 244, 245, 283, 302

P
Pakistan, 100
Paramilitary, 67
Parenthood, 9, 163, 168, 328
Paris Agreement, 68, 69, 79, 133, 206, 213
Participation, women, 233
Partnership, 20, 135, 142, 151, 156, 158, 183, 303
Paternalistic, 82

Patriarchal construct, 48
Patriarchal gender structure, 64
Patriarchal power relations, 18
Patriarchal social relations, 30
Patriarchy, 1, 10, 18, 272, 317, 325–326
Peacebuilding, 67
People-Centred Accelerator, 39
Permanent Sovereignty Act, 15
Persons with disabilities, 30
Petroleum, 15, 21, 23, 26, 66
Petroleum Act 2015, 15, 17, 21
Petroleum (Local Content) Regulations 2017, 21
PIGCCT Guajira, 77
Pink Electricity Zone, 50
Planes Integrales de Gestión del Cambio Climático Sectoriales (PIGCCS), 68, 69
Planes Integrales de Gestión del Cambio Climático Territorial (PIGCCT), 68, 75, 77
Plan Nacional de la Gestión del Riesgo de Desastres, 76
Plant processing, 18
Plataforma de Mujeres Negras, 78
PM SAUBHAGYA 2017, 45
Policy analysis, 63–85
Policy guidelines, 52, 83
Policy review, 5, 46
Political economy, 3, 317
Political participation, 2, 4, 82
Political science, 3, 317
Pollution, 17, 70, 107, 212, 214, 243, 269
Positionality, 75
Post-carbon future, 22
Poverty, 9, 37, 66, 70, 92, 100, 103, 109, 132, 167, 188, 193, 202, 204, 231, 263, 264, 266, 269, 271–273, 277, 282, 283, 286, 289, 301
Power infrastructure, 49
Power relations, 22, 41, 83, 152, 185, 250, 253, 319, 323, 327, 328
Power systems, 52
PRADAN, 50, 51
Procurement, 17, 20, 21, 28, 30, 31, 299, 323
Production Sharing Agreements, 21
Productive work, 8, 185, 199, 283
Productivity, 24, 49, 50, 96, 103–105, 116, 153, 163, 219, 263
Progression, 7, 142, 156
Protocols, 82, 300
Public engagement, 41
Public policy, 3, 16, 73, 80, 163, 186, 317
Public Procurement Regulatory Authority (PPRA), 30
Public resources, 8, 185, 187–188

Q

Quotas, 17, 18, 24, 25, 30, 72, 78, 165, 200, 203, 221, 236

R

Race, 8, 73, 75, 98, 153, 155, 169, 218, 264, 267, 270, 324, 327, 328
Rajasthan, 38
Raw materials, 20, 131
Refugee status, 9
Renewable energy, 38, 46, 115, 125–128, 130–133, 149, 150, 166, 167, 177–180, 182–185, 192, 196, 197, 199–201, 204, 205, 212–214, 218, 220–223, 226, 231, 234, 235, 244, 245, 247, 256, 299, 316, 324
Renewable energy development, 46

Representation, 2, 8, 24, 29, 45, 65, 72, 73, 81, 128, 132, 141, 150, 153, 154, 164, 165, 192, 193, 195, 197, 205, 206, 213, 218, 220, 225, 234, 235, 248, 253, 256, 258, 275, 301, 323
Reproductive role, 48, 75, 270
Resilience, climate, 6, 94, 96, 103–113, 115–117
Resilient energy systems, 111
Resource extraction, 16–18, 20, 24, 30, 67
Resource governance, 66
Resource misallocation, 25
Resource nationalism, 16
Risk analysis, 69
Risk management, 68, 83, 96
Rome, 39
Rural Electrification Policy 2006, 45
Rural employment, 46
Rural energy, 50
Rural Energy Policy 2006, 46
Rural households, 38
Rural people, 38
Rural women, 38, 75, 82, 201
Rwanda, 10, 299, 300, 302–304, 306, 308, 309

S

Safe-motherhood programme, 53
Sahel region, 105
Samudayeek Balbikash Tatha Mahila Jagaran, 53, 54
Science, technology, engineering, and mathematics (STEM), 28, 128n2, 132, 141, 149, 154, 157, 196, 203, 213, 247, 250, 251
Seasonal emergencies, 82
Security, 4, 42, 66, 77, 99, 104, 112, 114, 126, 137, 138, 160, 166, 179, 301, 308, 309, 316, 324
Self-Help Groups (SHGs), 47, 48, 50, 51
Senegal, 104, 105, 274
Service delivery, 38, 39, 46
Service industry, 23
Sex-disaggregated data, 76, 203, 248, 249
Sexual exploitation, 66
Sexual orientation, 8, 81, 85, 158, 264, 270, 328
Sexual violence, 9, 67, 159, 160, 301
Sierra Leone, 100
Single women, 46, 139, 140, 273
Skills, 16, 25, 28, 29, 56, 102, 127, 130, 137, 138, 142, 166, 185, 196, 197, 199, 206, 270
Skills development, 23, 26
Skills for Oil and Gas Africa (SOGA), 23, 26
Skills sector, 25
Small Industries Development Organisation (SIDO), 26
Small-scale mining, 16, 21, 28
Small scale producers, 82
Social capital, 42, 54, 55
Social conditions, 48, 56, 322, 323
Social conflicts, 65, 67
Social difference, 8, 264, 265, 267, 289
Social disruption, 16
Social enterprise, 50, 179
Social exclusion, 22, 272
Social inclusion, 39, 54, 154, 199
Socialist development, 22
Social justice, 5, 70, 127, 178, 193, 212, 214, 325
Social networks, 42, 56, 308
Social norms, 48, 57, 108, 110, 152, 155, 157, 189, 246, 273, 287, 323
Social power, 8, 264, 267
Social relations, 30, 64, 65, 76, 152, 168, 215, 216, 267, 315, 327, 328

Social sector, 56
Social Welfare Act 2060, 2003, 53
Social Welfare Council 1992, 53
Society, 1, 4, 6–8, 18, 19, 41, 46, 48, 50–52, 67, 68, 84, 95, 98, 116, 126–128, 134, 135, 137, 138, 140, 143, 154, 157, 162, 167, 168, 184–186, 190–193, 192n4, 195, 197–201, 204–206, 211, 214, 216, 218, 228, 232, 235, 236, 243, 244, 256, 258, 281, 300, 306, 307, 316–318, 324–327
Socio-cultural context, 43, 48, 52, 300, 324
Socioeconomic class, 8
Socio-environmental, 64, 70
Sociology, 3, 317
Socio-technical, 40–43, 48, 51, 300
Solar, 7, 8, 38, 47, 50, 51, 112, 114, 116, 126, 150–152, 154, 156, 157, 164, 168, 183, 194, 195, 199, 201, 204, 205, 223, 231, 274, 298, 304, 318
  energy, 7, 151–166, 168, 169, 183, 201, 227, 318
  mini-grids, 50–51
  products, 38, 56
Solar Energy Committee, 51
Solar energy, photovoltaic, 150
Solar energy systems, 38
Solar home systems (SHSs), 38, 47, 48
Solidarity, 19, 192, 234, 246, 325
South Africa, 26, 98
South American, 63
Spain, 8, 198–202, 223, 229, 230
Spatial planning, 76
Stand-alone energy, 47
Subcontracted work, 29
Sub-Saharan African, 9, 46, 265, 274, 288, 317, 328
Subsidy benefits, 46
Subsidy Policy for Renewable Energy 2013, 46
Supernormal patriarchy, 18, 20
Supervisors, 50
Supplier development, 26
Supply chain, 5, 18, 27, 45, 49, 51, 54–56, 130, 270
Supply system, 54, 113
Sustainable consumption, 2, 4, 322
Sustainable development, 2, 4, 8, 168, 178, 206, 226
Sustainable Development Goals (SDGs), 16, 31, 68, 79, 92, 103, 128, 138, 167, 168, 201, 202, 213, 263, 264, 275, 299
Sustainable energy, 22, 115, 130, 133, 134, 137, 138, 182, 212, 215, 226, 299, 322, 328, 329

T

Tanzania, 15–31, 274, 302, 318, 323
Tanzania Bureau of Standards (TBS), 26
Tanzania Chamber of Commerce, 26
Tanzania Engineering and Manufacturing Design Organisation (TEMDO), 26
Tanzania Entrepreneurship and Competitiveness Centre (TECC), 26
Tanzania Industrial Research Development Organisation (TIRDO), 26
Tanzania Investment Centre (TIC), 26
Tanzania Local Enterprise Development (TLED), 24
Tanzania Private Sector Foundation (TPSF), 26
Tanzania Trade Development Authority (TANTRADE), 26

Tanzania Women Miners Association (TAWOMA), 23
Tariff fixation, 52
Technical and Vocational Education and Training (TVET), 16, 20, 25–27, 30
Technological advancements, 27
Technology design, 27, 41
Tenth FYP (2002–2007), 46
Tertiary education system, 26
Third National Communication on Climate Change, 2015, 68
Transport, 9, 105, 128, 139, 140, 178, 186, 187, 263, 272, 283, 286, 288, 289
Turkey, 8, 198–202, 320

U
Udaypur, 53–54
Uganda, 8, 198–202, 302, 320
Ujamaa, 20
Underprivileged, 47
United Nations Climate Technology Centre and Network, 45
United Nations Economic Commission for Africa (UNECA), 16, 27
United Nations (UN), 4, 16, 40, 92, 103, 111, 128, 134, 158, 167, 213, 247, 273, 275
Universal electrification, 39
University of Pretoria, 26
University of the Witwatersrand, 26
User's Group Committee, 51, 52

V
Value chain, 7, 16, 20, 27, 28, 38, 167, 182, 192, 221, 231
Variabilidad climática, 81
Vidyut Sahayaks, 48–50, 55
Vigilance squad, 49
Village councils, 25, 116
Village institutions, 25
Village Land Council, 24
Violence, 9, 67, 81, 101, 103, 106, 107, 152, 157–161, 167, 168, 272, 274, 281, 282, 287, 289, 298–301, 303, 308
Visibility, 8, 157, 165, 193, 201, 213, 231, 235, 307
Vocational education, 25
Vocational Education and Training Authority (VETA), 16, 25, 26
Vulnerability, 30, 74, 76, 82, 92, 94–103, 105, 106, 110, 116, 117, 202, 205, 212, 214, 243, 271, 272, 289, 308, 318

W
Wayúu, 70
Welfare, 24, 39, 42, 43, 267, 282
Well-being, 6, 7, 66, 92, 94, 100, 103, 152, 157–163, 168, 178, 188, 219, 263, 267, 308, 316, 318, 321
Wind energy, 195, 199, 204–205, 220, 224
Wiwa, 70
Women entrepreneurs, 18, 28, 38
Women-led companies, 30
Women-led enterprises, 17, 57
Women's access, 99, 110, 283
Women's agency, 10, 43, 102, 270
Women's economic empowerment (WEE), 5, 15–31, 53, 323
Women's empowerment, 5, 37–57, 92, 93, 110, 115, 198, 202, 221, 232, 235, 283, 323
Women's empowerment, economic, 5, 15–31, 53, 323
Women's energy needs, 185, 273, 323

Women's involvement, 38, 45, 46, 115
Women's Network in Solar Energy, 151
Workforce, 6, 7, 18, 125–143, 150, 154, 163–168, 213, 219, 220, 247, 318
Working environment, 29
Working women, 57, 139, 140, 281
Workloads, 17, 139, 140, 202, 263
World Bank, 25, 66

Y
Young Graduate Program, 23
Youth, 23, 30, 307
Youth Business Tanzania, 23

The manufacturer's authorised representative in the EU is Springer Nature Customer Service Centre GmbH, Europaplatz 3, 69115 Heidelberg, Germany. If you have any concerns regarding our products, please contact ProductSafety@springernature.com

Printed and bound by CPI Group (UK) Ltd, Croydon, CR0 4YY
01/10/2026
02225894-0017